T0206127

An Introduction to Compressible Flow

An Introduction to Compressible Flow

Second Edition
Forrest E. Ames and Clement Tang

CRC Press
Taylor & Francis Group
Boca Raton London New York

CRC Press is an imprint of the
Taylor & Francis Group, an **informa** business

Second edition published 2022
by CRC Press
6000 Broken Sound Parkway NW, Suite 300, Boca Raton, FL 33487-2742

and by CRC Press
2 Park Square, Milton Park, Abingdon, Oxon, OX14 4RN

© 2022 Forrest E. Ames and Clement Tang

First edition published by Momentum Press 2018

CRC Press is an imprint of Taylor & Francis Group, LLC

ISBN: 978-0-367-89567-9 (hbk)
ISBN: 978-0-367-69779-2 (pbk)
ISBN: 978-1-003-04294-5 (ebk)

Typeset in Palatino
by Spi Global, India

Contents

Preface

The topic of compressible flow is an important foundation for a number of courses in the thermal fluid science area of mechanical and aerospace engineering. Often applied subjects such as gas turbines and aerodynamics require the integration of a section of compressible flow before many of the topics within each of these courses can be initiated. Many other subjects in the thermal fluid sciences and aerospace engineering would benefit from the subject area of compressible flow. This material is critical to the engineering fields related to energy conversion or aerospace. Unfortunately, a number of books which incorporate a chapter or a section on this material necessarily tend to be overly concise. In an effort to better support compressible flow material in gas turbine and aerodynamics courses, the first edition of *An Introduction to Compressible Flow* was written. This first edition included material on isentropic flow, normal shock waves, oblique shock waves, and Prandtl–Meyer expansion fans. Later, after the first edition of *An Introduction to Compressible Flow* was written, we began to use it in a compressible flow course taught every year. The material in this first edition comprised about 60% of the material in the course. Naturally, it made sense to extent this material to support a full semester-long course on compressible flow. The additional material added to the original volume includes a chapter on applications and chapters on linearized flow, flow with friction (Fanno-line flow), and flow with heat addition (Rayleigh-line flow). A section on supersonic flow over axisymmetric cones was also added to the oblique shock chapter.

This book is designed to cover sufficient material for a typical semester-long undergraduate course in compressible flow. Initially, this book begins with a brief review of thermodynamics and control volume fluid dynamics, which sets the entry point for the study of compressible flow. Some additional information about gases is also added to provide more background to the reader. The second chapter introduces static and stagnation conditions in moving and stationary reference frames, the speed of sound, and the development of the isentropic relations for one-dimensional compressible flow. This chapter also includes the applications of converging nozzles as well as converging–diverging nozzles. The third chapter introduces and develops relationships for normal shock waves. The application of traveling one-dimensional shock waves is introduced along with a very short section on shock tubes. The fourth chapter introduces oblique shock waves along with the application of oblique shock inlets for propulsion turbines. Additionally, the material on conical shocks was added to Chapter 4. The fifth chapter introduces Prandtl–Meyer flow with the application of supersonic nozzles and airfoils. In Chapter 6, this book applies the material covered from

Chapters 1–5 that includes, for example, applications of normal shocks for nozzles, and oblique shocks and Prandtl–Meyer expansion fans for diffusers and supersonic airfoils.

Chapter 7 introduces linearized flow, which is an approximation of flow over mild convex and concave corners which assumes flow is isentropic. This material is common in aerodynamics courses and fits well in a first course in compressible flow. The eighth chapter introduces the subject of flow with friction which is called Fanno-line flow. This material is critical to engineers in their understanding of compressible flow in pipes and the general engineering of natural gas pipe lines. This chapter also includes material on isothermal flow which provides a simple but powerful approximation to Fanno-line flow. Chapter 9 introduces flow with heat addition or Raleigh-line flow. This material is key to understanding how heat addition and cooling can influence compressible flows.

This book has been developed as a stand-alone resource that provides sufficient material for a semester-long undergraduate course on compressible flow. The initial few chapters of this book provide excellent supplementary material to support courses in gas turbines and aerodynamics. Engineering students who need an understanding of compressible flow as a foundation for more advanced classes will find this material helpful. In addition, this book will serve as a useful reference to engineers who need this material for their professional endeavors. All the chapters in this book have example problems and end-of-chapter problems; and a solution manual will be offered to instructors upon request.

The approach taken in developing this material is to maintain a focused and straightforward presentation. However, we have also attempted to provide enough discussion about the applications for this material to keep students interested and motivated.

Author Biographies

Forrest Ames has been Professor of Mechanical Engineering at the University of North Dakota (UND) for the last 23 years. Dr. Ames began his career at Allison Gas Turbine Div. of General Motors where he worked in the research laboratories. Dr. Ames has conducted research in the area of gas turbine heat transfer and aerodynamics for over 30 years. At UND, Dr. Ames is responsible for teaching in the thermal fluids area of mechanical engineering and regularly teaches classes on compressible flow, aerodynamics, gas turbines, thermodynamics, computational fluid dynamics, convective heat transfer, and fluid dynamics. Dr. Ames has been a member of the Heat Transfer Committee of the International Gas Turbine Institute for over 24 years. He has been a regular contributor to ASME Turbo Expo technical sessions as author, presenter, reviewer, and session organizer in the areas of turbine aerodynamics and heat transfer. He is a fellow of the ASME.

Clement Tang is an Associate Professor of Mechanical Engineering at the University of North Dakota (UND). He joined UND as a faculty member in 2011. Dr. Tang has research experience in the area of multiphase flow heat transfer and aerodynamics of thin flexible materials. He has been conducting experimental research in gas–liquid two-phase flow heat transfer for over 15 years. At UND, Dr. Tang has taught compressible flow, heat and mass transfer, heat conduction and radiation, HVAC, mechanical measurements, multiphase flow heat transfer, and thermodynamics.

1

Introduction

1.1 Background Information on Gases

Although a compressible flow can occur in any gas, it is most typically related to flows around aircraft and in aeroengines and gas turbines. Air is the most common gas that engineers deal with, therefore we begin by introducing some properties of air from a perspective of gas dynamics.

1.1.1 Air Composition and Air Molecules

Air is often described in terms of its properties, including its density, temperature, and pressure. A typical density given for air is about 1.2 kg/m³, which is almost exact for 21°C at 1 atmosphere. Consequently, the mass of the air in 1 m³ of volume is about 1.2 kg. From chemistry, we know that one kilogram mole (kmol) of any substance has 6.02252×10^{26} molecules. The typical composition for dry air is given as 78.08 percent nitrogen (N_2), 20.95 percent oxygen (O_2), 0.93 percent argon (Ar), and about 0.04 percent carbon dioxide (CO_2). If the percentages are multiplied by the individual molecular weights of these main four constituents, air is found to have a composite molecular weight of 28.97 kg/kmol. Consequently, 1 m³ of volume will typically contain nearly 2.5×10^{25} molecules of air, which is a nearly incomprehensible number. The root mean square of the speed of these molecules is around 503 m/s, and the average distance they travel before bumping into another molecule (the mean free path, λ) is about 66 nm. This means an average air molecule at 1 atmosphere and 21°C experiences about 7.6 billion collisions a second.

1.1.2 Temperature and Gases

Air is largely made up of diatomic molecules – molecules with two atoms. These air molecules can store energy not only through their kinetic energy of motion but also through rotational energies in two axes. The combination of kinetic energy through velocity or translation plus this energy of rotation is often called the internal energy of the gas. Generally, each of the directions of translation and each axis of rotation can store equal amounts of energy. This equivalence between energies of the three directions of translation and the two independent directions of rotation is called the equipartition of energy.

About 0.93 percent of air is Argon – a monotonic gas. The internal energy of this monotonic gas molecule is related to one-half the mean squared speed of the molecules $\frac{1}{2}\overline{C^2}$ times the mass, m, of the given molecule. In fact, the mean energy of translation for all the molecules in the gas can be quantified in a similar manner.

$$\tilde{e}_{tr} = \frac{1}{2} m \overline{C^2} \tag{1.1}$$

The energy of translation can also be estimated based on its temperature.

$$\tilde{e}_{tr} = \frac{3}{2} k T \tag{1.2}$$

Here, T is the absolute temperature and k is the Boltzman constant (k = 1.38054E−25) (J/molecule/K, note this term is only used in Chapter 1), which is essentially the gas constant per molecule. Consequently, absolute temperature is related directly to the mean squared speed of the molecules. This relationship can be put into more familiar terms by multiplying the right-hand side of both equations by Avogadro's number, $\hat{N}$. The new equality becomes:

$$\frac{1}{2} \hat{M} \overline{C^2} = \frac{3}{2} \hat{R} T \tag{1.3}$$

Here, $\hat{M}$ is the (composite) molecular weight of the gas, and $\hat{R}$ is the universal gas constant (8314.46 J/kmol/K). The square root of the mean squared speed of the molecules can be determined using Equation (1.3) yielding:

$$\sqrt{\overline{C^2}} = \sqrt{3\left(\hat{R}/\hat{M}\right)T} \tag{1.4}$$

From this relationship, it is apparent that both the square root of the mean squared speed of a molecular species or a composite gas like air is related to the molecular weight. At 21°C (294.15 K), using the composite molecular weight of air (28.97 kg/kmol), the composite velocity is determined to be around 503 m/s. Nitrogen, which is slightly lighter, will have a slightly higher speed on average, while oxygen and the other molecules in air, which are heavier, will move at slightly lower average speeds.

Later, the speed of sound of air will be derived from a macroscopic approach. This analysis will show that the speed of sound of air is directly related to the absolute temperature of air similar to the square root of the mean squared speed of a molecule. Note that the square root of the mean squared speed of a molecule is slightly different than the average or mean speed of molecular motion. However, this speed in Equation (1.4) can be determined easily from energy concepts related to molecular motion. In general, in thinking about molecular motion, realize that molecular speeds have a random distribution, which is described mathematically by the famous Maxwellian distribution. It turns out that the root mean square speed (Equation 1.4) is slightly higher than the average speed, which is slightly higher than the most probable speed. At a given total temperature, the maximum bulk velocity that a gas can achieve from expansion can be determined from energy concepts. Basically, if all the energy stored in a gas is applied to its expansion to maximize the bulk velocity, for a given total temperature for a gas, the maximum velocity will be:

$$V_{MAX} = \sqrt{2C_P T} \tag{1.5}$$

This equation means that all of the random and organized energy in the gas, including translational and rotational energies and flow work, is converted into kinetic energy. Here, C_P is the specific heat at constant pressure per mass for the composite gas (J/kg/K) and T is the absolute temperature (K).

1.1.3 Pressure and Gases

The concept of pressure can be related to the force that a gas exerts on a surface in a volume. A simple thought experiment would be to think of a gas as consisting of a very large number of molecules. However, instead of considering the huge number of collisions, which occur within a gas in a volume, let us think of a gas molecule as a point in space that has mass and velocity, but that takes up such a small volume that collisions do not need to be considered. When the molecule hits a wall, we can assume that it rebounds at an angle equal to the incoming angle. This assumption would mean that, every time a molecule hits a wall, it would impart a momentum equal to two times its mass times its normal velocity. The rate that it would hit that wall would be related to its normal velocity divided by two times the normal distance to the opposite wall. The pressure on a wall could then be calculated by the number of molecules times the momentum imparted per collision times the rate of collisions divided by the area of the wall. Consider a cubical volume of 1 m per side. Assume that the volume holds 1.2 kg of air at a temperature of 21°C (294.15 K). Based on our preceding estimate, the pressure could be calculated as:

$$P = \frac{M}{\hat{M}} \times \hat{N} \times 2mC_N \times C_N / (2l) / l^2 \tag{1.6}$$

This equation can be simplified by noting due to the equipartition of energy $\overline{C_N^2} = \overline{C^2} / 3$ and from internal energy considerations $\overline{C^2} = 3(\hat{R} / \hat{M})T$. Also noting that Avogadro's number $\hat{N}$ times the mean weight per molecule, m, is simply the molecular weight, $\hat{M}$, which cancels. Also the l^3 term in the denominator can be converted to volume, Ψ. Our resulting equation for pressure becomes:

$$P = M(\hat{R} / \hat{M})T / \Psi \tag{1.7}$$

This equation is our familiar ideal gas law, which is often written as $P \Psi = MRT$, with R being the gas constant per mass $\left(R = \hat{R} / \hat{M} \right)$ of the composite gas of interest. Using a mass of 1.2 kg, a gas constant for air of 287 J/kg/K at temperature of 294.15 K and a volume of 1 m³, we calculate our pressure to be 101,305 Pa. This value is essentially atmospheric pressure within round-off error. Two obvious issues that we omitted include the finite size of the molecules and attraction between molecules.

1.2 Control Volume Analysis and Fundamental Concepts

Compressible fluid flow is an engineering subject that is grounded in fluid mechanics and thermodynamics. Compressible flow is governed by the same principles as fluid mechanics. However, perhaps the key issue in compressible flow is that the change in density of the fluid is a natural feature of the flow. In the present treatment of one-dimensional compressible flow, skin friction due to fluid viscosity will not be considered. This and other simplifications will be discussed as we begin to address the governing equations in this chapter. Generally, four major principles and one key assumption will govern our development of the relationships, which we will use to understand and apply our knowledge of compressible flow. These principles include conservation of mass, the momentum principle, conservation of energy, and the second law of thermodynamics. In addition to these four principles, we will make the assumption that the gas we are considering can be treated as an ideal gas.

In this chapter, we will begin our study of compressible flow by reviewing the basic laws for a system, as well as the principles of conservation of mass, the momentum principle (or Newton's second law), and conservation of energy (or the first law of thermodynamics). Next, we will develop a generalized form for control volume equations and apply this to our first three principles. Subsequently, we will undertake a short review of thermodynamics and the ideal gas law.

1.2.1 Basic Laws for a System

Our initial goal is to understand the basic principles of a system and to apply those principles to solve problems in compressible flow. Good engineering solutions to many problems can be developed by applying the basic principles of conservation of mass, the momentum principal, and conservation of energy. In this section, we develop a basic framework of integral equations for control volume analysis, and we apply this framework to the principles of conservation of mass and the momentum principle. We will be looking at two types of systems in our analyses. One system is a fixed mass system, which can also be called a closed system, and another type of system is a control volume, which is also called an open system. In an open system, mass is allowed to flow into and out of the system. A classical approach to developing control volume equations is to initially look at an open system from the perspective of a fixed mass system. This approach will be used in this current chapter. However, before beginning to develop our control volume perspective, we will first review the basic laws for a system.

1.2.2 Conservation of Mass

One practical and consistent approach to understand and apply our basic principles is the production method. The basic principles can all be stated using this fundamental approach. Conservation of mass can be stated as follows:

$$\text{Production rate of}\left(\text{mass}\right) = 0 = \text{outflow rate} - \text{inflow rate}$$
$$+ \text{change in storage rate}$$

It can be written symbolically as:

$$\dot{P}_M = 0 = \dot{M}_{OUT} - \dot{M}_{IN} + \frac{dM_{SYS}}{dt} \tag{1.8}$$

For a fixed mass system, we have no outflow or inflow, so conservation of mass for a fixed mass system is simply:

$$\text{Production rate of}\left(\text{mass}\right) = 0 = \text{change in storage rate} = dM / dt$$

Or, symbolically:

$$\dot{P}_M = 0 = \frac{dM_{SYS}}{dt} \tag{1.9}$$

While this result seems trivial, it will have some utility when we apply this relationship to a control volume. For the case of conservation of mass, mass is the property of the system that we are interested in. *We can determine the property or the amount of mass in our system by integrating overall the differential mass in the system.* A similar but more practical approach to quantify our systems mass is to integrate the density or mass per unit volume over the entire volume of our system.

$$M_{SYS} = \int_{M_{SYS}} dm = \int_{V_{SYS}} \rho \, dV$$

1.2.3 Newton's Second Law

A basic statement on Newton's second law for a fluid system would be that a force applied to this system will cause an acceleration of the mass of the system. We can call this acceleration the time rate of change of system momentum, P.

$$F = \frac{dP}{dt} \tag{1.10}$$

The system momentum, P, can be quantified by integrating the momentum per unit mass, V, over the system volume similar to the quantification of the mass.

$$P_{SYS} = \int_{M_{SYS}} V \, dm = \int_{V_{SYS}} \rho V \, dV$$

Newton's second law has magnitude and direction, meaning it is a vector equation, and this equation is often rewritten in terms of its three Cartesian directions.

1.2.4 Energy Equation

The energy equation can be stated in a manner very similar to the conservation of mass principle.

Production rate of energy $= 0 =$ outflow rate $-$ inflow rate
                                    $+$ change in rate of storage

For a fixed mass system, we have no outflow or inflow of mass so for a process we can say:

$$P_E = 0 = W - Q + \Delta E \tag{1.11}$$

Outflow $-$ inflow $+$ change in storage

The simple energy equation for a simple fixed mass system can also be written on a time rate basis, where $\dot{Q}$ represents the heat transfer rate and $\dot{W}$ is the work transfer rate or power.

$$\dot{P}_E = 0 = \dot{W} - \dot{Q} + \frac{dE}{dt} \tag{1.12}$$

Outflow rate $-$ inflow rate $+$ change in rate of storage

As shown in our other examples, we can write the total system energy as an integral term.

$$E_{SYS} = \int_{M_{SYS}} e \, dm = \int_{V_{SYS}} \rho e \, dV$$

Here, e is taken as the sum of the specific internal, kinetic, and potential energies, $e = u + V^2/2g_C + gz/g_C$.

Note $g_C = 1$, SI units; $g_C = 32.174$ lbm-ft/lbf-s^2, English units.

1.2.5 Development of a Generalized Control Volume Equation

The principles of conservation of mass, Newton's second law, and the conservation of energy have been presented simply in the previous section. However, a method is needed to extend these equations to a general control volume or open system in a fluid field. These equations can be extended to their control volume form by rationalizing between a fixed mass system and an open system. This rationalization can be accomplished with a generalized approach using a general conserved property N as well as an intensive property, η. There is a substantial amount of similarity between control volume equations for these different principles. However, before we discuss the extensive and intensive properties of interest in developing control volume equations, the assumptions we are making should be reviewed.

The first assumption in the development of our control volume equations is the equations are taking a macroscopic view of the fluid field. Properties are assumed to be continuous throughout the fluid, and although they are allowed to change, they cannot change abruptly. By taking this continuum concept of properties, we can define a property at a point as it is based on sufficient matter so that any gradients or variations are well-behaved. As previously noted, we have some 2.5×10^{16} molecules in a cubic millimeter of air at ambient conditions, so this continuum assumption is not too challenging to accept in most compressible flow situations. This assumption could break down once our point of interest becomes on the order of a mean free path, which as noted previously is about 6.6×10^{-6} cm at ambient conditions.

For conservation of mass, our extensive property is the mass of the system or M, and our intensive property, mass per unit mass, can be represented by 1. Our extensive property for momentum is P, while the intensive form, momentum per unit mass is velocity, V, and these are vector quantities. Our extensive property for the energy of a system is E, while the intensive form for energy of a system is e taken as the sum of the internal, kinetic, and potential energies of the matter being considered. Our extensive property for the entropy of a system is given the symbol S, while its intensive form is written as s. The extensive property of a substance in a system, N_{SYS}, can be determined by integrating the intensive property, η, over all the differential mass of a system, dm. Alternatively, the same accounting for N_{SYS} can be accomplished by integrating the intensive property, η, times the local density, ρ, across all the differential volume, dV, of a system.

$$N_{SYS} = \int_{M_{SYS}} \eta \, dm = \int_{V_{SYS}} \rho \eta \, dV$$

The classical approach to develop a control volume form of equations from a fixed mass system is to take a fixed mass system in an arbitrary flow field. From the standpoint of the fixed mass system:

$$\left(\frac{dN}{dt}\right)_{SYS} = \lim_{\delta t \to 0} \frac{N_{t0+\delta t} - N_{t0}}{\delta t} \tag{1.13}$$

From the standpoint of an open system (or control volume), which is superimposed on the fixed mass system at time, t_0, as the flow field moves into or out of the control volume, it carries property, η. This intensive form of a property times the mass flow rate produces an inflow or outflow of the property N on a time rate basis. The development of the control volume form of equations can be derived in terms of this fixed mass system. As noted earlier at the initial time, t_0, our control volume is coincident with our fixed mass system. However, in a small increment of time, δt, our flow field of interest will move our fixed mass system partially out of the initial control volume as shown in Figure 1.1.

On the basis of the fixed mass system, in the time δt, the system has moved from regions I and II to regions II and III. We can use the regions to put these changes in the location of the fixed mass system in terms of the original control volume at time, $t_0 + \delta t$, plus the new region III less the old region I.

$$\frac{dN_{SYS}}{dt} = \lim_{\delta t \to 0} \left[\frac{N_{CV,t0+\delta t} + N_{III} - N_I - N_{CV,t0}}{\delta t} \right] \tag{1.14}$$

The system is located in regions I and II at time t_0, and this is coincident with the control volume at time t_0. The system moves to regions II and III over the time δt and occupies those regions at time $t_0 + \delta t$. We were able to construct this region using the control volume at time $t_0 + \delta t$ plus region III less region I. As a result, $N_{SYS,t0} = N_{CV,t0}$ and $N_{SYS,t0+\delta t} = N_{CV,t0+\delta t} + N_{III,t0+\delta t} - N_{I,t0+\delta t}$.

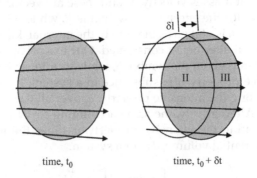

time, t_0 time, $t_0 + \delta t$

FIGURE 1.1
Flow of a fixed mass system through a control volume at two instants of time.

It is clear to see that the first and last term in the limit form the time derivative of the property N in the control volume, while the N_{III} term represents the rate of outflow from the control volume and N_I represents the rate of inflow.

$$\frac{dN_{CV}}{dt} = \lim_{\delta t \to 0} \left[\frac{N_{CV,t_0+\delta t} - N_{CV,t_0}}{\delta t} \right]$$

$$\dot{N}_{OUT} = \lim_{\delta t \to 0} \left[\frac{N_{III}}{\delta t} \right] (\text{outflow rate of } N)$$

$$\dot{N}_{IN} = \lim_{\delta t \to 0} \left[\frac{N_I}{\delta t} \right] (\text{inflow rate of } N)$$

The time rate of change of N in the control volume is best written in terms of the first time derivative of the integral of η around the control volume.

$$\frac{dN_{CV}}{dt} = \frac{d}{dt} \int_{CV} \eta \, dm = \frac{d}{dt} \int_{CV} \rho \eta \, d\Psi \tag{1.15}$$

In our production methodology, this term represents the time rate of change of extensive property N in our control volume. The second term, $\dot{N}_{III}$, represents the outflow of N from the control volume due to flow leaving the control volume. In the increment of time, δt, the surface of our system moves the distance, δl, from the control volume. We can represent the outflow of mass in terms of the density, ρ, times the volumetric outflow with time. This volume outflow $\delta \Psi$ can be further developed in terms of the dot product of δl with the incremental area normal $d\mathbf{A}$ integrated over the outlet surface.

$$\frac{dN_{III}}{dt} = \lim_{\delta t \to 0} \left\{ \eta \rho \, \delta\Psi_{III} / \delta t \right\} = \lim_{\delta t \to 0} \int_{A_{III}} \eta \rho \, \delta l / \delta t \cos \alpha \, dA$$

Note that $\lim_{\delta t \to 0} \frac{\delta l}{\delta t}$ is simply the velocity at the surface, V. The $\cos(\alpha)$ accounts for the angle between the local velocity vector and the area normal as only the normal velocity carries fluid across the system boundary and with this fluid, the mass, and other properties.

$$\frac{dN_{III}}{dt} = \int_{A_{III}} \eta \rho \, V \cdot d\mathbf{A}$$

This term represents the outflow rate of N leaving the control volume with the fluid. The inflow rate of N into the system can be described similarly

$$-\frac{dN_I}{dt} = \int_{A_I} \eta \rho \, V \cdot d\mathbf{A}$$

In the development, this third term was negative. However, the integral does not need to be written as a negative as the dot product between V and dA equates to $V\,dA\cos(\alpha)$. Note that for inflows into the system the velocity vector will have a trajectory into the surface. The angle α will, therefore, be between 90° and 270° for the inflow term, automatically subtracting the inflows. The dot product between the area normal essentially adds the outflows and subtracts the inflows. Consequently, these two terms can be put together into a single integral, which adds the outflow rate and subtracts the inflow rate.

$$\frac{dN_{III}}{dt} - \frac{dN_I}{dt} = \int_{A_{III}} \eta\rho\,V\cdot dA + \int_{A_I} \eta\rho\,V\cdot dA = \oint_{CS} \eta\rho\,V\cdot dA$$

While the dot product inside the integral allows us to combine the outflow rate minus the inflow rate into a single term, it is important conceptually to recognize that this term represents both terms.

Our development of our fixed mass system to our control volume perspective initially resulted in three separate terms, including a term representing the time rate of change of N storage in our control volume plus the outflow rate less the inflow rate. We can now combine these three terms into a single generalized control volume equation for N.

$$\frac{dN_{SYS}}{dt} = \frac{d}{dt}\int_{CV} \rho\eta\,d\Psi + \oint_{CS} \eta\rho\,V\cdot dA \qquad (1.16)$$

This result represents our rationalization between our fixed mass system approach and control volume approach, and we can now apply this result to the various principles that are used in compressible flow.

1.2.6 Conservation of Mass for a Control Volume

The conservation principle for mass in a fixed mass system initially seems trivial. However, rationalized to our control volume formulation, the relationship becomes highly useful in compressible flow.

$$\frac{dM_{SYS}}{dt} = 0 = \frac{d}{dt}\int_{CV} \rho\,d\Psi + \oint_{CS} \rho\,V\cdot dA$$

In the case of conservation of mass, our intensive property η is equal to 1 in the preceding equation. Typically, we reverse the two terms on the right side of the equal sign to move toward our production methodology.

$$\frac{dM_{SYS}}{dt} = 0 = \oint_{CS} \rho\,V\cdot dA + \frac{d}{dt}\int_{CV} \rho\,d\Psi$$

The preceding equation can now be easily written in the production methodology.

$$\dot{P}_M = 0 = \oint_{CS} \rho V \cdot dA + \frac{d}{dt} \int_{CV} \rho \, d\forall \tag{1.17}$$

Rate of production of mass = 0 = rate of outflow minus inflow plus change in storage rate.

This control volume form for conservation of mass is a key principle in the development of many one-dimensional flow relationships. Typically, the equation will be written in the form shown in Equation (1.17) before we simplify it and use it to help develop a new relationship in the application and understanding of compressible flow. However, Equation (1.17) has significant utility in conventional fluid mechanics. One application of this equation would be for the specialized case of incompressible flow where the density, ρ, is a constant. When density is a constant, we can take it out of the integrals and rewrite conservation of mass as:

$$\dot{P}_M = 0 = \rho \oint_{CS} V \cdot dA + \rho \frac{d}{dt} \int_{CV} d\forall$$

As the density is now a constant, we can divide it out of the equation and put it in the form of volumetric flow and the rate of expansion (contraction) of the control volume.

$$0 = \oint_{CS} V \cdot dA + \frac{d}{dt} \int_{CV} d\forall \tag{1.18}$$

If our control volume stays constant in time, then the last term of the equation goes to zero. In this case, the equation essentially states that the sum of the volumetric outflows is equal to the inflows.

$$0 = \sum_{i=1}^{n} \dot{\forall}_{OUT} - \sum_{i=1}^{n} \dot{\forall}_{IN} \tag{1.19}$$

Often, in compressible flow, we will make the assumption that the flow can be dealt with as a uniform and steady flow. In many internal flow situations, this is not the case. However, in many external flows, regions with significant effects of viscosity can be small, and this one-dimensional uniform and steady flow assumption has significant validity. Our steady-state control volume conservation of mass equation reduces to the following:

$$\dot{P}_M = 0 = \oint_{CS} \rho V \cdot dA \tag{1.20}$$

If we examine flow at one surface applying the uniform flow assumption:

$$\int_{SURF} \rho \overline{V} \cdot d\overline{A} = \rho V A_{SURF} \cos\alpha$$

Sketch:

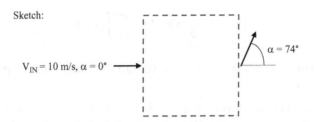

$V_{IN} = 10$ m/s, $\alpha = 0°$

$\alpha = 74°$

FIGURE 1.2
Sketch of control volume representing linear cascade.

Example 1.1: One-dimensional Uniform Flow

Flow in turbomachinery is often preliminarily analyzed as one-dimensional uniform flow. In this case, we have flow entering a linear cascade at 10 m/s at an angle perpendicular to the inlet surface as shown in Figure 1.2. At the exit, the flow leaves at an angle of 74°. Assuming one-dimensional uniform flow, determine the exit velocity. You can assume the flow is steady and incompressible.

Analysis: From our steady-state assumption, together with our uniform flow assumption, we can essentially write:

$$\dot{P}_M = 0 = \dot{M}_{OUT} - \dot{M}_{IN} = \rho\, V_{OUT}\, A_{OUT}\, \cos\alpha_{OUT} - \rho\, V_{IN}\, A_{IN}\, \cos\alpha_{IN}$$

The density can be eliminated from the equation with the incompressible flow assumption. Also, assuming flow enters from the left and leaves from the right, we can assume $A_{IN} = A_{OUT}$, and therefore cancel area from the equation. Additionally, $\alpha_{IN} = 0°$ so $\cos(\alpha_{IN}) = 1$. V_{IN} is known to be 10 m/s, which only leaves V_{OUT} as an unknown.

Solving:

$$V_{OUT} = \frac{V_{IN}}{\cos\alpha_{OUT}} = \frac{10\,\frac{m}{s}}{\cos 74°} = 36.28\,\frac{m}{s}.$$

1.2.7 Newton's Second Law for a Control Volume

Newton's second law or the momentum principle is another key relationship required to develop one-dimensional compressible flow relationships. As stated earlier, we can write this principle in terms of a force causing a time rate of change in momentum.

$$F = \frac{dP}{dt} \tag{1.21}$$

It is key to recognize that this equation is a vector equation and is typically broken into separate orthogonal relationships for each direction. The development of the relationship is similar to the development of conservation of mass. In this equation, the intensive property, momentum per unit mass is V. The momentum of a system, P, can be calculated by integrating the velocity, V, times the local density, ρ, over the volume of the system.

$$P = \int_{M_{SYS}} V\,dm = \int_{V_{SYS}} V\rho\,dV$$

In developing the control volume perspective, a generalized control volume equation was developed, which can be used together with the principle of the relationship and the particular intensive property to develop the control volume form for the principle. We can develop our control volume form of the momentum principle by making our substitutions for N and η.

$$\frac{dP_{SYS}}{dt} = \oint_{CS} V\rho V\cdot dA + \frac{d}{dt}\int_{CV} \rho V\,dV \tag{1.22}$$

At this point, we can put the preceding generalized form of the control volume formula together with the principle equation to generate the control volume form for the momentum equation.

$$F = \oint_{CS} V\rho V\cdot dA + \frac{d}{dt}\int_{CV} \rho V\,dV \tag{1.23}$$

We can interpret this to read the force vector on a system is equal to the outflow of momentum from minus the inflow of momentum to a system in addition to the rate of change of momentum in the system. The outflow minus inflow of momentum indicated by the first integral results from fluid movement from and to the system. Remember that the $V\cdot dA$ term automatically adds the outflows and subtracts the inflows and accounts for the angle that the velocity vector makes with the surface normal. The second term accounts for either the acceleration of the system or the acceleration of the fluid within the system. In general, the forces that can be applied to a system include the surface forces such as pressure and shear, the body forces such as gravity and solid body rotation, and the resultant forces, which need to be transmitted by something physical in the system.

$$F = F_S + F_B + R = \oint_{CS} V\rho V\cdot dA + \frac{d}{dt}\int_{CV} \rho V\,dV \tag{1.24}$$

This form of the equation is valid for analyses in SI units. However, the last two terms on the right-hand side should be divided by g_C for English units.

Now that we have written the control volume form of our momentum equation in vector form, it is convenient to breakdown this equation into three orthogonal directions. This new scalar form for the momentum equation in the X direction is given as follows:

$$\sum F_X = F_{S_X} + F_{B_X} + R_X = \oint_{CS} V_X \rho\, V \cdot dA + \frac{d}{dt} \int_{CV} \rho V_X\, d\mathcal{V} \tag{1.25}$$

We can easily extend this equation to the other two directions. We can state this equation simply as the summation of forces in the X direction is equal to the surface forces in the X direction plus the body forces in the X direction plus the reaction forces in the X direction. These forces are equal to the outflow less the inflow of X-momentum into the control volume with the mass flow plus the time rate of change of X-momentum within the control volume.

Example 1.2: Pelton Wheel

A Pelton wheel is essentially an impulse turbine that is designed to generate power from the kinetic energy of a water jet. Typically, a small dam will be developed to manage water from a high elevation region for use in the generation of power at a lower elevation region. We often call the gravitational potential energy generated by an elevation difference a fluid or water head. This water head is used to generate a jet of water. The Pelton wheel typically spins at approximately half of the speed of the water jet. The Pelton wheel bucket is shaped to turn the water through approximately a 160° turn. This change in direction together with the movement of the bucket allows the wheel to extract most of the jet's kinetic energy and covert this into power. The partial turn of the fluid allows the water to leave the bucket without interfering with the movement of the wheel.

Given: In this problem, we will assume that the water jet diameter is 0.0005 m² and the jet velocity is 10 m/s. We will determine the force that the jet puts on a single bucket moving at 5 m/s in the positive X direction. In this case, both the incoming jet and the outgoing flow are open to the atmosphere. We will assume that the movement of water over the bucket is frictionless. Assuming the Bernoulli equation along a streamline applies, the magnitude of the exiting flow velocity will be equal to the magnitude of the entering flow velocity.

Analysis: From our steady-state assumption together with our uniform flow assumption, we can write the integral around the control surface in terms of the mass flow rate out of the control volume times the X velocity of the outlet flow less the mass flow rate into the control volume times the X velocity of the inlet flow. In both cases, these velocities are with respect to the control volume, which moves with the Pelton wheel as indicated in Figure 1.3.

Sketch:

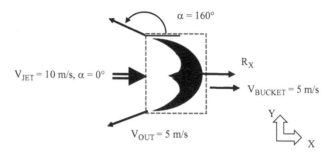

FIGURE 1.3
Sketch of moving control volume with cross-section of Pelton wheel bucket.

$$R_X = \dot{M}_{OUT} V_{X,OUT} - \dot{M}_{IN} V_{X,IN} = \rho\, V_{OUT}\, A_{OUT} V_{OUT}\, \cos\alpha_{OUT}$$
$$- \rho\, V_{IN}\, A_{IN}\, V_{IN}\, \cos\alpha_{IN}$$

The bucket moves in the positive X direction at a velocity of 5 m/s and the control volume, which is encompassed by the system boundary, moves with the bucket. This means that, relative to the control volume, the water moves into the bucket at $V_{IN} = V_{JET} - V_{BUCKET}$. Based on the conservation of mass, the flow rate moving into the bucket will also move out of the bucket.

Solving for the flow rate into and out of the control volume:

$$\dot{M}_{OUT} = \dot{M}_{IN} = \rho\, V_{IN}\, A_{IN} = 1000\,\frac{\text{kg}}{\text{m}^3} \times 5\,\frac{\text{m}}{\text{s}} \times 0.0005\ \text{m}^2 = 2.5\,\frac{\text{kg}}{\text{s}}$$

Solving for the reaction force in the simplest terms.

$$R_X = \dot{M}_{OUT}\, V_{X,OUT} - \dot{M}_{IN}\, V_{X,IN} = \dot{M}\left(V_{X,OUT} - V_{X,IN}\right)$$

Noting the outflow velocity in the X direction:

$$V_{X,OUT} = V_{IN}\cos\alpha_{OUT} = 5\,\frac{\text{m}}{\text{s}} \times \cos 160° = -4.698\,\frac{\text{m}}{\text{s}}$$

Solving for the force:

$$R_{X,OUT} = 2.5\,\frac{\text{kg}}{\text{s}}\left(-4.698\,\frac{\text{m}}{\text{s}} - 5\,\frac{\text{m}}{\text{s}}\right) = -24.24\ \text{N}$$

The negative sign shows us that the force, which pushes against the water is in the negative X direction, opposite from what was initially assumed.

1.2.8 Conservation of Energy for a Control Volume

The first law of thermodynamics has a critical but relatively simple impact on compressible flow. Introductory compressible flow phenomenon is largely an adiabatic process. Additionally, one-dimensional compressible flow will largely not be concerned with work addition or work extraction. Work addition or extraction does occur in turbomachinery. However, typically, when analyzing turbomachinery flows, the analysis is conducted in a relative velocity frame similar to the Pelton wheel analysis, and consequently, no work terms are part of the analysis in the relative velocity frame. The first law of thermodynamics largely involves a relationship between the local total temperature and the local static temperature and Mach number for most one-dimensional compressible flows. However, one situation where this adiabatic assumption does not apply occurs in flows with heat addition often referred to as Rayleigh line flow. The control volume form of the energy equation will be developed in a manner similar to conservation of mass and Newton's second law in our thermodynamics review.

We can begin our development of the control volume form of the first law of thermodynamics starting from the basic principle for the first law in a time-dependent form for a fixed mass system.

$$\dot{P}_E = 0 = \dot{W} - \dot{Q} + \frac{dE}{dt} \tag{1.26}$$

Outflow rate – inflow rate + change in rate of storage

Conventionally, work outflow is taken to be positive, and heat transfer inflow is taken to be positive. Starting with our generalized control volume form of our equation, we begin by substituting our intensive form for energy, e, for our generalized intensive property, η.

$$\frac{dE_{SYS}}{dt} = \oint_{CS} e\,\rho\,\mathbf{V} \cdot d\mathbf{A} + \frac{d}{dt} \int_{CV} \rho\,e\,d\forall$$

Here, the intensive form of energy, e, is taken to include the specific internal energy, u, the kinetic energy, $V^2/2g_C$, and the potential energy, gz/g_C, of the flow stream ($e = u + V^2/2g_C + gz/g_C$). Another aspect of the first law that we need to consider is with respect to the flow work. Conventionally, we suggest that the flow work term, the work the fluid does against pressure in moving into and out of the control volume is accounted for in the work term, $\dot{W}$. The intensive form of the flow work term is Pv or P/ρ. We can account for the outflow minus the inflow of the flow work by including it in the integral around the control volume.

$$\dot{W} = \oint_{CS} Pv\,\rho\,\mathbf{V} \cdot d\mathbf{A} + \dot{W}_{CV}$$

Now we have both the outflow minus the inflow of energy and the flow work accounted for in integrals around the control surface. We can add these terms together producing:

$$e + Pv = u + Pv + \frac{V^2}{2g_C} + \frac{g\,z}{g_C}$$

The property enthalpy, h, is a term, which is constructed by combining the specific internal energy, u, with the flow work, Pv ($h = u + Pv$). Enthalpy allows the accounting of both the internal energy and the flow work as a fluid moves into and out of a control volume. Understanding this significance is important, especially in the case where we have an incompressible flow and the flow work is the predominant term in the enthalpy, h. At this point, we can rewrite $e + Pv$ in the form of enthalpy combining the flow work integral with the energy transport integral and develop our control volume form of the first law of thermodynamics.

$$\dot{P}_E = 0 = \oint_{CS} \left(h + \frac{V^2}{2g_C} + \frac{g\,z}{g_C} \right) \rho \, \boldsymbol{V} \cdot d\boldsymbol{A} + \dot{W}_{CV} - \dot{Q}_{CV} + \frac{d}{dt} \int_{CV} \rho \left(u + \frac{V^2}{2g_C} + \frac{g\,z}{g_C} \right) d\boldsymbol{V}$$

$$(1.27)$$

The preceding equation can be interpreted as the production of energy is equal to zero and equal to the outflow rate minus the inflow rate of energy into the control volume (first integral term) plus the work outflow rate from the control volume less the heat transfer inflow rate to the control volume plus the change in rate of energy storage in the control volume (last integral term). However, at times, it is much easier to think about the first law in simpler terms for a single stream steady-state flow, this could be expressed simply as:

$$\dot{P}_E = 0 = \dot{M}_{OUT} \left(h + \frac{V^2}{2g_C} + \frac{g\,z}{g_C} \right) + \dot{W}_{CV} - \dot{M}_{IN} \left(h + \frac{V^2}{2g_C} + \frac{g\,z}{g_C} \right) - \dot{Q}_{CV} \quad (1.28)$$

The outflow of energy with the mass minus the inflow of the energy with the mass shown in the brackets in the preceding equation takes the place of the first integral shown. We also have the work outflow term as well as the heat transfer inflow term. It is often easier to visualize the first law for a control volume in the simplest essential terms. This form of the first law will be the starting point for the development of some very important equations in one-dimensional compressible flow.

1.2.9 The Second Law of Thermodynamics for a Control Volume

The second law of thermodynamics and the concept of entropy are often difficult concepts to master. Entropy can be thought of as a measure of unavailable energy. An example of the creation of unavailable energy is a waterfall. A waterfall takes the potential energy of water at a height and converts this height into kinetic energy through the force of gravity. As the water achieves its peak velocity, it simultaneously splashes down into the pool at the bottom of the waterfall where all this kinetic energy is converted into thermal energy. At the top of the dam, the water had the potential to do work. At the bottom of the dam, the conversion of kinetic energy into thermal energy raises the temperature of water by less than 1°F, for example, the Orville dam, which is 770 feet high. This slightly warmed water at the bottom of the dam no longer has the potential to do work. The energy content in the water is now unavailable to do work. Entropy and the second law of thermodynamics can be used to quantify the creation of unavailable energy or "exergy destruction."

A principle of the second law is that the entropy of an isolated system can stay the same or increase. An example might be taking a cup of hot water and a cube of ice, placing the ice in the cup, and insulating the system from the environment. As you know, in a short period of time, the ice will have melted and the water will have cooled and our "isolated system" will be in equilibrium. In the process, we will have lost the ability to do work, which may have occurred due to the temperature difference between the ice and water. You would never expect a cup of water at equilibrium, insulated from its surroundings to form an ice cube and warm water. Although, this later process would not violate the first law of thermodynamics, the probability of ice forming spontaneously in an isolated system out of lukewarm water would make winning the power ball lottery seem like a sure bet. It turns out we can form an isolated system by taking a system and the surroundings that it interacts with. This extended system is really the basis of the second law of thermodynamics.

The second law of thermodynamics can be put in terms of either the production methodology or in a control volume form. Fundamentally, the second law of thermodynamics is similar to the first law, except that no work terms or mechanical energy terms appear in the second law and the second law is an inequality. In an internally reversible system, the production of entropy in the system and its surroundings could, in theory, equal zero. However, in a real system, it is always greater than zero. In production form, we can write the second law for a control volume for a steady-state single stream as:

$$\dot{P}_S = \dot{M}_{OUT}s_{OUT} - \dot{M}_{IN}s_{IN} - \frac{\dot{Q}_{CV}}{T_B} \geq 0 \qquad (1.29)$$

Notice its similarities to the first law after taking into account the work and mechanical energy terms. In the second law, we use the specific entropy in place of the enthalpy or internal energy, and the heat transfer rate is divided by the temperature at the heat transfer boundary to the system. Our control volume form for the second law can be written with an interpretation similar to the single stream except for the time rate of change of storage term added at the end:

$$\dot{P}_S = \oint_{CS} s\rho \, \mathbf{V} \cdot d\mathbf{A} - \frac{\dot{Q}_{CV}}{T_B} + \frac{d}{dt} \int_{CV} \rho s \, d\mathcal{V} \geq 0 \qquad (1.30)$$

1.3 Review of Thermodynamics and the Ideal Gas Model

One-dimensional compressible flow analysis uses relationships developed from the principles of conservation of mass, the momentum principle, conservation of energy, and the second law to understand compressible flows and to solve related problems. Now that the first and second laws have been introduced, they can be used in our review of thermodynamics and the development of some useful relationships. However, before we develop some new relationships, it is useful to review the ideal gas law discussed in Section 1.1.

1.3.1 Ideal Gas Law

The conceptual basis for the ideal gas law discussed briefly in Section 1.1.3 was developed using a model of molecules imparting pressure to a wall due to their impact and resilient rebound. The rate of their collisions with the wall was based on the normal velocity of the molecules, and the distance between parallel walls and the momentum imparted was related to the normal velocity. Consequently, the pressure on a surface was related to the mean squared speed as is the energy of translation. The pressure in a given volume can be shown to be equal to:

$$P = \frac{MRT}{\cancel{V}} \tag{1.31}$$

The ideal gas law is commonly written as $P\cancel{V} = MRT$. The ideal gas model can be conceptualized based on gas molecules that take up zero volume and never collide except at walls where their rebound is spectral. Real gas molecules take up volume and, for given circumstances, exhibit a significant attraction between molecules. In the limit of decreasing pressure, the ideal gas law is highly accurate. However, at relatively low temperatures and high pressures, gases often exhibit a significant departure from the ideal gas model. Fortunately, for many engineering applications, including most compressible flow situations, air behaves very closely to an ideal gas. A generalized compressibility chart has been developed for checking the accuracy of the ideal gas law. The compressibility of a gas, Z, is given as:

$$Z = \frac{Pv}{RT} \tag{1.32}$$

Here, v is the specific volume, which can be determined for a system in equilibrium as $v = \cancel{V}/M$. Typical compressibility charts plot the compressibility, Z, as a function of the reduced pressure, P/P_C, and reduced temperature, T/T_C. Here, P_C and T_C are taken as the composite critical pressure and temperature of the gas. At relatively low temperatures and pressures around the critical pressure, the attraction forces between molecules can be substantial and Z can become much less than 1. While at relatively high pressures, the actual volume taken up by gas molecules can become increasingly large and Z can become significantly higher than 1. The composite critical temperature of air is 133 K, while the composite critical pressure is 37.7 bar. Consequently, at ambient conditions, air is a relatively low pressure moderately high temperature gas and the ideal gas model works very well.

Example 1.3: Ideal Gas Law

Wanted: Determine the density of air at 1 atmosphere and 298.15 K and at 20 atmospheres and 700 K.

Analysis:

$$Pv = RT \text{ so } v = 1/\rho \text{ and } \rho = P/RT$$

Gas constant for air:

$$R = \frac{\hat{R}}{\hat{M}} = \frac{8314.46 \dfrac{J}{kmol\,K}}{28.97 \dfrac{kg}{kmol}} = 287.0 \frac{J}{kg\,K}$$

The density of air at 1 atmosphere and 298.15 K is

$$\rho = \frac{P}{RT} = \frac{1\,atm \times 101325\,Pa}{287 \dfrac{J}{kg\,K} 298.15\,K} = 1.184 \frac{kg}{m^3}$$

The density of air at 20 atmospheres and 700 K is

$$\rho = \frac{P}{RT} = \frac{20\,atm \times 101325\,Pa}{287 \dfrac{J}{kg\,K} 700\,K} = 10.087 \frac{kg}{m^3}$$

The reduced pressure and temperature at condition one is

$$P_R = \frac{P}{P_C} = \frac{1.01325\,bar}{37.7\,bar} = 0.02688$$

$$T_R = \frac{T}{T_C} = \frac{298.15\,K}{133\,K} = 2.242$$

The generalized compressibility chart [2], which can be found in any engineering thermodynamics textbook, indicates that at these conditions, $Z \cong 1.0000$.

The reduced pressure and temperature at condition two is

$$P_R = \frac{P}{P_C} = \frac{20\,(1.01325\ \text{bar})}{37.7\ \text{bar}} = 0.5375$$

$$T_R = \frac{T}{T_C} = \frac{700\ \text{K}}{133\ \text{K}} = 5.263.$$

The generalized compressibility chart [2] indicates that at these conditions, $Z \cong 1.01$, which would mean that the density would be about 1 percent lower than what was calculated using the ideal gas law. These results would suggest that we could use the ideal gas law to accurately model air's equation of state at these conditions.

1.3.2 Specific Heats

The use of specific heats is important in one-dimensional compressible flow, as it allows a simple method to relate total temperature to static temperature and velocity. The specific heat ratio is an important parameter in calculating the speed of sound, as it relates the isentropic compressibility to the gas constant and the absolute static temperature. Specific heats at constant volume and at constant pressure relate the changes in internal energy and enthalpy to temperature.

$$C_v = \left(\frac{\partial u}{\partial T} \right)_v \tag{1.33}$$

$$C_p = \left(\frac{\partial h}{\partial T} \right)_p \tag{1.34}$$

Generally, the specific internal energy, u, is a function of both temperature and volume. However, for an ideal gas, u is a function of temperature alone [$u = u(T)$]. Also, the specific enthalpy, h, is a function of both temperature and pressure. However, for an ideal gas, h is a function of temperature alone [$h = h(T)$]. Consequently, for an ideal gas, the specific heats can be stated simply as:

$$C_v = \frac{du}{dT} \tag{1.35}$$

$$C_p = \frac{dh}{dT} \tag{1.36}$$

Also,

$$du = C_v(T)dT$$

$$dh = C_p(T)dT$$

These relationships will be very useful, as we relate velocity changes to changes in temperature.

1.3.3 Tds Equations

Relationships for the change in entropy can be developed from the first and second laws for a fixed mass system. A key assumption is that the process being looked at is internally reversible. This development begins by looking at an arbitrary fixed mass system, shown in Figure 1.4 in differential form.

The differentials, dq and dw, are inexact, as they do not lead directly to a change in a specific property. Also, largely in our development of the Tds equations, de can be taken to be du, assuming any changes in kinetic energy or potential energy are insignificant. The differential form of the first law can be written as:

$$P_E = 0 = dw - dq + du \tag{1.37}$$

The work term, dw, can be replaced with Pdv, assuming that the fluid is a simple compressible substance. Subsequently, solving for dq.

$$dq = du + Pdv$$

A second law balance around the same control volume will have terms associated with dq and du but, of course, there will be no work term.

$$P_S = -\frac{dq}{T_B} + ds \geq 0 \tag{1.38}$$

At this point in the development, the concept of an internally reversible process is put forth yielding a relationship between Tds and dq. Noting the subscript B has been omitted.

$$Tds = dq$$

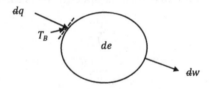

FIGURE 1.4.
Sketch of simple fixed mass thermodynamic system with differential analysis.

This equation essentially indicates that the heat transfer in a process plotted in T-s coordinates is the area under the curve. This relationship also forms the first of our two *Tds* equations when $du + Pdv$ is substituted for dq.

$$Tds = du + Pdv \qquad (1.39)$$

The differential for enthalpy provides a means to put the *Tds* equation in terms of enthalpy.

$$h = u + Pv \qquad (1.40)$$

Starting with the definition of enthalpy, the term can be differentiated using the chain rule.

$$dh = du + Pdv + vdP$$

The first two terms on the right-hand side of the equal sign are the same as the right-hand side of the initial *Tds* equation. Subtracting *vdP* from both sides an equivalent result which includes enthalpy is developed.

$$dh - vdP = du + Pdv$$

Substituting this expression into the *Tds* equation for the right-hand side, a second *Tds* equation is developed.

$$Tds = dh - vdP \qquad (1.41)$$

These two *Tds* equations that have been developed form the basis for evaluating the change in entropy for substances. These equations can be transformed into equations for ideal gases noting the respective equations for *du* and *dh*.

$$Tds = C_v(T)dT + Pdv$$

$$Tds = C_p(T)dT - vdP$$

By dividing each equation by *T* and then evaluating the second term in terms of the ideal gas law $(Pv = RT)$:

$$ds = \frac{C_v(T)dT}{T} + \frac{Rdv}{v}$$

$$ds = \frac{C_p(T)dT}{T} - \frac{RdP}{P}$$

Assuming constant specific heats and integrating compact relationships for the change of entropy for an ideal gas with a constant specific heat result:

$$s_2 - s_1 = C_v \ln\left(\frac{T_2}{T_1}\right) + R \ln\left(\frac{v_2}{v_1}\right) \qquad (1.42)$$

$$s_2 - s_1 = C_p \ln\left(\frac{T_2}{T_1}\right) - R \ln\left(\frac{P_2}{P_1}\right) \tag{1.43}$$

These relationships also lead to the isentropic relationships that will be used in many aspects of compressible flow. Taking the first equation for entropy change, a relationship for an isentropic process can be developed by setting the left-hand side to zero, $s_1 = s_2$. Taking the exponential of this new equation and solving for T_2/T_1:

$$\frac{T_2}{T_1} = \left(\frac{v_1}{v_2}\right)^{R/C_v} \tag{1.44}$$

This equation gives us a relationship for temperature change in an isentropic compression or expansion as a function of the specific volume ratio. This equation will typically be used with a fixed mass system, so the ratio of the specific volumes can be directly related to the ratio of the total volumes because the mass in a fixed mass system will stay constant. The second equation for entropy change in an ideal gas with constant specific heats can also be developed similarly to provide a relationship between pressure ratio and temperature ratio in an isentropic process.

$$\frac{T_2}{T_1} = \left(\frac{P_2}{P_1}\right)^{R/C_p} \tag{1.45}$$

It is common and useful to develop these equations in terms of the specific heat ratio, k ($k = C_p/C_v$). Taking the definition of enthalpy and replacing Pv with RT, the following relationship for enthalpy is developed.

$$h = u + Pv = u + RT$$

Taking the differential of enthalpy and the right-hand side results in:

$$dh = du + RdT$$

Using the definition of C_p and C_v for an ideal gas and making this substitution:

$$C_p dT = C_v dT + RdT$$

Eliminating the dT in this equation, a relationship for C_p and C_v and R is developed.

$$C_p = C_v + R$$

Using the definition of k and by dividing through alternatively by C_p, then C_v, the following relationships are developed.

$$\frac{R}{C_v} = k - 1 \tag{1.46}$$

$$\frac{R}{C_p} = \frac{k-1}{k} \tag{1.47}$$

Consequently, our temperature relationships, which use specific heat ratios and R, can be converted to the following relationships with k.

$$\frac{T_2}{T_1} = \left(\frac{v_1}{v_2}\right)^{k-1} \tag{1.48}$$

$$\frac{T_2}{T_1} = \left(\frac{P_2}{P_1}\right)^{k-\frac{1}{k}} \tag{1.49}$$

This last isentropic relationship will be used often in compressible flow. The two relationships can also be extended to an isentropic relationship between pressure and specific volume.

$$\left(\frac{P_2}{P_1}\right) = \left(\frac{v_1}{v_2}\right)^k \tag{1.50}$$

Another way to write this relationship is:

$$Pv^k = \text{constant} \tag{1.51}$$

These relationships will be used in the development of a number of compressible flow relationships.

Example 1.4: Nozzle Flow

A reservoir of warm high-pressure air is generated for a compressible flow experiment. The air is heated to 800 K and pressurized to 1 MPa absolute, and it has a velocity of 100 m/s. Determine the maximum velocity that can be reached if the air is expanded isentropically to atmospheric pressure.

Analysis: The most straightforward approach to solving this problem is to use the first law of thermodynamics for a steady flow single stream as written next. Flow through a nozzle will preclude any work addition and any heat transfer for an isentropic process. The potential energy terms can be eliminated for a typical gas flow and for a steady single stream the outflow and inflow of mass will be the same.

$$\dot{P}_E = 0 = \dot{M}_{OUT}\left(h + \frac{V^2}{2g_C} + \frac{g\,z}{g_C}\right) + \dot{W}_{CV} - \dot{M}_{IN}\left(h + \frac{V^2}{2g_C} + \frac{g\,z}{g_C}\right) - \dot{Q}_{CV}$$

The simplifications results in the following relationship:

$$h_1 + \frac{V_1^2}{2g_C} = h_2 + \frac{V_2^2}{2g_C}$$

Solving for the exit velocity:

$$V_2 = \left[2g_c \left(h_1 - h_2 \right) + V_1{}^2 \right]^{1/2}$$

At this point, the problem could be solved using the change in pressure to find the change in enthalpy using the ideal gas air tables. Another approach is to assume that an average C_p can be used then the equation can be rewritten as:

$$V_2 = \left[2g_c\, C_p \left(T_1 - T_2 \right) + V_1{}^2 \right]^{1/2}$$

The specific heat at constant pressure, C_p, has a value of 1.051 kJ/kg/K at 600 K, which might be an average temperature for this process. The specific heat ratio, k, at this temperature is 1.376.

$$T_2 = T_1 \left(\frac{P_2}{P_1} \right)^{k-\frac{1}{k}}$$

$$T_2 = 800\,K \left(\frac{101\,\text{kPa}}{1000\,\text{kPa}} \right)^{.376/1.376} = 427.6\,K$$

The single unknown has now been found and the exit velocity of the nozzle can be determined. However, when converting energy to velocity, it is useful to remember that J/kg has the units of m²/s². Solving for velocity:

$$V_2 = \left[2(1)1051\frac{J}{kg\,K} \left(800\,K - 427.6\,K \right) + \left(100\,m/s \right)^2 \right]^{1/2} = 890.4\,m/s$$

This problem was solved by using the first law of thermodynamics to develop a relationship for the exit velocity based on the change in temperature of the expanded gas. The change in temperature of the expanded gas was determined using one of the isentropic relationships developed earlier. The exit velocity was determined using an averaged specific heat and the corresponding specific heat ratio taken at a mean temperature.

References

[1] Fox, R.W., A.T. McDonald, and P.J. Pritchard. 2004. *Introduction to Fluid Mechanics*, 6th ed. John Wiley & Sons.

[2] Moran, M.J., H.N. Shapiro, D.D. Boettner, and M.B. Bailey. 2011. *Fundamentals of Engineering Thermodynamics*, 7th ed. John Wiley & Sons.

[3] Vincenti, W.G., and C.H. Kruger. 1965. *Introduction to Physical Gas Dynamics*. New York: Wiley.

Solved Problems

1. A steady airflow enters a duct at 50 m/s with a density of 1.2 kg/m³ through an area of 0.35 m². The duct shown in Figure 1.5 has two outlets, one outlet is 0.2 m² and has air leaving at a density of 1.1 kg/m³ at a velocity of 40 m/s. The other outlet is also 0.2 m² in area and based on the outlet pressure and temperature the density of the flow is 1.15 kg/m³. (Conservation of mass)

Wanted: Determine the velocity of the third airflow

Analysis: This problem can be solved using conservation of mass and the 1D uniform flow assumption.

$$\dot{P}_M = 0 = \oint_{CS} \rho V \cdot dA + \frac{d}{dt} \int_{CV} \rho \, dV$$

Rate of production of mass = 0 = rate of outflow minus inflow plus change in storage rate.

Because the flow is steady state, the transient terms disappear and based on the 1D flow assumption, the problem can be solved as the sum of outflow rates minus the inflow rate.

$$\dot{P}_M = 0 = \Sigma \dot{M}_{OUT} - \Sigma \dot{M}_{IN} = \rho_2 V_2 A_2 + \rho_3 V_3 A_3 - \rho_1 V_1 A_1$$

The only unknown is V_3, so solving for V_3:

$$V_3 = \frac{(\rho_1 V_1 A_1 - \rho_2 V_2 A_2)}{\rho_3 A_3} = \frac{1.2 \frac{kg}{m^3} * 50 \frac{m}{s} \times 0.35 \, m^2 - 1.1 \frac{kg}{m^3} \times 40 \frac{m}{s} \times 0.2 \, m^2}{1.15 \frac{kg}{m^3} \times 0.2 \, m^2}$$

$$= 53.04 \frac{m}{s}$$

Sketch:

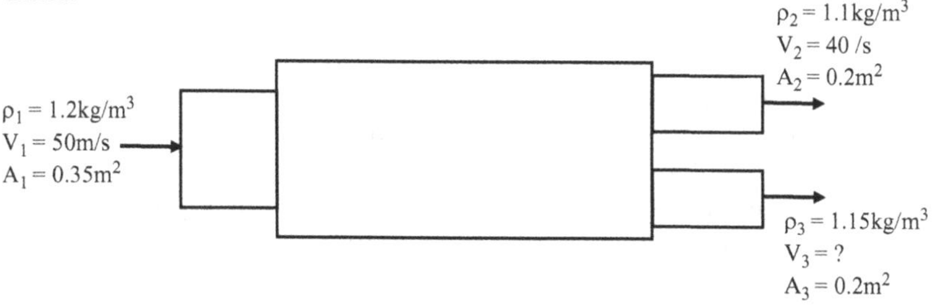

$\rho_2 = 1.1 kg/m^3$
$V_2 = 40 \, /s$
$A_2 = 0.2 m^2$

$\rho_1 = 1.2 kg/m^3$
$V_1 = 50 m/s$
$A_1 = 0.35 m^2$

$\rho_3 = 1.15 kg/m^3$
$V_3 = ?$
$A_3 = 0.2 m^2$

FIGURE 1.5
Sketch of a control volume with one inlet and two outlets.

2. A turbofan engine shown in Figure 1.6 hangs from the aft fuselage of a regional jet. The pilot holds the brakes on, so the plane will not move. The engine is set to full power just prior to the start of take-off. This problem can be considered to be steady state. (Momentum principle)

Given: The core of the engine entrains air at a rate of 45 kg/s. The turbofan's bypass ratio is 5. The gas exiting from the core jet is traveling at a velocity of 444 m/s, while the velocity of the bypass jet is 308 m/s.

Find: Determine the thrust produced by the jet.

Sketch:

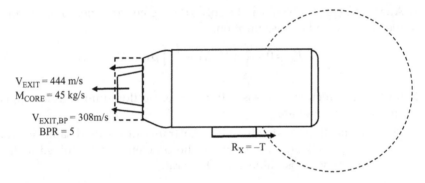

FIGURE 1.6
Sketch of an unmixed exit turbofan engine for momentum analysis.

Analysis: Momentum principle: Steady state

$$\sum F_X = F_{S_X} + F_{B_X} + R_X = \oint_{CS} V_X \, \rho \, \mathbf{V} \cdot dA + \frac{d}{dt} \int_{CV} \rho V_X \, dV$$

Evaluation: First, the problem is steady state and the transient term is dropped. Secondly, the body force is typically negligible in dealing with gas flows and can be dropped. Also, the inlet to the turbofan engine is driven by the fan. It creates a low-pressure region in front of the engine causing the surrounding air to be entrained into the engine. In order to avoid needing to evaluate both the inlet velocity contribution and the pressure force contribution to momentum, the inlet control surface is drawn out well away from the inlet where the velocity is very low and the pressure is very nearly ambient. Based on this representation, the sum of the two influences is estimated to be negligible. Consequently, the remaining items include two exhaust streams with specified velocities and flow rates. These terms can be written in simplified terms, neglecting the integrals. Also, both of these nozzles are converging, but expected to operate below a Mach number of 1, so the exit pressure can be considered to be equal to the backplane pressure. Consequently, the surface or pressure forces are negligible.

$$R_X = \Sigma \dot{M}_{OUT} V_{X,OUT} - \Sigma \dot{M}_{IN} V_{X,IN} = \dot{M}_{CORE}\left(V_{X,HOT} - V_{X,IN} + BPR \times \left(V_{X,COLD} - V_{X,IN}\right)\right)$$

Additionally, as the control volume was drawn to a point where the inlet velocity was negligible, it can be deleted leaving:

$$R_X = \dot{M}_{CORE}\left(V_{X,HOT} + BPR \times \left(V_{X,COLD}\right)\right) = 45\frac{kg}{s} \times \left(-444\frac{m}{s} + 5 \times -308\frac{m}{s}\right)$$
$$= -89280\ N = -T$$

The velocities are exiting to the left in the negative X direction, so negative signs need to be added to the velocities. This produces a negative reaction force to hold the engine in place or a positive thrust of 89,290 N or 20,071 lbf (1 lbf = 0.22481 N).

3. A regional jet flying at altitude experiences a leak and begins to lose pressure. The fuselage can be modeled as a 30- m-long cylinder of 2.74 m in diameter. The leak occurs abruptly due to a 2-cm-diameter hole. If the cabin is originally held at 80,000 Pa (P_T), how quickly would the cabin pressure drop to 57,000 Pa, the minimum cabin pressure allowed for aircraft without pressurization?

Given: The cabin temperature can be assumed to be a steady 298K, and the exit velocity of the flow leaving the cabin can be assumed to be 316 m/s. The static temperature is based on this velocity.

Wanted: Find the time for the cabin pressure to drop to 57,000 Pa given the size of the leak and the size of the fuselage. The pressure and temperature in the cabin will be the unsteady total conditions.

Analysis: This problem requires the use of unsteady conservation of mass, isentropic relations, and energy. Unsteady conservation of mass can be used to determine the time required for this depressurization. The pressure and temperature of the gas leaving will be at the static conditions.

$$\dot{P}_M = \oint_{CS} \rho V \cdot dA + \frac{d}{dt}\int_{CV} \rho\, dV\, 0 =$$

Noting that the volume of the fuselage stays constant the equation can be rewritten:

$$V\frac{d\rho}{dt} = -\rho(t) V A_{HOLE}$$

Substituting the ideal gas law:

$$V \frac{dP_T}{RT_T dt} = \frac{-P(t)}{RT} V A_{HOLE}$$

Separating variables and simplifying:

$$\frac{dP_T}{P(t)} = \frac{-T_T}{T} \frac{V A_{HOLE} dt}{V}$$

Assuming sonic flow out of the hole, there will be a set ratio between P_T and P at the exit of the hole.

$$\frac{dP_T}{P_T(t)} = \frac{-T_T}{T} \frac{P}{P_T} \frac{V A_{HOLE} dt}{V}$$

Integrating:

$$Ln\left(\frac{P_T(t)}{P_{T0}}\right) = \frac{-T_T}{T} \frac{P}{P_T} \frac{V A_{HOLE}(t-t_0)}{V}$$

Solving:

$$(t-t_0) = -\left(\frac{T}{T_T}\right)\frac{P_T}{P}\frac{V}{V A_{HOLE}} Ln\left(\frac{P_T(final)}{P_{T0}}\right)$$

The ratio T/T_T can be found from energy, noting the total enthalpy stays constant:

$$h_T = h_1 + \frac{V_1^2}{2g_c} = h_2 + \frac{V_2^2}{2g_c} = h + \frac{V^2}{2g_c}$$

Writing enthalpy in terms of $h = C_pT$

$$C_pT_T = C_pT + \frac{V^2}{2g_c}$$

Or

$$T_T = T + \frac{V^2}{2g_cC_p} \text{ yielding } T = T_T - \frac{V^2}{2g_cC_p} = 298K - \frac{\left(316\frac{m}{s}\right)^2}{2\times1005\frac{J}{kg\,K}} = 248.3K$$

The pressure ratio, P/P_T, can be related to T/T_T using our isentropic relation, Equation (1.49) with the exponent reversed:

$$\left(\frac{T_2}{T_1}\right)^{\frac{k}{k-1}} = \left(\frac{P_2}{P_1}\right) \text{ and the ratios can be combined } \frac{T}{T_T}\left(\frac{T_T}{T}\right)^{\frac{k}{k-1}} = \left(\frac{T_T}{T}\right)^{\frac{1}{k-1}}$$

The fuselage volume:

$$V = \frac{\pi}{4}D^2 L = 0.7854 \times (2.74\,\text{m})^2 \times 30\,\text{m} = 176.9\,\text{m}^3$$

The resulting time to reach the minimum pressure becomes:

$$\left(t - t_0\right) = -\left(\frac{298\,\text{K}}{248.3\,\text{K}}\right)^{\frac{1.0}{0.4}} \frac{176.9\,\text{m}^3}{316\,\frac{\text{m}}{\text{s}} \times \frac{\pi}{4} \times (0.02\text{m})^2} \text{Ln}\left(\frac{57,000\,\text{Pa}}{80,000\,\text{Pa}}\right) = 953\,\text{sec}$$

Discussion: This analysis indicates that, if the pressure inside a regional jet was allowed to decay due to a leak from a 2 cm diameter hole that the pressure could drop from 80,000 Pa to 57,000 Pa in about 16 min. However, typical jets use bleed air from one or both of the compressors on the jet engines for cooling. Consequently, as long and the thrust engines are operating, the cabin would be continuously supplied with pressurized air. Further, this problem required the use of unsteady continuity, energy, and the ideal gas law for solution.

Chapter 1 Problems

1. Determine the number of air molecules in a cubic volume with sides of 0.001 inch at standard conditions (Note 1 inch = 0.0254 m). How does this size compare with the mean free path of air?

2. Determine the maximum velocity that air at a temperature of 2000K can be expanded to. Compare this to helium at 2000K, noting $C_{P,He} = 5/2\,R$, $R = R/M$ and $M_{He} = 4.003$ kg/kmol.

3. A blower can drive an airflow of up to 5000 CFM (ft^3/min). If the duct flow area is 150 inch^2, determine the velocity in ft/s through the duct. Here, air can be treated as an incompressible flow.

4. The UND water tower tank is about 40 feet in diameter. Assume that the typical campus water usage (outflow) is 250 gallons/min. Assume that the water tank is down 10 feet. If the pump capacity

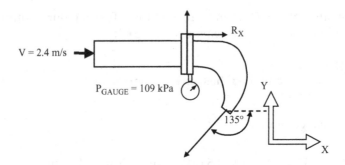

FIGURE 1.7
Schematic of 135° turning nozzle for momentum and pressure force analysis.

must be high enough to fill the tank in 8 hours under typical water usage, how large does the pump need to be in gallons per minute. Note 7.48 gallons $\cong$ 1 ft³.

5. A rocket engine exhausts gas at 1350 m/s at a mass flow rate of 250 kg/s. (a) Determine the thrust if the nozzle is perfectly expanded ($P_{EXIT} = P_{ATM}$). (b) Determine the thrust if $P_{EXIT} = 100,000$ Pa, $P_{ATM} = 50,000$ Pa, and $A_{EXIT} = 0.21$ m².

6. A pipe carrying water includes an elbow and nozzle system that turns the flow 135° as shown in Figure 1.7. The pipe flow area is 0.0125 m², while the jet area is 0.002 m². The pipe inlet velocity is 2.4 m/s and the pressure at the inlet is 109,000 Pa gauge. Find the force in the X and Y directions on the flange. You can assume that the density of water is 992 kg/m³.

7. A large centrifugal pump is used to supply 500 gallons/min of water over a pressure rise of 250,000 Pa. You can assume that the water has a density of 992 kg/m³. Note that 268.2 gallons $\cong$ 1 m³. Determine the input power to the pump required if the pump efficiency is 80 percent.

8. A large thruster expands air with a total temperature of 2400K and a total pressure of 6.61 MPa to an ambient pressure of 100 kPa. You can assume that the expansion process is isentropic and steady state and the average Cp for air is 1200 J/kg/K. Determine the exit static temperature and the exit velocity. Also if the flow rate is 200 kg/s, size the exit of the nozzle.

9. Calculate the mass of air in a room that is 3 m by 4 m by 2.5 m high. You can assume that the pressure in the room is 1 bar (100,000 Pa) and the temperature is 295K.

10. Hydrogen, H_2, is stored at 135 bar and 298K. At these conditions, the generalized compressibility chart shows that Z = 1.095. Note that Z = Pv/RT. Find the mass of H_2 in a 1 m³ tank.

2

Isentropic Flow

This chapter will introduce the fundamentals of isentropic flow. In many situations, air and other gases behave closely to an isentropic flow. Isentropic flow is, by nature, frictionless, so this treatment will not deal with boundary layers. This assumption is largely valid when boundary layers are very thin compared to the macroscopic dimensions of the flow. Isentropic flow does not include flows with any form of shock waves nor flows with any heat addition or work extraction. In this chapter, the difference between stagnation or total conditions and static conditions will be examined in both stationary and moving reference frames. The dynamics of sound waves through compressible media will be explored with the development of a relationship for the speed of sound. The conservation of energy will be the starting point in the development of one-dimensional isentropic equations, which will also rely on the isentropic relationship established earlier. At this point, the relationships that have been developed for isentropic flow will be applied to flow in converging nozzles and then flow in converging–diverging nozzles.

2.1 Stagnation and Static Conditions

This section will introduce the difference between static and stagnation conditions as well as stationary and moving reference frames. Stagnation conditions in enthalpy, temperature, and pressure are also called total conditions. One framework used in the application of these conditions is a stationary reference frame. In a stationary reference frame, there is often a compressor, blower, or fan, which generates the total condition and some type of duct or test section where the static condition is applicable. Consider a wind tunnel in a blow down configuration shown in Figure 2.1. The air is initially entrained through a blower. The purpose of the blower is to do work on the air to increase its pressure in order to generate the required velocity in the duct or test section. This situation can be examined in a conceptual wind tunnel. In our conceptual wind tunnel, air has been entrained into a blower. Downstream from the blower, air is directed through a series of diffusers and flow conditioning. The analysis begins at the entrance of the nozzle where the flow is examined between the low-speed entrance and the high-speed test section. At location 1, the flow is stagnated and the inlet total temperature is acquired as is the inlet total pressure. At location 2, the exit static pressure is acquired.

Sketch:

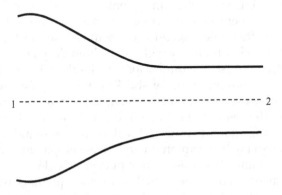

FIGURE 2.1
Schematic of nozzle and test section of a wind tunnel.

Example 2.1: Calculation of Velocity Using the First Law and Isentropic Relationships

Given:

$$P_{TOTAL} = 102,500 \, Pa, P_{STATIC} = 100,000 \, Pa, T_{TOTAL} = 301 \, K$$

Analysis: Flow leaves the flow conditioning section and enters the nozzle at a low velocity. At the entrance to the nozzle, the inlet total pressure is measured with a total pressure probe, which stagnates the flow (brings it to a stop) and senses the total pressure. At the entrance to the nozzle, a total temperature probe is used to acquire in the inlet total temperature as well. This probe also stagnates the flow over the probe in acquiring the total temperature. Downstream, as the flow is forced through the contraction by the pressure difference between the inlet total and the test section static pressure, the flow accelerates then maintains its velocity through the test section. The first law can be applied between points 1 and 2, assuming a steady uniform flow with a single stream.

$$\dot{P}_E = 0 = \dot{M}_{OUT}\left(h_2 + \frac{V_2^2}{2g_c} + \frac{g\,z_2}{g_c}\right) + \dot{W}_{CV} - \dot{M}_{IN}\left(h_1 + \frac{V_1^2}{2g_c} + \frac{g\,z_1}{g_c}\right) - \dot{Q}_{CV}$$

The equations can be simplified by recognizing there is no work extraction or heat addition and the potential energy terms in this gas flow are negligible. The mass flow rate terms can also be eliminated recognizing that the flow is a steady single stream in this present problem. The steady-state control volume energy equation reduces to the following:

$$h_2 + \frac{V_2^2}{2g_C} = h_1 + \frac{V_1^2}{2g_C} = h_T \tag{2.1}$$

This equation essentially indicates that the total energy, in terms of the enthalpy plus kinetic energy of the flow stream, remains constant. In other words, the total enthalpy (h_T) is a constant when the flow remains isentropic (no friction, shocks, heat addition, or work extraction). Typically, in air, the flow can be taken as an ideal gas with a constant specific heat. Using a constant specific heat at constant pressure (C_P), the preceding equation can be rewritten in terms of temperatures.

$$T_2 + \frac{V_2^2}{2g_C C_P} = T_1 + \frac{V_1^2}{2g_C C_P} = T_T$$

In this problem, the inlet total temperature is given, so the static temperature in the test section is unknown. However, the problem indicates that the inlet total pressure and test section static pressures are known. Consequently, assuming an isentropic flow, the inlet total pressure to test section static pressure can be related to the inlet total temperature to test section static temperature through the isentropic relationship.

$$T_2 = T_T \left(\frac{P_2}{P_T} \right)^{k-\frac{1}{k}} \tag{2.2}$$

As the inlet total temperature is 301 K, the specific heat ratio, k, will assumed to be 1.4.

$$T_2 = 301\,\text{K} \left(\frac{100\,\text{kPa}}{102.5\,\text{kPa}} \right)^{0.4/1.4} = 298.884\,\text{K}$$

In the present case, three significant digits are used past the decimal point in the temperature. However, in the current case, the relative precision in the temperature needs to be related to the relative precision in the pressure. A typical resolution of static pressure relative to the total pressure in the current case would be about 2.5 Pa. This precision in pressure makes the third digit in temperature significant. Velocity is solved from the energy equation. Note that it is important to put C_P in terms of J/kg/K.

$$V_2 = \left[2g_C\, C_P \left(T_T - T_2 \right) \right]^{1/2} \tag{2.3}$$

$$V_2 = \left[2(1)1005\frac{J}{\text{kg K}} \left(301\,\text{K} - 298.884\,\text{K} \right) \right]^{1/2} = 65.2\,\text{m/s}$$

Discussion: The present example highlights stagnation and static conditions in a stationary control volume. The stagnation or total pressure and total temperature were determined upstream using a total pressure probe and a total temperature probe that stagnated (stopped) the flow and read the total conditions. The change in velocity can be determined from the first law for a closed system. The velocity change can be related to the temperature change. However, measuring a true static temperature can be challenging. Measuring a static pressure is typically straightforward, and the change in the static temperature can be related to the change in the static pressure for an isentropic flow. This stationary control volume is common and important for flows in wind tunnels, flows in the static sections of turbomachinery, flows in ducting, and many other industrial settings. Yet, at times, our datum of interest is moving like a car, an airplane, or even a rotating section in a turbomachine, and a moving reference frame offers a better approach for analysis.

The most common moving reference frame that will be examined in compressible flow problem-solving is a moving reference frame around a moving vehicle. In this type of moving reference frame, the air is typically at rest with the surroundings and the relative speed results from the movement of the vehicle through the air. However, an aircraft can encounter winds aloft, and automobiles can encounter head and tail winds, which can complicate flow in moving reference frames. Rotating reference frames in turbomachinery most typically have an absolute velocity moving into the rotating framework, which needs to be considered. In this treatment of a moving reference frame, the air will be assumed to be at rest with the surroundings.

A simple example of a moving reference frame is an automobile moving through air. The blunt front of the car accelerates the air up to the speed of the car and creates a stagnation region in the front portion of the automobile. The air at the stagnation region is at the total condition relative to the car. The resulting high-pressure air accelerates around the car and attempts to fill the region behind the car. This acceleration of air around the car creates a low-pressure region as the air speeds up to move around the car. A passenger in the car might hold their hand out and feel the stagnation pressure of the air as their hand accelerates the air moving by up to the speed of the automobile. A stationary observer along the side of the road initially feels the push of higher pressure air in the leading edge region of the car and then the suction of the air filling around the back of the car as it moves by.

The focus of the moving reference frame example will be an airplane, like a Cirrus SR22 flying at 180 knots on a standard day at 5000 m in altitude. At 5000 m on a standard day, the pressure is estimated to be 54,050 Pa and the local temperature is estimated at 255.68 K. This velocity is equal to 92.6 m/s. Determine the stagnation temperature and pressure experienced by a stagnation region such as the leading edge of a wing.

Example 2.2: Calculation of Total Pressure Using a Moving Reference Frame

Given:

$$P_{STATIC} = 54,050 \text{ Pa}, T_{STATIC} = 255.68 \text{ K}, V = 92.6 \text{ m / s}$$

Wanted: Determine the total temperature and pressure experienced by the leading edge of the wing.

Analysis: The stagnation temperature relative to the moving reference frame can be calculated from the energy equation.

$$T_T = T_{STATIC} + \frac{V_1^2}{2g_C C_p}$$

$$T_T = 255.68 \text{ K} + \frac{\left(92.6 \dfrac{\text{m}}{\text{s}}\right)^2}{2(1)1005 \dfrac{\text{J}}{\text{kg K}}} = 259.95 \text{ K}$$

The total pressure can be determined from our isentropic relationship, but revised from the version, which yields total temperature.

$$P_T = P_S \left(\frac{T_T}{T_S}\right)^{k/k-1} \tag{2.4}$$

$$P_T = 54,050 \text{ Pa} \left(\frac{259.95 \text{ K}}{255.68 \text{ K}}\right)^{1.4/0.4} = 57,273 \text{ Pa}$$

Although, in this case, the total temperature and pressure were determined from static conditions rather than the other way around. The first law, conservation of energy, and the isentropic relationships were used to solve this problem.

A key idea to remember regarding moving and stationary reference frames is that the static condition is always taken as the condition at rest with the air. This static condition can be visualized as an observer who is at rest with the air, whether it is flowing in a wind tunnel similar to a stationary reference frame or if it is not moving, as is sometimes the case with a moving reference frame. The total condition occurs when the flow is isentropically stagnated on the object of interest or with a total pressure or temperature probe.

This section reviewed conservation of energy in the context of static and stagnation or total properties. The total condition was found when the fluid was stagnated with respect to the leading edge of an object in the reference frame. The first law requires that energy be conserved, and in the absence of heat transfer and work interactions, the total enthalpy of a stream flow will remain constant. This result is very useful in an isentropic flow, which allows relating total to static temperature changes to total to static pressure changes and vice versa. Isentropic Mach number relationships will be developed from these results.

2.2 The Speed of Sound in a Gas or Compressible Media

Pressure waves in gases move at the speed of sound, which makes sense because sound waves are really pressure waves. If you have ever observed lightning, then counted one thousand one, one thousand two, and so forth, to determine the distance to the lightning strike, you know that sound travels about a mile in 5 s. Earlier, the speed of molecules in air was discussed, and the mean square of the molecular velocity was related to the absolute temperature. You might have wondered if the speed of sound in air is related to the velocity of the air molecules and as a result, the temperature. In this section, pressure waves in a compressible media will be analyzed generally and then applied to an ideal gas.

The analysis of the speed of sound or the velocity of a pressure wave in a compressible media will begin using a long constant area tube filled with gas at rest as shown in Figure 2.2. At the start of the tube on the left-hand side of the following drawing, there is a piston contained within the tube with a diameter essentially equal to the inside of the tube. At some instant, the piston is abruptly given a constant velocity to the right and a compression wave propagates before the piston. The pressure wave moves into the undisturbed gas at the speed of sound. The gas on the left side of the compression wave moves at the speed of the piston. In this region, the gas is at a slightly higher pressure, $P + dP$, resulting in a slightly higher density, $\rho + d\rho$, than the undisturbed gas to the right at conditions, P and ρ.

The pressure or compression wave, which is propagating to the right at the speed of sound, a, can be analyzed using a moving reference frame. Initially, the control volume form of conservation of mass can be written to give guidance on the terms needed to conduct the analysis. The tube along which the compression wave propagates has a constant area, A. At this point, boundaries for a simple moving reference frame are placed around the wave and conservation of mass is applied.

$$\dot{P}_M = 0 = \oint_{CS} \rho \mathbf{V} \cdot d\mathbf{A} + \frac{d}{dt} \int_{CV} \rho \, d\mathcal{V} \tag{1.17}$$

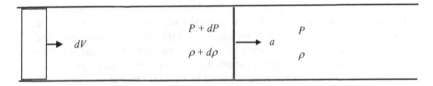

FIGURE 2.2
Schematic of a constant area tube with piston and compression wave propagating at the speed of sound.

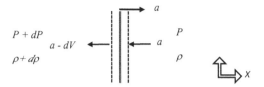

FIGURE 2.3
Schematic of control volume around compression wave propagating in X.

The time-dependent term is zero, as this analysis is assumed to be steady state as flow enters and leaves the moving control volume. The remaining integral term represents the outflow minus the inflow of mass out of and into the control volume. The integral term can handle variations in density and velocity. However, for this analysis, the assumption will be that the flow is uniform and one dimensional at the inlet and the outlet (Figure 2.3).

The compression wave moves into the still gas at the speed of sound, a, and the control volume shown with dashed lines moves with it. The gas at rest moves into the control volume at the initial condition at the speed of sound, a, and leaves the control volume at the new conditions at a velocity equal to the speed of sound, a, less the incremental velocity of the gas, dV.

$$\dot{P}_M = 0 = \dot{M}_{OUT} - \dot{M}_{IN} = (\rho + d\rho)(a - dV)A - \rho a A$$

After cancelling similar terms, a relationship can be developed between the differential velocity and the differential density.

$$\rho dV = d\rho a \tag{2.5}$$

In addition to conservation of mass, the momentum principle is also applied in this development to analyze the balance between pressure forces and inertial forces. Writing down the control volume form of the momentum principle, the time-dependent term (last term on the right) can be eliminated in this steady-state problem. Additionally, there is nothing physical that an attachment can be made to, so no reaction force is possible. The body forces in one-dimensional compressible flow are typically negligible. This equation can also be written in terms of one-dimensional uniform flow, as no variation in velocity or other properties is expected.

$$\sum F_X = F_{Sx} + F_{Bx} + R_X = \oint_{CS} V_X \rho V \cdot dA/g_c + \frac{d}{dt} \int_{CV} \rho V_X dV/g_c \tag{1.25}$$

Accounting for the simplifications discussed, the momentum equation applied to the control volume surrounding the moving compression wave becomes:

$$(P + dP)A - PA = -(a - dV)\rho a A/g_c - (-a)\rho a A/g_c$$

This equation is arrived at by noting that the mass flow into and out of the control volume is equal to $(\rho \, a \, A)$. Canceling duplicate terms and area the momentum, the equation becomes:

$$dP = \rho a dV / g_c$$

If the relationship from conservation of mass is substituted and rearranged, the resulting relationship for speed of sound can be written as:

$$a^2 = \frac{dP}{d\rho} g_c \qquad (2.6)$$

This relationship is applicable to other media in addition to ideal gases. This relationship can be applied to finding the speed of sound in liquids and solids. The bulk modulus, β_S, is defined as:

$$\beta_S = \rho \left(\frac{dP}{d\rho} \right)_S \qquad (2.7)$$

Consequently, the speed of sound of a liquid or solid could be determined from:

$$a = \sqrt{\frac{\beta_S}{\rho} g_c} \qquad (2.8)$$

The bulk modulus could be reported in terms of the isentropic value or the isothermal value. However, in solids the values are reasonably similar. The bulk modulus of structural steel is 160 GPa and its density is 7860 kg/m^3. The speed of sound in structural steel can be calculated to be:

$$a = \sqrt{\frac{160\,\text{GPa}}{7860 \dfrac{\text{kg}}{\text{m}^3}} (1)} = 4512 \frac{\text{m}}{\text{s}}$$

The speed of sound in a gas can be developed in a manner similar to the speed of sound in a solid. However, there is a significant difference between the derivative of pressure with respect to density at isothermal conditions (T) versus isentropic conditions (S). The relationship for the speed of sound in a gas can be developed from the general relationship for speed of sound in a compressible media.

$$a = \sqrt{\frac{dP}{d\rho} g_c} \qquad (2.9)$$

Sound waves tend to persist well and are not very dissipative. Computational methods used for the prediction of sound need to use non-dissipative calculation methods such as central differencing methods in order to accurately capture the level of noise. The persistence of sound waves is also very apparent if you have ever used parabolic sound reflectors. In a large room with parabolic sound reflectors at the two ends, a person at one end can talk in a normal voice and be heard clearly by another person nearly a football field in distance away. These observations suggest that the derivative of pressure with respect to density is isentropic. Previously, when isentropic relationships were developed, the following equation was developed:

$$Pv^k = constant$$

The specific volume, v, is the reciprocal of density, ρ, so the same relationship can be written as:

$$P = C\rho^k \tag{2.10}$$

The isentropic derivative of pressure with respect to density becomes:

$$\frac{dP}{d\rho} = kC\rho^{k-1} = k\frac{P}{\rho}$$

This relationship can be simplified using the ideal gas law. Noting $\rho = P/RT$, the derivative above can be replaced with kRT. The speed of sound in a gas can be written simply as:

$$a = \sqrt{kg_cRT} \tag{2.11}$$

This relationship can be used to determine the speed of sound of air at standard conditions ($T = 288.15$ K):

$$a = \sqrt{1.4(1)287\frac{J}{kg\,K}288.15K} = 340.3\frac{m}{s}$$

The speed of sound in air at standard conditions can also be determined in English units:

$$a = \sqrt{1.4\ 32.174\frac{lbm\ ft}{lbf\ s^2}53.343\frac{ft\ lbf}{lbm°R}518.67°R} = 1116.4\frac{ft}{s} = 340.3\frac{m}{s}$$

2.3 One-Dimensional Isentropic Mach Number Relationships

Many compressible flow problems can be solved relatively easily with dimensional variables. However, many situations in compressible flow are more easily understood and solved in terms of the Mach number. The Mach number, M, is equal to the local gas velocity over the local speed of sound ($M = V/a$). The Mach number is typically based on local absolute conditions, but in certain situations, it can depend on the reference frame that is being used. The development of isentropic Mach number relationships can begin by converting the relationship between total and static temperature.

$$T_T = T + \frac{V^2}{2 g_c \, C_p} \tag{2.2}$$

This statement is derived from energy, and it can be interpreted as a statement that the total temperature of a compressible flow stays constant as long as work is not being added or extracted and as long as thermal energy is not being added or extracted. Earlier, a relationship for R/C_p was developed (Equation 1.46). Solving for C_p:

$$C_p = \frac{k R}{k-1}$$

Substituting this relationship into the total temperature equation and factoring out static temperature.

$$T_T = T \left[1 + \frac{(k-1)V^2}{2 g_c \, k R T} \right]$$

The definition of the speed of sound ($a = [k \, g_c \, R \, T]^{1/2}$) in a gas is evident in the denominator. Making this substitution, the relationship becomes:

$$T_T = T \left[1 + \frac{(k-1)V^2}{2 a^2} \right] = T \left[1 + \frac{(k-1)}{2} M^2 \right] \tag{2.12}$$

Although this relationship is now in a different form, which suggests that the ratio of T_T/T is a function of Mach number and specific heat ratio, the relationship originates in the energy equation. The energy equation shows that, in a flow, the total enthalpy or similarly the total temperature stays constant as long as work or thermal energy is not extracted nor added. This equation provides a useful means to determine the static temperature from total temperature when the local Mach number is known.

This equation shows that there will be a drop in the static temperature with an increase in Mach number or equivalently an increase in velocity. The properties of a high-speed flow are dependent on the static conditions. Static temperature and pressure are needed to find the density. Static temperature is needed to find the speed of sound, the absolute viscosity, and the thermal conductivity of a gas. However, determining the static temperature of a flow directly is much more difficult. The most common method to determine the temperature of a gas is through the use of an immersion probe. An immersion probe will sense the recovery temperature. The recovery temperature of a probe in a high-speed gas is typically between the total temperature and the static temperature. The actual local temperature a probe will sense is dependent on the design of the probe and the flow over the probe. Temperature probes designed for high-speed flows typically slow down the flow to immerse the probe in gas that is closer to the total temperature. Consequently, the preceding equation is often used to determine the static temperature from the total temperature and the local Mach number.

The local Mach number can be related to the local total pressure and local static pressure as long as the flow is isentropic. An equation that relates total pressure to static pressure and Mach number can be developed using the relationship for temperature and the isentropic relationships developed earlier.

$$\frac{T_T}{T} = \left[1 + \frac{(k-1)}{2}M^2\right] \tag{2.13}$$

$$\frac{P_T}{P} = \left[\frac{T_T}{T}\right]^{k/k-1} = \left[1 + \frac{(k-1)}{2}M^2\right]^{k/k-1} \tag{2.14}$$

The preceding equation allows either total pressure to be determined from static pressure, or static pressure can be determined from total pressure if the local Mach number is known. The experimental determination of both total and static pressures is typically much easier than the experimental determination of static or total temperature. The static pressure of a flow adjacent to a surface can typically be sensed using a static pressure port. A static pressure port is typically a small hole with sharp edges, which is normal to a surface. A total pressure is typically sensed by a probe that stagnates the flow and thereby senses the stagnation pressure (Figure 2.4). A simple installation is shown as follows.

The preceding equation that relates the total to static pressure ratio to a relationship for the Mach number can be inverted to determine Mach number from the total to static pressure ratio.

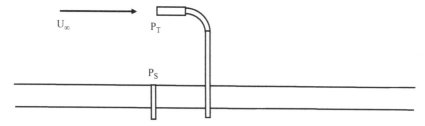

FIGURE 2.4
Schematic of total pressure probe and static pressure tap used to sense dynamic pressure.

$$M = \left\{ \left[\left(\frac{P_T}{P} \right)^{k-\frac{1}{k}} - 1 \right] \frac{2}{k-1} \right\}^{\frac{1}{2}} \tag{2.15}$$

The preceding equation might be used for finding the local Mach number. Assume that a pressure reading has been taken of the total to static pressure difference as measured by the total pressure probe and static pressure tap pictured. The reading in this case is 66,700 Pa. The local static pressure and temperature are assumed to be standard day conditions at sea level with the static temperature (T) of 288.15 K and the local static pressure (P) of 101,325 Pa. The use of the *Bernoulli equation* will result in a notable *inaccuracy in high-speed flows*. The simple constant density Bernoulli equation can be given by:

$$P_T = P + \frac{\rho V^2}{2g_c} \tag{2.16}$$

The velocity can be determined solving for V.

$$V = \sqrt{\frac{2g_c (P_T - P)}{\rho}} \tag{2.17}$$

The density needs to be determined from the ideal gas law.

$$\rho = \frac{P}{RT} = \frac{101,325\,\text{Pa}}{287 \frac{\text{J}}{\text{kg K}} \, 288.15\,\text{K}} = 1.2252 \frac{\text{kg}}{\text{m}^3}$$

Solving for velocity.

$$V = \sqrt{\frac{2g_c (P_T - P)}{\rho}} = \sqrt{\frac{2(1)66,700\,\text{Pa}}{1.2252 \frac{\text{kg}}{\text{m}^3}}} = 330.0\,\text{m}/\text{s}$$

The use of the Mach number relationship that was previously developed is an accurate approach in high-speed flows. Initially, Mach number can be found from the total to static pressure ratio. The specific heat ratio, k, is 1.4 for air at these conditions.

$$M = \left\{ \left[\left(\frac{P_T}{P} \right)^{k-\frac{1}{k}} - 1 \right] \frac{2}{k-1} \right\}^{\frac{1}{2}} = \left\{ \left[\left(\frac{101,325\,\text{Pa} + 66,700\,\text{Pa}}{101325\,\text{Pa}} \right)^{\frac{0.4}{1.4}} - 1 \right] \frac{2}{0.4} \right\}^{\frac{1}{2}} = 0.882$$

The speed of sound needs to be determined next.

$$a = \sqrt{kg_c RT} = \sqrt{1.4(1)287 \frac{J}{kg\,K} 288.15\,K} = 340.3 \frac{m}{s}$$

The velocity will simply be the Mach number times the speed of sound.

$$V = Ma = 0.882 * 340.3 \frac{m}{s} = 300\,m/s$$

Determining the velocity from the Bernoulli equation resulted in a 10 percent error, and this error will become larger at higher velocities.

2.4 Converging Nozzles

The application of compressible flow equations in solving flow through nozzles is one of the most direct and useful applications of compressible flow. Whether or not we are calculating flow through a rocket nozzle, the annular stages of a gas turbine engine or assessing compressible flow through a valve in a compressed air system, these equations are very useful. In applying these compressible flow equations to a nozzle, a one-dimensional uniform flow is assumed. In many situations, this assumption is valid as boundary layers are thin and flow has been accelerated enough to be essentially uniform. One issue with flow in a converging nozzle is the maximum Mach number that can be reached is a Mach number of one. Consider flow through a nozzle shown next in Figure 2.5. Initially, the back pressure (P_B) at a, and the inlet total pressure, P_T, are the same and no flow is entrained through the nozzle. As the back pressure at the exit of the nozzle is dropped, flow is entrained through the nozzle and pressure drops, b, as the velocity rises. As the backplane pressure is further lowered, c, more flow is entrained through the nozzle, the velocity rises in the nozzle, and the static pressure within the nozzle further lowers. At some back pressure, d, the flow at the minimum area of the nozzle will become choked or sonic. At this point, further reductions in the back pressure will not increase the flow rate through the nozzle. As sonic flow is reached, pressure waves cannot propagate upstream to enhance a faster flow through the nozzle. As pressure is lowered further, e, the flow remains choked and the velocity and pressure distribution through the nozzle remains fixed. At the exit of the nozzle, a phenomenon called expansion fans develops, allowing the flow to expand further to the back pressure of the surroundings.

The flow through the nozzle can also be visualized in terms of the flow rate, as P_B/P_T decreases, the mass flow rate, $\dot{m}$, increases, as presented in

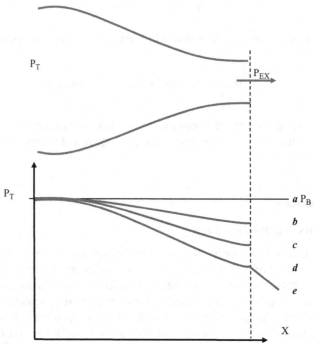

FIGURE 2.5
Sketch of converging nozzle with figure of related static pressure distributions.

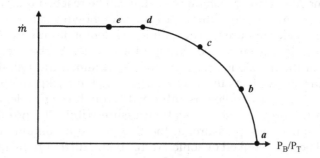

FIGURE 2.6
Plot of mass flow rate versus P_B/P_T for both choked and non-choked conditions.

Figure 2.6. You can correlate the increase in the mass flow rate with the decrease in the back pressure to total pressure ratio. The mass flow rate continues to increase for a given total pressure until choking conditions occur at *d*. At this point, any further decrease in the back pressure will not affect the mass flow rate.

Example 2.3: Converging Nozzle

There are two basic situations for a converging nozzle. The nozzle can be choked, which means flow is sonic at the exit of the nozzle where the area is a minimum, or flow is subsonic at the exit. Consequently, in dealing with a compressible flow problem, in a converging nozzle, one of the first steps is to check to see whether the nozzle is choked or whether it is subsonic.

Given: A converging nozzle shown in Figure 2.7 with a 1.25 cm² exit area has an absolute inlet total pressure of 800 kPa. Initially, the back pressure on the nozzle is 500 kPa. Later, the back pressure is dropped to atmospheric pressure 101.325 kPa. The inlet total temperature to the nozzle is 325 K.

Wanted: Find the exit Mach number, the exit velocity, and the mass flow rate for the air leaving the nozzle for the two conditions.

Sketch:

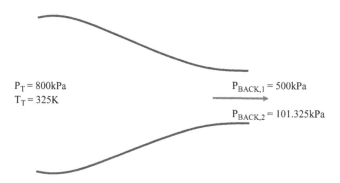

$P_T = 800\text{kPa}$
$T_T = 325\text{K}$

$P_{BACK,1} = 500\text{kPa}$

$P_{BACK,2} = 101.325\text{kPa}$

FIGURE 2.7
Schematic of a converging nozzle.

Analysis: Initially, the nozzle flow must be determined to be subsonic or choked. Based on the isentropic relationship between total to static pressure ratio and Mach number, the back pressure to choke the flow can be determined. k will be assumed to be 1.4.
At a Mach number of 1.

$$\frac{P_T}{P} = \left[1 + \frac{(k-1)}{2} M^2\right]^{\frac{k}{k-1}} = \left[1 + \frac{(.4)}{2} 1^2\right]^{\frac{1.4}{0.4}} = 1.8929$$

$$P_{CHOKED} = \frac{P_T}{\dfrac{P_T}{P}} = \frac{800 \text{ kPa}}{1.8929} = 422.6 \text{ kPa}$$

This means that, in the initial case, the flow is subsonic. However, in the second case, the flow will be choked. In the second case, the velocity will be sonic at the exit, and this will restrict the minimum pressure to condition d in the pressure versus distance figure aforementioned. In the present case, the exit pressure will be held at 422.6 kPa and the velocity and mass flow rate will be based on the choking conditions at the exit of the nozzle.

CASE 1

P_{BACK} = 500 kPa

The flow with a back pressure of 500 kPa was determined to be subsonic. The analysis can begin by determining the exit Mach number based on the total to static, or in this case, exit pressure ratio.

$$M = \left\{ \left[\left(\frac{P_T}{P} \right)^{k-\frac{1}{k}} - 1 \right] \frac{2}{k-1} \right\}^{\frac{1}{2}} = \left\{ \left[\left(\frac{800 \text{ kPa}}{500 \text{ kPa}} \right)^{0.4/1.4} - 1 \right] \frac{2}{0.4} \right\}^{\frac{1}{2}} = 0.8477$$

The exit velocity must also be found. The exit velocity can be determined by multiplying the Mach number by the speed of sound. The speed of sound depends on the local static temperature. The static temperature can be determined from the Mach number relationship determined from the energy equation.

$$\frac{T_T}{T} = \left[1 + \frac{(k-1)}{2} M^2 \right]$$

$$T = \frac{T_T}{\left[1 + \frac{(k-1)}{2} M^2 \right]} = \frac{325 \text{ K}}{\left[1 + \frac{0.4}{2} 0.8477^2 \right]} = 284.2 \text{ K}$$

The speed of sound at this temperature is:

$$a = \sqrt{kg_c RT} = \sqrt{1.4 (1) 287 \frac{J}{\text{kg K}} 284.2 \text{ K}} = 337.9 \frac{\text{m}}{\text{s}}$$

The velocity is determined to be:

$$V = M a = 0.8477 \times 337.9 \frac{\text{m}}{\text{s}} = 286.4 \text{ m / s}$$

At this point, the mass flow rate can be calculated by assuming a one-dimensional uniform flow.

$$\dot{m} = \rho V A$$

The area of the nozzle was given and the velocity has been determined. The remaining variable that needs to be calculated is the density, which can be determined accurately for these conditions using the ideal gas law. However, the density is a local property that is based on the local static conditions.

$$\rho = \frac{P}{RT} = \frac{500,000 \text{ Pa}}{287 \frac{J}{kg \text{ K}} \, 284.2 \text{ K}} = 6.130 \frac{kg}{m^3}$$

The resulting mass flow rate is determined to be:

$$\dot{m} = \rho V A = 6.13 \frac{kg}{m^3} \times 286.4 \frac{m}{s} \times 0.000125 \text{ m}^2 = 0.219 \frac{kg}{s}$$

CASE 2

$P_T = 101.325$ kPa

This case has already been established to be choked with $M = 1$ and with an exit pressure of 422.6 kPa. The exit velocity will also be the speed of sound since the nozzle is choked at this time. Consequently, the static temperature must be determined in order to calculate the speed of sound.

$$T = \frac{T_T}{\left[1 + \frac{(k-1)}{2} M^2\right]} = \frac{325 \text{ K}}{\left[1 + \frac{0.4}{2} 1^2\right]} = \frac{325 \text{ K}}{1.2} = 270.8 \text{ K}$$

$$a = \sqrt{kg_c RT} = \sqrt{1.4\,(1)\,287 \frac{J}{kg \text{ K}} \, 270.8 \text{ K}} = 329.9 \frac{m}{s} = V$$

The mass flow rate can be determined similarly with Case 1 by first calculating the density and then the mass flow rate using the one-dimensional flow assumption. Note that the static properties at the exit are used to determine both the speed of sound and the density. The backplane pressure is not used.

$$\rho = \frac{P}{RT} = \frac{422{,}600\,\text{Pa}}{287\,\dfrac{J}{\text{kg K}}\,270.8\,\text{K}} = 5.437\,\frac{\text{kg}}{\text{m}^3}$$

$$\dot{m} = \rho\,V\,A = 5.437\,\frac{\text{kg}}{\text{m}^3} \times 329.9\,\frac{\text{m}}{\text{s}} \times 0.000125\,\text{m}^2 = 0.224\,\frac{\text{kg}}{\text{s}}$$

Discussion: Based on this calculation, the mass flow rate has increased. However, the increase in the mass flow rate is only a little more than 2 percent, even though the Mach number has increased from 0.8477 to 1.0. This very small change in flow rate in spite of a moderately large change in Mach number is an interesting aspect of flow in the range near the speed of sound. Another aspect of this calculation was the assumption of one-dimensional uniform flow. How accurate is this assumption? The assumption in this particular case is quite reasonable. Due to the strong acceleration expected in the nozzle, the high pressure, and the high velocity through the nozzle, the boundary layer inside the nozzle can be expected to be very thin. This assumption is often valid at moderate to higher Reynolds numbers, but at lower Reynolds numbers, the displacement thickness of local boundary layers should be checked.

2.5 Flow in Varying Area Ducts

When a flow is incompressible, that is, when density stays constant or nearly constant like it does with water and other fluids, flow in varying area ducts is straightforward to understand. Based on a one-dimensional uniform flow assumption, the relationship between mass flow rate, density, and velocity is simply:

$$\dot{m} = \rho\,V\,A \text{ or } \rho_1\,V_1\,A_1 = \rho_2\,V_2\,A_2$$

As density is constant in an incompressible flow, the velocity of a uniform flow will vary inversely with area. However, when we deal with compressible flow, the variation of velocity with area is not so clearly defined. Understanding the effects of area change in compressible flow is best guided through the development of relationships for the variation of velocity and area in compressible flow. Relating the influence of area change on velocity and pressure can be developed using the momentum principle and conservation of mass. Useful relationships that guide our understanding will be developed next and initially shown in Figure 2.8.

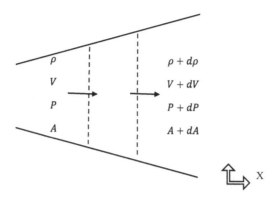

FIGURE 2.8
Schematic of diverging duct for analysis of compressible flow with area change.

This analysis can be started using the control volume formulation of conservations of mass assuming a steady flow. Basically, this relationship can interpreted as the outflow rate less the inflow rate equals zero.

$$\dot{P}_M = 0 = \oint_{CS} \rho V \cdot dA = (\rho + d\rho)(V + dV)(A + dA) - \rho VA$$

Multiplying each term keeping only first-order terms or lower then dividing through by ρVA, the following relationship is obtained:

$$\frac{d\rho}{\rho} + \frac{dV}{V} + \frac{dA}{A} = 0 \tag{2.18}$$

This relationship will be used later in the development. Next, the steady-state X momentum equation can be used to develop a balance between pressure forces and inertial forces.

$$F_{Sx} + F_{Bx} + R_X = \oint_{CS} V_X \rho V \cdot dA / g_c$$

Based on the momentum equations, there will be pressure forces in the positive X direction and pressure forces in the negative X direction. These pressure forces must balance the outflow minus inflow of X momentum. This change in momentum can be most simply written as the mass flow rate, ρVA, times the change in X momentum.

$$PA + \left(P + \frac{dP}{2}\right)dA - (P + dP)(A + dA) = \frac{\rho AV(V + dV - V)}{g_c}$$

Multiplying each side out, dropping second-order terms, and canceling similar terms, X momentum reduces to:

$$\cancel{PA} + \cancel{PdA} - \cancel{PA} - \cancel{PdA} - dPA = \frac{\rho A V dV}{g_c}$$

Canceling area, the following simple relationship results:

$$-dP = \frac{\rho V dV}{g_c} \tag{2.19}$$

This simple relationship is very useful in helping to interpret isentropic compressible flows. Essentially, when velocity increases, pressure must decrease, and when velocity decreases, pressure must increase. A similar method to interpret this relationship symbolically is $V \uparrow P \downarrow$ *and* $V \downarrow P \uparrow$. However, another substitution is needed to relate the change in area with the change in pressure. Starting with the result from continuity for a varying area duct, a relationship for the change in velocity can be developed to eliminate it from the preceding equation.

$$dV = -V\left(\frac{d\rho}{\rho} + \frac{dA}{A}\right)$$

Inserting this into the relationship developed from X momentum, a new relationship with area changes results.

$$-dP = -\frac{\rho V^2}{g_c}\left(\frac{d\rho}{\rho} + \frac{dA}{A}\right)$$

The relationship can be further developed by solving for $d\rho$ in the speed of sound equation.

$$d\rho = \frac{dP}{a^2} g_c$$

Substituting for $d\rho$ and factoring out dP, a relationship is developed, which guides our understanding of how pressure changes for situations of changing area with compressible flow.

$$dP\left(1 - \frac{V^2}{a^2}\right) = \frac{\rho V^2}{g_c}\frac{dA}{A}$$

The term on the left-hand side can be written to put the equation in terms of Mach number.

$$dP\left(1 - M^2\right) = \frac{\rho V^2}{g_c}\frac{dA}{A} \tag{2.20}$$

The Mach number can be seen to effectively flip the sign between the pressure and area relationships. For subsonic flow, $M < 1$, increasing area results in rising pressure and decreasing area results in dropping pressure.

For supersonic flow, $M > 1$, increasing area results in dropping pressure, while decreasing area results in rising pressure. These latter results are not intuitive, as they are opposite to what is found in incompressible flow. Based on the earlier observation between pressure and velocity, supersonic flow produces increasing velocity with increasing area and decreasing velocity for decreasing area. It is worthwhile to review changing area effects for subsonic and supersonic flow.

The top two ducts with area change shown in Figure 2.9 symbolize subsonic flow, which suggests that, for a converging duct, pressure decreases and velocity rises, while in a diverging duct, pressure increases and velocity decreases. The bottom two ducts symbolize supersonic flow, suggesting that, for a converging duct, pressure increases and velocity decreases, while in a diverging duct, pressure drops and velocity increases. The two equations that are used to guide our understanding are provided next.

$$-dP = \frac{\rho V dV}{g_c} \tag{2.19}$$

$$dP\left(1 - M^2\right) = \frac{\rho V^2}{g_c}\frac{dA}{A} \tag{2.20}$$

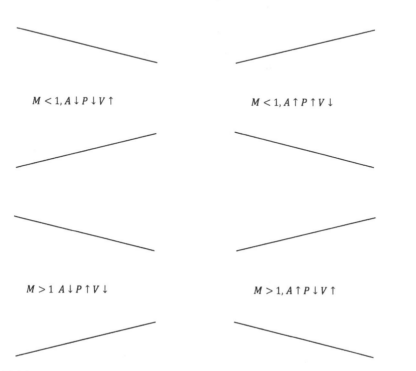

$$M < 1, A \downarrow P \downarrow V \uparrow \qquad\qquad M < 1, A \uparrow P \uparrow V \downarrow$$

$$M > 1\ A \downarrow P \uparrow V \downarrow \qquad\qquad M > 1, A \uparrow P \downarrow V \uparrow$$

FIGURE 2.9
Sketch of converging and diverging ducts for both subsonic and supersonic flow.

The Mach number relationship for temperature, which comes from energy, and the isentropic Mach number relationship for pressure are not restricted to converging ducts. The preceding exercise provided guidance in understanding how supersonic flows behaved in varying area ducts. Some of the principles that have been presented include that, to accelerate subsonic flow, a converging duct is needed, and to accelerate supersonic flow, a diverging duct is needed. Additionally, from the analysis and discussion for converging ducts, the maximum velocity possible to accelerate a subsonic flow to in a converging duct is sonic flow or flow with a Mach number of 1. If these different guiding principles are integrated, the result would suggest that, to achieve supersonic flow from subsonic flow, the gas must be accelerated through a converging–diverging duct. However, before converging–diverging ducts are addressed, the development of the last equation for isentropic flow tables, the A/A^* relationship will be presented. This analysis will show that there is a one-to-one relationship between area and Mach number in a compressible isentropic flow.

The A/A^* equation can be developed from the principles of continuity or the one-dimensional uniform flow assumption, energy, and an isentropic flow assumption. The ideal gas law is also used as is the relationship for the speed of sound. Starting from the one-dimensional uniform flow assumption, the mass flow rate can be calculated:

$$\dot{m} = \rho V A$$

Using the ideal gas law, the concept of the Mach number with the speed of sound, this equation can be converted to more basic variables.

$$\dot{m} = \frac{P}{RT} M \sqrt{k g_C R T}\, A \tag{2.21}$$

The problem of using basic static variables, such as pressure (P), temperature (T), Mach number (M), and area (A), is that all of these variables change in a variable area duct. Static pressure and temperature can be related to the total pressure and temperature for an isentropic flow with the following relationships.

$$\frac{T_T}{T} = \left[1 + \frac{(k-1)}{2} M^2 \right]$$

$$\frac{P_T}{P} = \left[1 + \frac{(k-1)}{2} M^2 \right]^{k/k-1}$$

As the relationship for the ratio of total to static pressure was developed from an isentropic relationship with temperature, the relationship contains the same relationship in brackets. This means that the equation for mass flow rate can be developed from the previous relationship using total pressure and temperature with only one additional relationship.

$$\dot{m} = \frac{P_T}{RT_T} M A \sqrt{k g_C R T_T} \left[1 + \frac{(k-1)}{2} M^2 \right]^{-k/_{k-1} + 1/_2} \tag{2.22}$$

The exponent for the relationship in the bracket can be simplified by developing a common denominator.

$$\frac{-2k}{2(k-1)} + \frac{(k-1)}{2(k-1)} = \frac{-(k+1)}{2(k-1)}$$

Our final relationship for the mass flow rate can be expressed as:

$$\dot{m} = \frac{P_T}{RT_T} M A \sqrt{k g_C R T_T} \left[1 + \frac{(k-1)}{2} M^2 \right]^{\frac{-(k+1)}{2(k-1)}} \tag{2.23}$$

This relationship basically suggests that there is a one-to-one relationship between the Mach number (M) and the area (A) for an isentropic flow with fixed total conditions and mass flow rate. This equation is also very useful in finding the mass flow rate if the total conditions, local Mach number, and the area are known. The area can also be found with this equation if the mass flow rate is known. However, the observation of critical interest is the functional relationship between Mach number and area. The development of an equation that could be used to correlate area with Mach number would be very useful. However, this development would be simplified with the specification of a reference condition. One possible reference condition would be the area and conditions at choked flow. Typically, this condition is given the designation "*." Consequently, the area at the choked or sonic condition is A^*, and this designation can be given to other properties at the choked condition. Note that, for a given mass flow rate and total conditions, there is only one A^*. At the A^* condition, $M = 1$, and the mass flow rate equation can be expressed as:

$$\dot{m} = \frac{P_T}{RT_T} A^* \sqrt{k g_C R T_T} \left[1 + \frac{(k-1)}{2} \right]^{\frac{-(k+1)}{2(k-1)}} \tag{2.24}$$

This equation is very useful in determining the mass flow rate through choked nozzles, including both converging nozzles and converging diverging nozzles. An equation that relates Mach number to the ratio of A/A^* can be determined by equating the two relationships for mass flow rate.

$$\frac{P_T}{RT_T} M A \sqrt{k g_C R T_T} \left[1 + \frac{(k-1)}{2} M^2 \right]^{\frac{-(k+1)}{2(k-1)}} = \frac{P_T}{RT_T} A^* \sqrt{k g_C R T_T} \left[1 + \frac{(k-1)}{2} \right]^{\frac{-(k+1)}{2(k-1)}}$$

The total pressure, P_T, total temperature, T_T, the gas constant, R, and everything in the square root sign can be eliminated. Solving for A/A^*, the following relationship is arrived at.

$$\frac{A}{A^*} = \frac{\dfrac{1}{M}\left[1+\dfrac{(k-1)}{2}\right]^{\frac{-(k+1)}{2(k-1)}}}{\left[1+\dfrac{(k-1)}{2}M^2\right]^{\frac{-(k+1)}{2(k-1)}}} \tag{2.25}$$

This equation is the third and final equation, which comprises the isentropic Mach number equations for compressible flow. The other two are the relationships for static to total pressure ratio and static to total temperature ratio, which are given next in the form that they appear in the isentropic tables, which appears in Appendix A.1.

$$\frac{T}{T_T} = \frac{1}{\left[1+\dfrac{(k-1)}{2}M^2\right]} \tag{2.13}$$

$$\frac{P}{P_T} = \frac{1}{\left[1+\dfrac{(k-1)}{2}M^2\right]^{k/k-1}} \tag{2.14}$$

The tables that give P/P_T, T/T_T, and A/A^* as a function of Mach number provide a means to rapidly solve and also understand compressible flow equations. Examining the equations for T/T_T and P/P_T, it is apparent that both of these relationships monotonically decrease with increasing Mach number. Noting that a typical value of the specific heat ratio, k, is 1.4 for air at ambient conditions, P/P_T clearly decreases much faster than T/T_T. However, how does A/A^* vary with Mach number, and is there any observation that can be noted, which can help our understanding of compressible flow? A/A^* is plotted versus Mach number in Figure 2.10. The figure indicates for a given mass flow rate that the minimum area occurs at a Mach number of 1. This means that the mass flux is a maximum. Additionally, the previous varying area relationships indicated a converging duct is needed to accelerate subsonic flow, while a diverging duct is needed to accelerate supersonic flow. This curve is equivalent to the relationship between area and Mach number for a choked flow in a converging or diverging duct. The figure also indicates that small changes in area equate to large changes in Mach number near $M = 1$. The equations that govern isentropic Mach number flow have been developed and are provided in the preceding section. Isentropic Mach number tables for

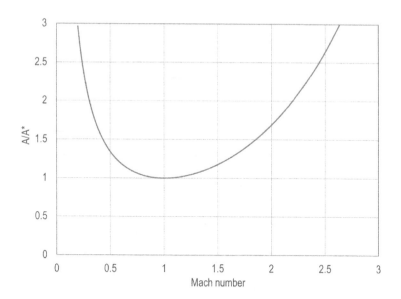

FIGURE 2.10
Variation of A/A^* with Mach number for $k = 1.4$.

$k = 1.4$ and $k = 1.3$ are tabulated in Appendix A.1. However, applying these relationships to a flow situation is a very useful tool in developing our understanding of compressible flow.

Example 2.4: Varying Area Duct

Given: The air flow entering a variable area duct has an inlet Mach number of 0.5, an inlet static pressure of 120 kPa, and inlet static temperature of 300 K, as pictured in Figure 2.11. The area of the duct at location 1 is 0.1 m². The duct converges to an area of 0.08 m² at location 2.

Wanted: Determine the mass flow rate through the duct and determine the exit Mach number of the flow at $A_2 = 0.08$ m² and the static conditions at location 2.

Sketch:

Analysis: The mass flow rate needs to be determined and the exit conditions at location 2 need to be calculated. The mass flow rate can be determined from the one-dimensional uniform flow assumption. The conditions for the calculation can be taken at either end, but because the conditions are given at location 1, using these conditions is easiest and will be accurate.

$$\dot{m} = \rho A V = \frac{P_1}{R T_1} A_1 M_1 a_1$$

Sketch:

$M_1 = 0.5$ $M_2 = ?$

$A_1 = 0.1\ m^2$ $A_2 = 0.08\ m^2$

$P_1 = 120\ kPa$ $P_2 = ?$

$T_1 = 300K$ $T_2 = ?$

FIGURE 2.11
Schematic of converging duct with compressible flow.

The speed of sound is determined from the static temperature and the gas constant for air.

$$a_1 = \sqrt{k\, g_c\, R\, T_1} = \sqrt{1.4\,(1)\,287\,\frac{J}{kg\ K}\,300\ K} = 347.2\ m\,/\,s$$

Solving for the mass flow rate:

$$\dot{m} = \frac{120,000\ Pa}{287\,\dfrac{J}{kg\ K}\,300\ K}\,0.1\ m^2\,0.5 \times 347.2\,\frac{m}{s} = 24.19\,\frac{kg}{s}$$

The values in the appendix can be used to determine the unknowns at location 2. The state at location 2 can be found using the A/A^* ratios.

$$\frac{A_2}{A^*} = \frac{A_1}{A^*} \times \frac{A_2}{A_1}$$

A_1/A^* can be found in the Appendix A.1 using the Mach number, M_1. Useful ratios for P/P_T and T/T_T can also be found here.

$M_1 = 0.5 : A_1\,/\,A* = 1.339844, T_1\,/\,T_T = 0.952381, P_1\,/\,P_T = 0.843019$

Solving for A_2/A^*

$$\frac{A_2}{A^*} = 1.339844 \times \frac{0.08\ m^2}{0.1\ m^2} = 1.11875$$

Looking at the isentropic Mach number table for $k = 1.4$, the Mach number at location 2, M_2, is close to 0.67. The table gives the following values: $T_2/T_T = 0.917616$, $P_2/P_T = 0.74014$. The static conditions at location 2 can be determined from condition one in the following manner.

$$T_2 = T_1 \times \frac{T_T}{T_1} \frac{T_2}{T_T} = 300 \, \text{K} \times \frac{0.917616}{0.952381} = 289.0 \, \text{K}$$

$$P_2 = P_1 \times \frac{P_T}{P_1} \frac{P_2}{P_T} = 120 \, \text{kPa} \times \frac{0.74014}{0.843019} = 105.4 \, \text{kPa}$$

In the relationship for temperature, the local static temperature at 1 is divided by the ratio of static to total temperature at that location. This provides the local total temperature, which is a constant in these isentropic flows. The total temperature is then multiplied by the ratio of the static to total temperature for the Mach number found at location 2. This product determines the local static temperature. Note that T_1 and T_T cancel from the equation. Over the 20 percent reduction in area, a 34 percent increase in Mach number occurs, indicating the necessity of using compressible flow equations.

2.6 Converging–Diverging Nozzles

The analysis of flow in varying area ducts showed that to accelerate subsonic flow, a converging duct was required, and to accelerate supersonic flow, a diverging duct was required. A relationship for A/A^* was developed, which had a minimum value $(A/A^* = 1)$ for choked flow or where M = 1. This A/A^* relationship suggested that the path for a flow to go from subsonic to supersonic flow was through a duct that first converged then, at the choking point, began to diverge. In this section, we will examine flow in a converging–diverging nozzle in a manner similar to the approach taken in Section 2.4 for converging nozzles.

Flow in a converging–diverging nozzle is discussed here with help from Figures 2.12 and 2.13 shown next. This conceptual problem is started with a sketch of a converging–diverging nozzle. The nozzle has an inlet plane total pressure of P_T and an exit plane static pressure of P_B. At the beginning of the conceptual test, the back pressure P_B is the same as the inlet total pressure P_T. At this point, there is no flow entrained through the nozzle. Analytically, either the Bernoulli equation could be used or our isentropic Mach number relationship could be used to show that when $P_B = P_T$, then our velocity must be zero. This condition is equivalent to condition, *a*, on the pressure versus axial distance, X, chart. The chart shows that the static pressure in the nozzle does not vary. The bottom chart represents the mass flow rate as a function of P_B/P_T. When P_B/P_T is equal to 1, then $P_B = P_T$, then our mass flow rate for condition, *a*, will be zero.

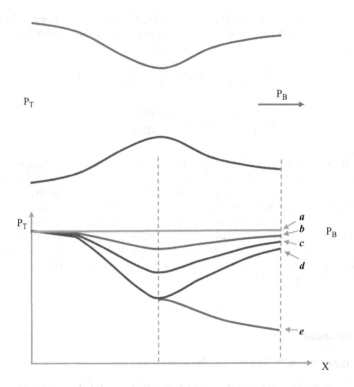

FIGURE 2.12
Sketch of converging–diverging nozzle with figure of related static pressure distributions.

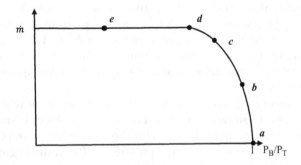

FIGURE 2.13
Plot of mass flow rate vs. P_B/P_T for both choked and non-choked conditions.

When the back plane pressure is initially lowered below the inlet total pressure, flow is entrained through the nozzle. As subsonic flow enters the converging section, the velocity increases as the area decreases. The increasing velocity will result in a decrease in the local static pressure, which is depicted in curve *b*. At this point, the flow is subsonic throughout the nozzle. As subsonic flow enters the diverging section, the velocity of the flow decreases with the increasing area and the static pressure rises toward the exit of the nozzle. The pressure versus distance along the nozzle chart shows an initially decreasing pressure to the minimum area of the nozzle and then an increasing pressure to the exit of the nozzle.

Later, the backplane pressure is lowered below the back pressure of P_{Bb} to the back plane pressure of curve P_{Bc}. The lower back pressure induces more flow and correspondingly higher velocities throughout the nozzle, lowering the pressure in all regions of the nozzle. However, as the flow is subsonic throughout the nozzle, when the flow reaches the minimum area and the nozzle begins to diverge, the subsonic flow begins to decrease in velocity and the pressure within the nozzle begins to rise. However, with the higher flow rate, all pressures throughout the nozzle are lower than those of case *b*.

The backplane pressure is now lowered to P_{Bd} where the flow finally reaches the sonic condition at the minimum area in the nozzle. At this point, the flow in the minimum area or throat is now choked. This means that the flow at the throat is moving at the speed of sound and pressure waves cannot travel upstream into the converging portion of the nozzle, signaling to the flow that the backplane pressure has been further lowered. However, the back plane pressure is the maximum value that can choke the nozzle. The flow velocity in the diverging portion of the nozzle decreases and the pressure rises to the exit plane pressure P_{Bd}. Any further decrease in the backplane pressure will not change the flow rate within the nozzle and the pressure distribution upstream from the throat will not change.

The backplane pressure can be lowered to P_{Be} where the flow in the nozzle is supersonic. Note that upstream of the throat, pressure did not change and the flow rate through the nozzle did not change. If the backplane pressure was lowered below P_{Be}, then the pressure through the entire nozzle would be fixed. As noted previously, for a given mass flow rate, there is a one-to-one correlation between area and Mach number or equivalently the pressure distribution. In Figure 2.12, for any given A/A^*, there were two solutions, a subsonic solution and a supersonic solution. Curves *d* and *e* represent the subsonic and supersonic solutions for this current choked converging–diverging nozzle at the given inlet total pressure and temperature conditions, respectively, which produce a choked nozzle.

The mass flow rate curve, presented in Figure 2.13, shows that once condition *d* is reached, any further decrease in the pressure ratio, P_B/P_T, results in no further change in the mass flow rate. Note that this curve has a more rapid rise in flow rate with backplane pressure compared with the converging nozzle. Note that these results are restricted to an isentropic flow with an ideal gas.

Example 2.5: Converging–Diverging Nozzle

There are three basic conditions that we can examine currently in our
converging–diverging nozzle. These include flows that are (1) subsonic
but not choked, (2) subsonic but choked, and (3) supersonic and choked.
Subsonic flows that are not choked are represented by curves *b* and *c*.
Subsonic flow that is choked is represented by curve *d*, and supersonic
flow is represented by curve *e*. Subsonic choked flow is a critical point
in the flow curve, as for pressures at or below P_{Bd}, the flow in the nozzle
is choked, and there cannot be an increase in the flow rate without the
change in one or both of the inlet total pressure or temperature conditions.

Given: A converging–diverging nozzle with an 11.00 cm² exit area and a
throat area of 10 cm² has an absolute inlet total pressure of 320 kPa. This
nozzle is shown in Figure 2.14. Initially, the back pressure on the nozzle
is 250 kPa. Later, the back pressure is dropped to atmospheric pressure
105 kPa. The inlet total temperature to the nozzle is 360 K.

Sketch:

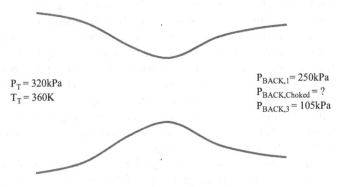

$P_T = 320\text{kPa}$
$T_T = 360\text{K}$

$P_{BACK,1} = 250\text{kPa}$
$P_{BACK,Choked} = ?$
$P_{BACK,3} = 105\text{kPa}$

FIGURE 2.14
Sketch of converging–diverging nozzle with two different back pressure conditions.

Wanted: (a) Find the exit Mach number, the exit velocity, and the mass
flow rate for the air leaving the nozzle for the initial conditions. (b) Find
the maximum back pressure to choke the nozzle and determine the exit
Mach number and mass flow rate at this condition. (c) Find the exit Mach
number and static conditions for the case with the back plane pressure
equal to 105 kPa. Has the mass flow rate changed?

Analysis: Initially, the nozzle flow must be determined to be subsonic or
choked. A converging–diverging nozzle will have a higher choking pressure
than a simple converging nozzle. In order to determine the maximum choking pressure, the subsonic back pressure for the condition $A_{EXIT}/A_{THROAT} =
A/A*$ must be determined. If the entire flow is subsonic, then the mass flow
rate and temperature can be determined simply from the exit pressure. If the
flow in the nozzle is supersonic, then the isentropic solution will occur at the
condition $A_{EXIT}/A_{THROAT} = A/A*$ for the supersonic solution.

$$A_{EXIT} / A_{THROAT} = A / A* = 11\,\text{cm}^2 / 10\,\text{cm}^2 = 1.1$$

$A / A* = 1.1$ for subsonic flow : $M = 0.69, A / A* = 1.101822,$
$$T / T_T = 0.913059, P / P_T = 0.727353$$

The subsonic choking point is

$$P_{EX} = P_T \times \frac{P}{P_T} = 320 \text{ kPa} \times 0.727353 = 232.8 \text{ kPa}$$

As $P_{BACK} = 250$ kPa $> P_{EX}$, the initial condition is not choked. At this pressure, the local Mach number can be determined and the exit static temperature can also be determined.

$$M = \left\{ \left[\left(\frac{P_T}{P} \right)^{k-\frac{1}{k}} - 1 \right] \frac{2}{k-1} \right\}^{\frac{1}{2}} = \left\{ \left[\left(\frac{320 \text{ kPa}}{250 \text{ kPa}} \right)^{0.4/1.4} - 1 \right] \frac{2}{0.4} \right\}^{\frac{1}{2}} = 0.6045$$

The static temperature can be determined from the inlet total temperature and the local Mach number.

$$T = \frac{T_T}{\left[1 + \frac{(k-1)}{2} M^2 \right]} = \frac{360 \text{ K}}{\left[1 + \frac{0.4}{2} 0.6045^2 \right]} = 335.5 \text{ K}$$

The speed of sound at this temperature is

$$a = \sqrt{k g_C R T} = \sqrt{1.4 (1) 287 \frac{J}{\text{kg K}} 335.5 \text{ K}} = 367.1 \frac{\text{m}}{\text{s}}$$

The velocity is determined to be

$$V = M a = 0.6045 \times 367.1 \frac{\text{m}}{\text{s}} = 221.9 \text{ m / s}$$

At this point, the mass flow rate can be calculated by assuming a one-dimensional uniform flow.

$$\dot{m} = \rho V A$$

The area of the nozzle was given and the velocity has been determined. The remaining variable that needs to be calculated is the density, which can be determined accurately for these conditions using the ideal gas law. However, the density is a local property that is based on the local static conditions.

$$\rho = \frac{P}{R T} = \frac{250,000 \text{ Pa}}{287 \frac{J}{\text{kg K}} 335.5 \text{ K}} = 2.596 \frac{\text{kg}}{\text{m}^3}$$

The resulting mass flow rate is determined to be

$$\dot{m} = \rho\, V\, A = 2.596\frac{kg}{m^3} \times 221.9\frac{m}{s} \times 0.0011\, m^2 = 0.634\frac{kg}{s}$$

The exit Mach number and exit static pressure at the maximum (subsonic solution) exit pressure was previously found at $M = 0.69$ and $P_{EX} = 232.8$ kPa.

The static temperature at this point is

$$T_{EX} = T_T \times \frac{T}{T_T} = 360\, K \times 0.913059 = 328.7\, K$$

The speed of sound at this temperature is

$$a = \sqrt{kg_c RT} = \sqrt{1.4\,(1)\,287\frac{J}{kg\, K}\,328.7\, K} = 363.4\frac{m}{s}$$

The velocity is determined to be

$$V = M\, a = 0.69 \times 363.4\frac{m}{s} = 250.8\, m\,/\,s$$

The mass flow rate can be calculated by assuming a one-dimensional uniform flow, $\dot{m} = \rho V A$.

The area of the nozzle was given and the velocity has been determined. The remaining variable that needs to be calculated is the density, which can be determined using the ideal gas law.

$$\rho = \frac{P}{R\,T} = \frac{232{,}800\, Pa}{287\dfrac{J}{kg\, K}\,328.7\, K} = 2.468\frac{kg}{m^3}$$

The resulting mass flow rate is determined to be

$$\dot{m} = \rho\, V\, A = 2.468\frac{kg}{m^3} \times 250.8\frac{m}{s} \times 0.0011\, m^2 = 0.681\frac{kg}{s}$$

The resulting increase in the mass flow rate is about 7 percent. At this point, the nozzle is choked and any further drop in back pressure will not influence the mass flow rate. The last case involves finding the exit Mach number and static conditions when the back pressure is 105 kPa. First, the Mach number can be determined.

$$M = \left\{ \left[\left(\frac{P_T}{P} \right)^{k-1/k} - 1 \right] \frac{2}{k-1} \right\}^{1/2} = \left\{ \left[\left(\frac{320\, kPa}{105\, kPa} \right)^{0.4/1.4} - 1 \right] \frac{2}{0.4} \right\}^{1/2} = 1.369$$

This exit Mach number is essentially consistent with an exit Mach number of 1.37, which has an $A/A^* = 1.099$, which is very close to 1.10, the design A/A^*. This result can be interpreted as the flow at the supersonic isentropic condition. The total to static temperature ratio is given as $T/T_T = 0.727072$. The back pressure is equivalent to the isentropic supersonic back pressure so $P = 105$ kPa. The static temperature needs to be found:

$$T_{EX} = T_T \times \frac{T}{T_T} = 360 \text{ K} \times 0.727072 = 239.0 \text{ K}$$

As the nozzle is choked, the mass flow rate has not changed.

References

John, J.E.A., and T.G. Keith. 2006. *Gas Dynamics*, 3rd ed. Prentice Hall.
Moran, M.J., H.N. Shapiro, D.D. Boettner, and M.B. Bailey. 2011. *Fundamentals of Engineering Thermodynamics*, 7th ed. John Wiley & Sons.
Saad, M.A. 1993. *Compressible Fluid Flow*, 2nd ed. Prentice Hall.

Chapter 2 Problems

1. A very light jet is flying at 155 m/s at an altitude of 5000 m where the static temperature is 260 K. The static pressure at that altitude is 54 kPa. Determine the total pressure and temperature that is sensed by the leading edge of the fuselage and the instrument probe. Solve this problem using the energy approach and compare the answer to the Bernoulli equation.

2. The static pressure in a compressible flow wind tunnel is measured to be 30,000 Pa and the inlet total pressure is 48,000 Pa. The inlet total temperature has been measured to be 325 K. Determine the local Mach number, velocity, and density for the flow. What is the local static temperature?

3. On a warm summer evening, the air temperature is 303 K. You see a flash of lightning and immediately start to count one thousand one, one thousand two, and so forth, and at a value of 8, you hear the thunder clap. Calculate the speed of sound in the air and estimate the distance of the lightning strike.

4. Calculate the speed of sound in copper that has a density of 8900 kg/m³ and a bulk modulus. β_S of 121.8 GPa (1.218E11 Pa).

5. A converging nozzle has a back pressure of 150 kPa and a reservoir pressure of 250 kPa. The reservoir temperature (total temperature) is 350 K. The nozzle has an exit area of 0.001 m². Determine the exit Mach number, the static pressure, and temperature at the exit, and the mass flow rate of the air leaving the nozzle. Perform the calculation again for a back pressure of 100 kPa.

6. A converging–diverging nozzle has an exit to throat area ratio of 1.25. The reservoir or total pressure supplied to the nozzle is 370 kPa and the total temperature is 400 K. Determine the maximum back pressure that can still choke the nozzle. Determine the pressure needed for the isentropic supersonic solution. Determine the mass flow rate for the air if the exit area is 0.0125 m².

7. An air flow enters a diverging channel at a Mach number of 0.89 where the area is 0.01 m². At the exit of the duct, the area is 0.012 m². If the inlet static pressure is 80 kPa, find the exit static pressure and the exit Mach number.

8. Flow enters a diverging channel at a Mach number of 1.11 where the area is 0.01 m². At the exit, the duct area is 0.012 m². If the inlet static pressure 62 kPa, find the exit static pressure and the exit Mach number.

3

Normal Shock Waves

A normal shock wave occurs when supersonic flow abruptly adjusts to a change in pressure. In a normal shock wave, supersonic flow enters going into the wave and subsonic flow leaves the wave. Across the shock wave, the static pressure abruptly rises from the supersonic side to the subsonic side. Supersonic flow behaves differently from subsonic flow. In a subsonic flow, like flow around a typical commercial aircraft, the subsonic speed of the aircraft allows pressure waves to proceed upstream of the aircraft enabling the upstream flow to adjust to the presence of the aircraft as it flies through the air. This potential for the flow to adjust to the presence of the oncoming aircraft allows the use of streamlining. When a supersonic aircraft moves through the air after takeoff and before landing, it travels faster than the speed of sound. Sound waves cannot travel upstream of the aircraft, so the air cannot adjust to its presence as it moves through the air. Consequently, the air must abruptly adjust to the presence of the aircraft, and this occurs through shock waves. Normal shock waves are examined in this chapter. Initially, subsonic and supersonic flows are briefly discussed. Next, equations for normal shock waves are developed and stationary shock waves are analyzed. Normal shock waves are then examined in a moving reference frame. This moving reference frame includes normal shock reflections. Finally, flow in shock tubes and other normal shock applications are discussed and analyzed.

3.1 Subsonic and Supersonic Flow

Pressure waves proceed a body in subsonic flow. Figure 3.1 depicts a subsonic aircraft or projectile at a Mach number of about 0.6. Initially, a projectile is shown at time $t = 0$ when it begins to send out a sound wave. Later, the same projectile is shown at time $t = 1$, and at this time, it also sends out a sound wave. Later, when the projectile arrives at time $t = 2$, the figure also shows a sound wave that has traveled from that point. In the figure, the projectile is shown at times $t = 0, 1, 2$, and 3. Additionally, sound waves that originated at time $t = 0, 1$, and 2 are shown, but time $t = 3$ is the current time in the figure. By time $t = 3$, the sound wave originating from the location of $t = 0$ has traveled 3a, where **a** is the speed of sound. The sound wave is shown by a circle that has a radius of 3a. Next, a sound wave originating from the projectile at $t = 1$ is shown at time $t = 3$. Similar to the initial sound wave, this sound

Subsonic flow:

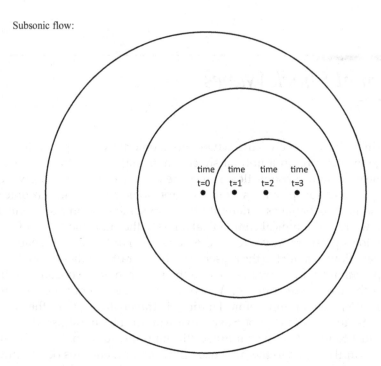

FIGURE 3.1
Visualization of subsonic projectile with sound wave propagating at different times.

wave originates at the projectile at time $t = 1$ and has a radius of 2**a**. Finally, a sound wave originating from the projectile at time $t = 2$ is shown at time $t = 3$ with a radius of 1**a**. This simplified view of sound waves, which have originated from the projectile's position at various times, shows that the sound waves proceed the projectile and allow the air to sense and move around the projectile. Consequently, this allows streamlining to be possible. Streamlining is a means to reduce aerodynamic losses, which typically are maximized when flow separates from the surface of an aircraft or projectile resulting in a significant pressure drag.

Flow adjusts abruptly to the presence of a supersonic projectile. Figure 3.2 is a representation of a projectile or aircraft moving at supersonic speed. The position of the object is initially shown at time $t = 0$. A sound wave is presented that was initiated from that position at time $t = 0$, but shown at time $t = 3$. The object's position is also shown at time $t = 1$. A sound wave originating at that time but shown at time $t = 3$ is also present. The object is also shown at time $t = 2$ with a sound wave originating at that time but shown at time $t = 3$. Finally, the object is shown at time $t = 3$. In the figure, the sound waves are propagating at the speed of sound, so the initial wave has traveled

Supersonic flow:

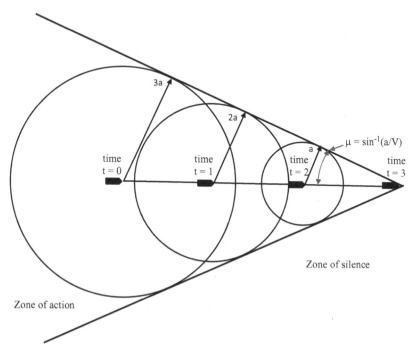

FIGURE 3.2
Visualization of supersonic projectile with sound wave propagating at different times.

3a from time $t = 0$ to time $t = 3$. This sound wave is represented by a circle with a radius of 3a. A similar sound wave originating from time $t = 1$ is also shown with a radius of 2a. A third sound wave originating at time $t = 2$ is also shown with a radius of 1a. In this supersonic flow, the object is traveling faster than the speed of sound, so $V > a$ or $M > 1$. In this case, based on the geometry of the figure, the Mach number of the object is about 2.5. As the object steadily passes through air, it creates a conical sound wave. Outside of this wave, the presence of the object cannot be sensed acoustically, as the pressure wave is limited by the velocity of sound. This region outside of the wave is sometimes referred to as the zone of silence. Inside the cone, the presence of the object can be sensed, and this region is sometimes referred to as the zone of action.

The surface of the cone made by the sound waves is perpendicular to the original source of the sound wave generated by the object. The angle of the cone, μ, is therefore the arcsin(a/V) or equivalently $\sin^{-1}(1/M)$. Here,

M is the Mach number, V/a. As the object is moving faster than the speed of sound, the air it moves into does not have time enough to adjust to the presence of the object before it arrives. This means that the adjustment process is necessarily abrupt as the object passes through the air and the general concept of streamlining is not effective. However, there are approaches to reduce the aircraft and projectile drag associated with supersonic flow.

3.2 Normal Shock Wave Equations

Normal shock waves are one of the ways that supersonic airflow adjusts abruptly to changes in the local boundary conditions. The fundamental relationship governing a normal shock wave is based on the momentum principle. However, conservation of energy and conservation of mass are two other relationships, which are used in the development of the normal shock equations. An analogous behavior in fluids to a normal shock is a hydraulic jump where flow abruptly moves from supercritical flow to subcritical flow. However, unlike a hydraulic jump, the thickness of a normal shock wave is very thin. Typically, the thickness of a normal shock wave is on the order of a few mean free paths of air, which is very small at ambient conditions. A hydraulic jump is much thicker as it depends on mixing with length scales on the order of the change in height. However, in both normal shock waves and hydraulic jumps, as the speed of the fluid is traded for a pressure rise in the air or a rise in the head of the water, there is a loss in the total pressure or the ability of the fluid to do work. Therefore, both a normal shock wave and a hydraulic jump are irreversible processes.

The flow entering a normal shock wave is always supersonic while the flow leaving a normal shock wave is always subsonic. The static pressure always abruptly increases from the supersonic side of a normal shock wave to the subsonic side. The possibility of an expansion shock cannot exist as this will violate the second law of thermodynamics. The following section discusses the reinforcement of compression waves and the physical inability of expansion waves to produce a similar reinforcement.

3.2.1 Reinforcement of Compression Waves into a Normal Shock Wave

Normal shock wave equations will be developed based on the principles discussed previously. However, before we begin the development, it is useful to perform some thought experiments on a series of compression waves and expansion waves. This physical discussion on compression and expansion waves was previously presented by John [1].

Consider a piston in a long cylindrical tube as shown in Figure 3.3 at time t_0. At some point in time, the piston is given an incremental velocity to the right

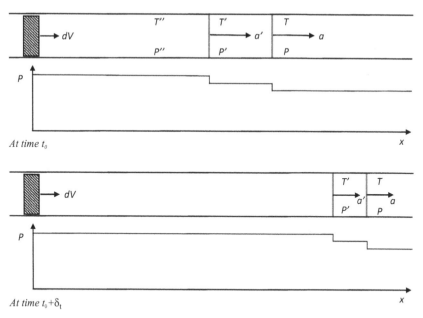

FIGURE 3.3
Schematic of two compression waves moving down a constant area duct at time t_0 and time $t_0+\delta t$.

forming a compression wave which moves into the still gas media within the tube at the speed of sound. Then a very short time later a second wave is formed by again giving that piston a second incremental velocity to the right forming a second compression wave. As the first wave moves into the still media it produces a small rise in pressure to P' as shown in the plot below the sketch. At the same time, the gas experiences a small increase in temperature to T'. The second compression wave moves into the gas downstream from the first compression wave moving at the speed of sound relative to the gas in this slightly compressed media. The compression wave produces a similar rise in pressure compared to the first. Since T' is slightly higher than T, the second wave has a slightly higher speed of sound than the first. If we look at the pressure distribution at some time along the tube, we see a slight rise in pressure after the passing of the first wave and a second slight rise in pressure after the passing of the second wave. If we look at these two compression waves in this tube a short time later as shown in Figure 3.3 at time $t_0+\delta t$, we can see that the second wave, which is traveling into slightly higher temperature gas than the first, has a higher speed of sound, and begins to catch up with the first wave.

This phenomena of the second wave moving faster than the first can be further explored by imagining a rapid series of compression waves being formed by a series of differential velocities imposed on the cylinder moving toward the right. As each wave moves through the media, the gas experiences

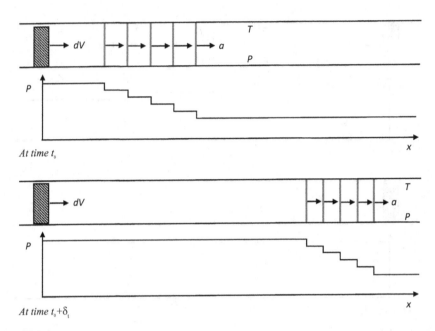

FIGURE 3.4
Schematic of a series of compression waves moving down a constant area duct at time t_0 and time $t_0+\delta t$.

a slight rise in pressure and temperature as each compression wave passes. This sketch of the piston cylinder and pressure waves is shown in Figure 3.4 at time t_0. The expected pressure distribution along the tube at that snapshot in time is shown below the sketch showing a stair like rise in pressure.

Imagine the system a short time later. The trailing waves begin to catch the preceding waves as shown in Figure 3.4 at time $t_0+\delta t$. The pressure distribution in the pipe at this snapshot of time is shown below the sketch. Now the stair-like pressure distribution is noticeably steeper than before. This series of trailing compression waves clearly catch up to the preceding waves and can reinforce one another to form a discrete shock wave. What happens in the case of an expansion wave?

Consider again this piston in a long cylindrical tube as shown in Figure 3.5 at time t_0. Imagine at some point in time that the piston is rapidly accelerated to a differential velocity to the left forming an expansion wave. A very short time later, imagine that another differential velocity is imposed on the piston in the same direction forming a second expansion wave. The expected pressure distribution can be visualized in the plot below. Note that as each expansion wave passes along the tube, the pressure in the tube incrementally drops as does the temperature due to the expansion. Consequently, the second wave which follows the first wave moves into a gas media which is lower in temperature and consequently has a lower speed of sound. The second wave will follow the first but fall further and further behind as shown in Figure 3.5 at

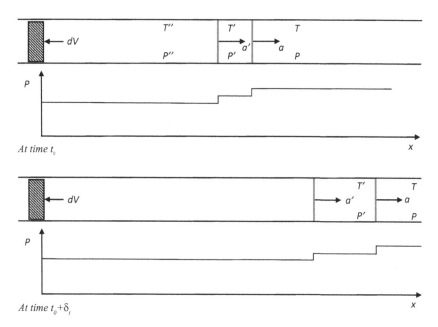

FIGURE 3.5
Schematic of two expansion waves moving down a constant area duct at time t_0 and time $t_0+\delta t$.

time $t_0+\delta t$. Therefore, a series of expansion waves formed by giving the piston a series of incremental velocities to the left will spread out over time as
they move further down the tube. This thought experiment helps in understanding that expansion waves can never reinforce each other to form a discrete expansion shock wave. Additionally, we will see a compression shock
wave with a high-speed supersonic flow entering and a subsonic flow leaving result in a static pressure rise and a total pressure loss. However, a conceptual expansion shock wave, where a high-speed supersonic flow emerges
from a discrete subsonic flow, would produce an increase in total pressure
violating the second law of thermodynamics.

3.2.2 Normal Shock Wave Analysis

A control volume for a normal shock wave in a varying area duct is shown
in Figure 3.6. The conditions of the flow upstream of the shock wave are
given in terms of condition 1 for density, ρ; velocity, V; area, A; and pressure,
P. As a shock wave is very thin, a good assumption is that there is no area
change across the shock wave, $A_1 = A_2$. The analysis for the development of
normal shock wave equations will be based on the idea that this is a one-
dimensional uniform, steady flow with an ideal gas. The analysis can begin
by applying conservation of mass. Starting with our control volume formula,
the integral around the test surface can be interpreted as the outflow rate of

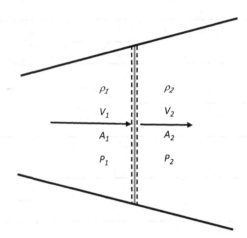

FIGURE 3.6
Schematic of normal shock wave sitting in a duct.

mass from minus the inflow rate of mass to the control volume and written in terms of a one-dimensional uniform flow.

$$\dot{P}_M = 0 = \oint_{CS} \rho V \cdot dA = \rho_2 V_2 A_2 - \rho_1 V_1 A_1 \qquad (1.20)$$

Noting that the shock wave is very thin and $A_1 = A_2$, this simplifies to:

$$\rho_2 V_2 = \rho_1 V_1$$

This relationship will be used to integrate results from conservation of energy and the momentum principle. Conservation of energy is a straightforward but important relationship to apply across a shock wave.

$$T_T = T\left[1 + \frac{(k-1)}{2}M^2\right] = T_1\left[1 + \frac{(k-1)}{2}M_1^2\right] = T_2\left[1 + \frac{(k-1)}{2}M_2^2\right] \qquad (2.12)$$

Essentially, this is a statement that total temperature is constant across a shock wave, as there is no method to extract or add work or thermal energy across a shock wave.

Applying the momentum principle across a shock wave provides the physical understanding needed to comprehend shock waves. The analysis can begin using the control volume form of the momentum principle with the steady-state assumption and with the flow in the X direction.

$$F_{Sx} + F_{Bx} + R_X = \oint_{CS} V_X \, \rho V \cdot dA / g_c \tag{1.25}$$

In this analysis, there are no physical attachments in the control volume, so reaction forces can be neglected. Also, as this is a gas assumed to be flowing in the horizontal direction, body forces will be assumed to be negligible. These simplifications leave a balance between the change in pressure forces equaling the outflow minus inflow of momentum. This balance between pressure forces and momentum or inertial forces is the fundamental principle of shock waves.

$$P_1 A_1 - P_2 A_2 = \frac{\rho_2 V_2^2 A_2}{g_c} - \frac{\rho_1 V_1^2 A_1}{g_c} \tag{3.1}$$

Again, noting the very thin nature of shocks, the area can be eliminated and the pressure and momentum terms from the two sides can be equated.

$$P_1 + \frac{\rho_1 V_1^2}{g_c} = P_2 + \frac{\rho_2 V_2^2}{g_c} \tag{3.2}$$

If the ideal gas law is substituted for density, a relationship for pressure change can be developed.

$$P_1 + \frac{P_1 V_1^2}{R T_1 g_c} = P_2 + \frac{P_2 V_2^2}{R T_2 g_c} \tag{3.3}$$

Factoring out pressure from each side:

$$P_1 \left(1 + \frac{V_1^2}{R T_1 g_c} \right) = P_2 \left(1 + \frac{V_2^2}{R T_2 g_c} \right) \tag{3.4}$$

Note that the denominator multiplied by the specific heat ratio k is equal to a^2. Multiplying the velocity terms on each side by (k/k), the relationship for static pressure change across a shock wave can be written in terms of the Mach number and the specific heat ratio.

$$P_1 \left(1 + k M_1^2 \right) = P_2 \left(1 + k M_2^2 \right) \tag{3.5}$$

This relationship is more often given in terms of the static pressure ratio across a shock wave.

$$\frac{P_2}{P_1} = \frac{\left(1 + k M_1^2 \right)}{\left(1 + k M_2^2 \right)} \tag{3.6}$$

In a similar manner to the pressure ratio, the static temperature change can be developed in terms of a ratio noting the underlying principle is total temperature remains constant.

$$\frac{T_2}{T_1} = \frac{\left[1 + \dfrac{(k-1)}{2} M_1^2\right]}{\left[1 + \dfrac{(k-1)}{2} M_2^2\right]} \tag{3.7}$$

At this point, the result from continuity can be developed in terms of a change in Mach number. Using the ideal gas law, Mach number, and the relationship for the speed of sound, continuity can be written in terms of static pressures, static temperatures, Mach number, the gas constant, and specific heat ratio.

$$\frac{P_2}{RT_2} M_2 \sqrt{k g_C R T_2} = \frac{P_1}{RT_1} M_1 \sqrt{k g_C R T_1} \tag{3.8}$$

Putting this relationship in terms of ratios and canceling similar terms on each side:

$$\frac{P_2}{P_1} \frac{M_2}{M_1} = \sqrt{\frac{T_2}{T_1}} \tag{3.9}$$

Squaring both sides and substituting our relationships for the static pressure and temperature ratio:

$$\left[\frac{\left(1 + k M_1^2\right)}{\left(1 + k M_2^2\right)}\right]^2 \frac{M_2^2}{M_1^2} = \frac{\left[1 + \dfrac{(k-1)}{2} M_1^2\right]}{\left[1 + \dfrac{(k-1)}{2} M_2^2\right]} \tag{3.10}$$

At this point, this equation can be multiplied out and developed into a quadratic equation for M_2^2, which can be solved with the following result.

$$M_2 = \sqrt{\frac{M_1^2 + \dfrac{2}{k-1}}{M_1^2 \dfrac{2k}{k-1} - 1}} \tag{3.11}$$

Equations have been developed for P_2/P_1, T_2/T_1, and M_2 across a shock wave. Additionally, most normal shock tables provide P_{T2}/P_{T1} to help determine the total pressure loss across a shock wave and ρ_2/ρ_1, which is related to the velocity ratio, V_1/V_2, across a shock wave. The total pressure loss across a shock can be developed from P_2/P_1 and the isentropic relationships for P_1/P_{T1} and P_2/P_{T2}.

$$\frac{P_{T2}}{P_{T1}} = \frac{P_2}{P_1} \frac{P_1}{P_{T1}} \frac{P_{T2}}{P_2} \tag{3.12}$$

The density ratio can be developed from the ideal gas law and then the pressure and temperature ratios from the normal shock tables.

$$\frac{\rho_2}{\rho_1} = \frac{P_2}{RT_2} \frac{RT_1}{P_1} = \frac{P_2}{P_1} \frac{T_1}{T_2} \tag{3.13}$$

The equations and relationships developed for M_2, P_2/P_1, T_2/T_1, P_{T2}/P_{T1}, and ρ_2/ρ_1 are the essential equations, which can be used to generate a normal shock table. Normal shock tables for $k = 1.4$ and $k = 1.3$ are provided in Appendix A.2.

Example 3.1: Stationary Normal Shock in a Duct

Given: A high-speed flow with a Mach number of 1.83, P_1 = 50 kPa, T_1 = 250 K undergoes a normal shock as shown in Figure 3.7.

Wanted: Determine the Mach number downstream from the shock, M_2, as well as the static pressure, P_2, Static temperature, T_2, and the total pressure loss. Also determine the velocity of the flow just upstream and just downstream from the shock.

Sketch:

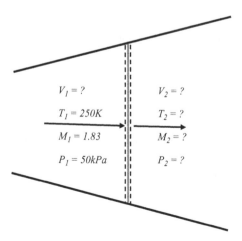

$V_1 = ?$ $V_2 = ?$

$T_1 = 250K$ $T_2 = ?$

$M_1 = 1.83$ $M_2 = ?$

$P_1 = 50kPa$ $P_2 = ?$

FIGURE 3.7
Schematic of normal shock wave sitting in a duct with defined upstream conditions.

Analysis: Initially, the Mach number relationship can be used to find M_2 and then the relationships for P_2/P_1, T_2/T_1 can be used to find P_2 and T_2. Later, P_{T2}/P_{T1} can be found using P_2/P_1 and the isentropic relationships. V_1 and V_2 can be determined from the Mach number and speed of sound. Later, these values will be checked using the normal shock table for $k = 1.4$.

$$M_2 = \sqrt{\frac{M_1^2 + \frac{2}{k-1}}{M_1^2 \frac{2k}{k-1} - 1}} = \sqrt{\frac{1.83^2 + \frac{2}{0.4}}{1.83^2 \frac{2 \times 1.4}{0.4} - 1}} = 0.610$$

With M_2 and M_1 known, the pressure ratio then P_2 can be determined.

$$\frac{P_2}{P_1} = \frac{\left(1 + k M_1^2\right)}{\left(1 + k M_2^2\right)} = \frac{1 + 1.4 * 1.83^2}{1 + 1.4 * 0.61^2} = 3.740$$

$$P_2 = P_1 * \frac{P_2}{P_1} = 50 \text{ kPa} * 3.740 = 187.0 \text{ kPa}$$

The temperature ratio and downstream temperature can also be determined.

$$\frac{T_2}{T_1} = \frac{\left[1 + \frac{(k-1)}{2} M_1^2\right]}{\left[1 + \frac{(k-1)}{2} M_2^2\right]} = \frac{\left[1 + \frac{0.4}{2} 1.83^2\right]}{\left[1 + \frac{0.4}{2} 0.61^2\right]} = 1.554$$

$$T_2 = T_1 * \frac{T_2}{T_1} = 250 \text{ K} * 1.554 = 388.5 \text{ K}$$

The inlet total pressure and exit total pressures can be determined using our isentropic relationships. Starting with our total to static pressure ratio, the inlet total pressure can be determined.

$$\frac{P_{T1}}{P_1} = \left[1 + \frac{(k-1)}{2} M_1^2\right]^{\frac{k}{k-1}} = \left[1 + \frac{0.4}{2} 1.83^2\right]^{\frac{1.4}{0.4}} = 6.016$$

$$P_{T1} = P_1 * \frac{P_{T1}}{P_1} = 50 \text{ kPa} * 6.016 = 300.75 \text{ kPa}$$

The exit total pressure can be determined using a similar manner.

$$\frac{P_{T2}}{P_2} = \left[1 + \frac{(k-1)}{2} M_2^2\right]^{\frac{k}{k-1}} = \left[1 + \frac{0.4}{2} 0.61^2\right]^{\frac{1.4}{0.4}} = 1.2856$$

$$P_{T2} = P_2 * \frac{P_{T2}}{P_2} = 187.0 \text{ kPa} * 1.2856 = 240.4 \text{ kPa}$$

Consequently, our total pressure loss is 60.35 kPa or just over 20 percent.
Finally, the velocity of the flow before and after the shock wave can be determined using the speed of sound times the Mach number.

$$V_1 = M_1 a_1 = M_1 \sqrt{kg_C RT_1} = 1.83 \sqrt{1.4\,(1)\,287\,\frac{J}{kg\,K}\,250\,K} = 1.83 * 316.9\,\frac{m}{s}$$
$$= 580.0\,m\,/\,s$$

$$V_2 = M_2 a_2 = M_2 \sqrt{kg_C RT_2} = 0.61 \sqrt{1.4\,(1)\,287\,\frac{J}{kg\,K}\,388.5\,K} = 0.61 * 395.1\,\frac{m}{s}$$
$$= 241.0\,m\,/\,s$$

The resulting velocity ratio, V_1/V_2, is equal to 2.4066. This value can be checked using the density ratio.

$$\frac{\rho_2}{\rho_1} = \frac{P_2}{P_1}\frac{T_1}{T_2} = \frac{3.740}{1.554} = 2.4067$$

Consequently, the two calculations for velocity ratio are consistent within round off error. The other values can be checked using the normal shock table.

Normal shock table: $M_1 = 1.83, M_2 = 0.6099, P_2 / P_1 = 3.7404,$
$T_2 / T_1 = 1.5541, P_{T2} / P_{T1} = .7993, \rho_2/\rho_1 = 2.4067.$

The preceding calculations are consistent with the tables, which can often be used to streamline these types of compressible flow analyses.

Another application for normal shocks might be at the entrance or exit of the diverging duct. This type of problem requires that information related to both isentropic flow and normal shock waves is used.

Example 3.2: Supersonic Flow with Area Change and Shock

Given: An airflow enters a diverging duct at a Mach number, $M_1 = 1.33$, a static pressure, $P_1 = 45$ kPa, and a static temperature, $T_1 = 270$ K. The area of the inlet is 0.1 m². The duct diverges to an area of 0.15 m². Subsequently, a normal shock occurs at the outlet to the duct, as pictured in Figure 3.8.

Wanted: Find the supersonic Mach number, M_2, at the end of the duct and the subsonic Mach number after the shock, M_3. Determine the static pressures just before, P_2, and just after, P_3 the shock. Also, find the static temperatures before and after the shock. Determine the total pressure before and after the shock.

Sketch:

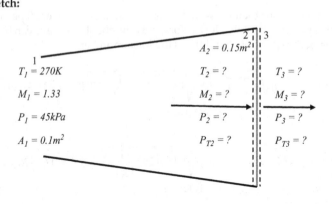

FIGURE 3.8

Schematic of supersonic flow entering a diverging channel with a normal shock at the exit.

Analysis: At the inlet of the nozzle, supersonic flow is entering a diverging duct. The flow upstream of the shock can be considered isentropic, so the flow from the inlet to the exit can be evaluated using our A/A^* ratios as the flow upstream from the shock can be assumed to have a single virtual throat at $M = 1$.

Isentropic tables, $k = 1.4$: $M_1 = 1.33$, $A_1/A^* = 1.07956$, $P_1/P_{T1} = 0.34640$, $T_1/T_{T1} = 0.73867$

$$\frac{A_2}{A^*} = \frac{A_1}{A^*}\frac{A_2}{A_1} = 1.07956\frac{0.15\,\text{m}^2}{0.1\,\text{m}^2} = 1.6193$$

Isentropic tables, $k = 1.4$: $M_2 = 1.95$, $A_2/A^* = 1.6193$, $P_2/P_{T2} = 0.13813$, $T_2/T_{T2} = 0.56802$

$$P_2 = P_1 * \frac{P_{T1}}{P_1} * \frac{P_2}{P_{T2}} = 45\,\text{kPa} * \frac{0.13813}{0.34640} = 17.94\,\text{kPa}$$

$$T_2 = T_1 * \frac{T_{T1}}{T_1} * \frac{T_2}{T_{T2}} = 270\,\text{K} * \frac{0.56802}{0.73867} = 207.6\,\text{K}$$

Having found the conditions of the flow upstream from the shock, normal shock tables can be used to determine the change in conditions across the shock.

Normal shock table: $k = 1.4$: $M_2 = 1.95$, $M_3 = 0.5862$, $P_3/P_2 = 4.2696$, $T_3/T_2 = 1.6473$, $P_{T3}/P_{T2} = 0.7442$, $\rho_3/\rho_2 = 2.5919$

$$P_3 = P_2 * \frac{P_3}{P_2} = 17.94\,\text{kPa} * 4.2696 = 76.6\,\text{kPa}$$

$$T_3 = T_2 * \frac{T_3}{T_2} = 207.6 \text{ K} * 1.6473 = 342.0 \text{ K}$$

$$P_{T2} = P_{T1} = \left. P_1 \middle/ \frac{P_1}{P_{T1}} \right. = \frac{45 \text{ kPa}}{0.3464} = 129.9 \text{ kPa}$$

$$P_{T3} = P_{T2} \frac{P_{T3}}{P_{T2}} = 129.9 \text{ kPa} * 0.7442 = 96.7 \text{ kPa}$$

In this example problem, supersonic flow entered a diverging duct. Based on the area change relationships developed for supersonic flow, increasing area causes decreased pressure, resulting in increased velocity. Inside the duct, the Mach number rose from 1.33 to 1.95, while pressure dropped from 45 kPa to 17.94 kPa. At the exit of the duct, the flow encountered a normal shock due to an elevated back pressure. This produced a rise in pressure from 17.94 kPa to 96.7 kPa. The Mach number also dropped to 0.586. The isentropic Mach number table provided a means to solve the flow variation upstream of the shock, and the normal shock table was used to find the change in conditions across the shock. One aspect of the total pressure loss across the shock is the resulting increase in size of the virtual throat, A^*. Equating the mass flow rates through the virtual throats (where $M = 1$) on the two sides of the shock wave in the previous problem.

$$\frac{P_{T2}}{RT_{T2}} A_2^* \sqrt{kg_c RT_{T2}} \left[1 + \frac{(k-1)}{2} \right]^{\frac{-(k+1)}{2(k-1)}} = \frac{P_{T3}}{RT_{T3}} A_3^* \sqrt{kg_c RT_{T3}} \left[1 + \frac{(k-1)}{2} \right]^{\frac{-(k+1)}{2(k-1)}}$$

Noting that the total temperature does not change, this relationship simplifies to:

$$P_{T2} A_2^* = P_{T3} A_3^*$$

In terms of the A^* ratio:

$$\frac{A_2^*}{A_3^*} = \frac{P_{T3}}{P_{T2}} \tag{3.14}$$

The increase in the size of the virtual throat across the shock is related to the reciprocal in the total pressure ratio across the shock. This requirement for an increased flow area downstream from a shock is an important constraint of many supersonic flow systems.

Normal shock waves can also appear within converging–diverging nozzles, as shown in Figure 3.9. Taking the previous solution from the nozzle in Chapter 2 where the backplane pressure was reduced from *a–d* until sonic flow occurred in the throat of the nozzle but with a subsonic solution

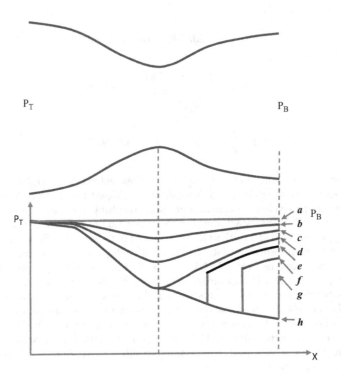

FIGURE 3.9
Sketch of converging diverging nozzle static pressure distributions with normal shocks.

for the exit area (A/A^*), the pressure is now lowered to e. At this pressure, supersonic flow is initially established inside the nozzle. However, the backplane pressure cannot support it, and the flow shocks up through a normal shock. The static pressure rises across the shock and the flow is now subsonic. Moving through a diverging duct, the flow velocity slows and the pressure rises to the exit. As the backplane pressure is further lowered to f, supersonic flow moves further into the nozzle, but again shocks up to subsonic flow to meet the back pressure condition at the exit of the nozzle. Here, the pressure initially drops as the flow goes supersonic in the diverging portion of the nozzle before shocking up to subsonic flow. The subsonic flow moving through the diverging section of the nozzle decreases in velocity and rises in pressure to the exit matching the backplane pressure. As the pressure is further lowered at the nozzle backplane to g, supersonic flow moves further into the nozzle until a normal shock resides at the exit of the nozzle. Any further lowering of the backplane pressure would be expected to produce an oblique shock wave at the nozzle exit. Also, shown on this figure is a representation of the isentropic solution for supersonic flow through the nozzle given by pressure h.

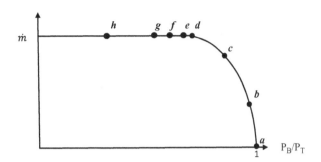

FIGURE 3.10
Plot of mass flow rate versus P_B/P_T for converging–diverging nozzle with shocks.

The mass flow also shown schematically is essentially similar to the previous solution. Once the nozzle is choked, the mass flow rate through the nozzle becomes fixed without regard to the decreasing backplane pressure, as shown in Figure 3.10.

3.3 Moving Shock Waves and Shock Reflections

Stationary shock waves are encountered in the analysis of flows in wind tunnels and in the inlets and exits of supersonic engines at certain points in the cycle. Stationary shock waves can also be encountered in compressible flows in piping systems under certain circumstances. Moving shock waves are typically related to transient events like the abrupt closing of a valve, a shock tube, or even an explosion. The present section provides a concise introduction to moving shock waves, providing some insight into this typically transient phenomenon, as well as the differences between stationary and moving reference frames for normal shocks.

In the current treatment of moving shock waves, the shock wave is initially moving into an undisturbed gas at some given conditions such as an ambient pressure and temperature. As the undisturbed gas is assumed to be stationary and a shock wave requires supersonic flow at its inlet, the shock wave moving into undisturbed gas must be moving at a supersonic velocity with respect to the undisturbed gas. Based on the treatment of stationary normal shock waves, flow enters a shock at a supersonic velocity and leaves at a subsonic velocity. However, if the phenomenon is looked at in the absolute reference, there is an immediate and significant rise in static pressure, as the shock passes a region along with a gas velocity following the shock. Think about a stationary shock wave, flow enters at a very high, supersonic velocity and leaves at a lower, subsonic velocity. *A shock wave moving along at a supersonic velocity leaves a gas velocity, V_G equal to $|V_1-V_2|$.*

Example 3.3: Moving Shock Wave

Given: A moving shock wave enters undisturbed air at a velocity of 650 m/s, as shown in Figure 3.11. The local ambient air pressure is 100,000 kPa and the ambient temperature is 21 °C (294.15 K).

Wanted: Determine the static pressure and temperature behind the shock wave. Also, calculate the gas velocity of the moving air behind the shock wave and the resulting total temperature and pressure in the absolute reference frame.

Sketch:

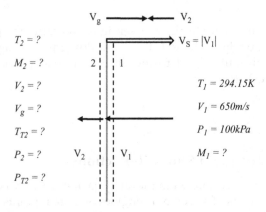

FIGURE 3.11
Control volume for a shock wave moving at V_S.

Analysis: A moving shock wave travels into still air at a velocity of 650 m/s. Behind the shock wave, the static pressure and temperature rise, and this higher pressure and temperature gas travels behind the shock wave at V_{gas}. In the moving reference frame, the moving shock wave can be analyzed like a normal shock wave. However, to determine the total properties of the gas in the absolute reference frame, the system must be put back into an absolute reference frame.

First, the Mach number of the entering flow relative to the moving wave will be determined.

$$M_1 = \frac{V_1}{a_1} = \frac{V_1}{\sqrt{k g_c R T_1}} = \frac{650\,\frac{m}{s}}{\sqrt{1.4(1)287\,\frac{J}{kg\,K}\,294.15\,K}} = \frac{650\,\frac{m}{s}}{343.8\,\frac{m}{s}} = 1.89$$

Now, the normal shock wave table can be used to determine M_2 and the static conditions behind the wave.

Normal shock table, $k = 1.4$: $M_1 = 1.89$, $M_2 = 0.5976$, $P_2/P_1 = 4.0008$, $T_2/T_1 = 1.6001$, $P_{T2}/P_{T1} = 0.7720$, $\rho_2/\rho_1 = 2.500$

$$P_2 = P_1 * \frac{P_2}{P_1} = 100\,\text{kPa} * 4.0008 = 400.08\,\text{kPa}$$

$$T_2 = T_1 * \frac{T_2}{T_1} = 294.15\,\text{K} * 1.6002 = 470.7\,\text{K}$$

The velocity leaving the shock wave can be determined from the density ratio, ρ_2/ρ_1.

$$V_2 = V_1 * \frac{V_2}{V_1} = \frac{V_1}{\dfrac{\rho_2}{\rho_1}} = \frac{650\,\dfrac{\text{m}}{\text{s}}}{2.500} = 260\,\frac{\text{m}}{\text{s}}$$

The gas velocity traveling behind the shock, $\vec{V_G}$, can be determined by the vector addition of the velocity of the moving shock wave, $\vec{V_S}$, plus the velocity of the gas leaving the shock, $\vec{V_2}$. This value is also equal to the absolute value of $|\vec{V_1}|$ less the absolute velocity of $|\vec{V_2}|$.

$$V_G = |\vec{V_S} + \vec{V_2}| = |\vec{V_1}| - |\vec{V_2}| = 650\,\frac{\text{m}}{\text{s}} - 260\,\frac{\text{m}}{\text{s}} = 390\,\frac{\text{m}}{\text{s}}$$

The total temperature and pressure in the absolute reference frame can be determined if the Mach number of the gas is first determined.

$$M_G = \frac{V_G}{a_G} = \frac{V_G}{\sqrt{k g_c R T_G}} = \frac{390\,\dfrac{\text{m}}{\text{s}}}{\sqrt{1.4(1)287\,\dfrac{\text{J}}{\text{kg K}}\,470.7\,\text{K}}} = \frac{390\,\dfrac{\text{m}}{\text{s}}}{434.9\,\dfrac{\text{m}}{\text{s}}} = 0.897$$

Isentropic tables, $k = 1.4$: $M_G = 0.90$, $P_G/P_{TG} = 0.5913$, $T_G/T_{TG} = 0.8606$

$$P_{TG} = P_G \frac{P_{TG}}{P_G} = \frac{400.08\,\text{kPa}}{0.5913} = 676.6\,\text{kPa}$$

$$T_{TG} = T_G \frac{T_{TG}}{T_G} = \frac{470.7\,\text{K}}{0.8606} = 546.9\,\text{K}$$

The moving shock wave produces a substantial rise not only in static conditions behind the shock, but also in the total conditions in the absolute reference frame. The static conditions on the lower velocity side of the shock are also the static conditions of the moving gas.

Example 3.4: Projectile moving in a barrel

A moving shock wave will also proceed a projectile in a gun barrel as the projectile moves through the barrel. If the velocity of the shock wave is known, then the velocity of the projectile can be calculated. However, if only the velocity of the projectile is known, the determination of the velocity of the shock wave is much more complicated. Fortunately, this general problem has been solved for an ideal gas with a constant specific heat, and the resulting equation is provided next [1, 2]:

$$V_S = V_G \left(\frac{k+1}{4} \right) + \sqrt{V_G^2 \left(\frac{k+1}{4} \right)^2 + a_1^2} \tag{3.15}$$

In this equation, V_S is the velocity of the shock wave traveling into a still gas, V_G is the velocity of the gas behind the shock wave, and this velocity is also the velocity of the projectile. The speed of sound in the undisturbed gas is a_1. This equation will be further highlighted in an example problem.

Given: A projectile moves down a barrel at 600 m/s into still air at the temperature of 23 °C and a pressure of 100,000 Pa as shown in Figure 3.12.

Wanted: Find the velocity of the moving shock wave, as well as the static temperature and pressure of the gas following the shock wave.

Sketch:

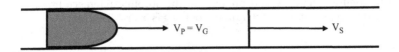

FIGURE 3.12
Sketch of shock wave proceeding projectile in a gun barrel.

Analysis: The projectile has a velocity of 600 m/s, and it also pushes the compressed gas in front of it at 600 m/s. The shock wave moving in front of the shock moves faster noting that the shock wave velocity is the sum of the projectile speed, $V_P = V_G$ and the velocity leaving the shock, V_2. In order to calculate the velocity of the shock wave, the speed of sound through the undisturbed gas must be determined.

$$a_1 = \sqrt{k g_c R T_1} = \sqrt{1.4(1)287 \frac{J}{kg\,K} 296.15\,K} = 345.0 \frac{m}{s}$$

Now, the velocity of the shock wave can be determined.

$$V_S = 600 \frac{m}{s} \left(\frac{1.4+1}{4} \right) + \sqrt{\left(600 \frac{m}{s} \right)^2 \left(\frac{1.4+1}{4} \right)^2 + \left(345 \frac{m}{s} \right)^2} = 858.6 \frac{m}{s}$$

The Mach number of the shock wave is

$$M_S = \frac{V_S}{a_1} = \frac{858.6\,\frac{m}{s}}{345\,\frac{m}{s}} = 2.49$$

Using the normal shock table or the normal shock equations, the static pressure and temperature of the gas downstream from the shock can be found.

Normal shock table: $k = 1.4$: $M_1 = 2.49$, $M_2 = 0.514$, $P_2/P_1 = 7.0668$, $T_2/T_1 = 2.1276$, $P_{T2}/P_{T1} = 0.5030$, $\rho_2/\rho_1 = 3.3214$

The pressure and temperature ratios can be used directly to find the conditions after the shock.

$$P_G = P_1 * \frac{P_G}{P_1} = 100\,\text{kPa} * 7.0668 = 706.7\,\text{kPa}$$

$$T_G = T_1 * \frac{T_G}{T_1} = 296.15\,\text{K} * 2.1276 = 630.1\,\text{K}$$

The total pressure ratio has no useful meaning in current reference frame, and the density ratio could be used to calculate or verify V_2.

There are some cases in engineering practice, research, and explosive phenomena when a moving gas impinges on a surface. One example is when a gas moving at a high velocity in a compressible flow system suddenly encounters a closing valve. Another situation is the reflection of a shock wave in the shock tube. A third situation occurs when a moving shock wave due to an explosion encounters a wall. The third situation will be chosen as an example, and the conditions from the initial moving shock wave example will be used for this situation.

Example 3.5: Reflected shock wave

In Example 3.3, the velocity of the gas behind the moving shock wave, V_G, was traveling at 390 m/s. The static pressure behind the wave was determined to be 400.08 kPa and the static temperature was 470.7 K. In this situation, when a shock wave reflects off a surface, a new shock wave is formed, which travels in the direction opposite to the original shock and the gas moving behind the shock wave toward the wall. This reflected shock wave moves into this gas moving in the opposite direction. Now, the gas leaving the moving reflected shock wave must equal the speed of the shock wave as the gas behind the reflected shock wave has to be stationary, $V_2 = V_{S/R}$. The velocity of the shock wave relative to the incoming gas is equal to the velocity of the reflected shock wave plus the velocity of the moving gas, $V_{S/G} = V_{S/R} + V_G$. Our relationship in calculating the velocity of the shock wave relative to the gas is given in Equation 3.16:

Given: Gas properties behind the initial moving shock wave, $V_G = 390 \, m/s$, $P_G = 400.08 \, kPa$, $T_G = 470.7 \, K$.

Wanted: Find velocity of the reflected shock wave, shown in Figure 3.13, and the static temperature and pressure behind this wave:

Sketch:

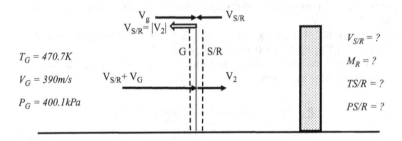

FIGURE 3.13
Schematic of moving normal shock wave reflecting off of a wall.

Analysis:

$$V_{S/R} + V_G = V_G \left(\frac{k+1}{4}\right) + \sqrt{V_G^2 \left(\frac{k+1}{4}\right)^2 + a_G^2} \qquad (3.16)$$

Based on the conditions taken from the initial moving shock wave:

$$V_{S/R} + V_G = 390\frac{m}{s}\left(\frac{1.4+1}{4}\right) + \sqrt{\left(390\frac{m}{s}\right)^2\left(\frac{1.4+1}{4}\right)^2 + \left(434.9\frac{m}{s}\right)^2} = 727.9\frac{m}{s}$$

$$V_{S/R} = \left(V_{S/R} + V_G\right) - V_G = 727.9\frac{m}{s} - 390\frac{m}{s} = 337.9\frac{m}{s}$$

The resulting Mach number across the reflected shock wave becomes:

$$M_R = \frac{V_{S/R} + V_G}{a_G} = \frac{727.9\frac{m}{s}}{434.9\frac{m}{s}} = 1.674$$

Interpolating from our normal shock tables: $k = 1.4$: $M_1 = 1.674$, $M_2 = 0.6474$, $P_2/P_1 = 3.1027$, $T_2/T_1 = 1.4389$, $P_{T2}/P_{T1} = 0.8664$, $\rho_2/\rho_1 = 2.155$

$$P_{S/R} = P_G * \frac{P_{S/R}}{P_G} = 400.08 \, kPa * 3.1027 = 1241.3 \, kPa$$

$$T_{S/R} = T_G * \frac{T_{S/R}}{T_G} = 470.7 \, K * 1.4389 = 677.3 \, K$$

This analysis shows that the properties behind a reflected shock wave can be substantially changed. In the initial problem, the shock wave was moving toward the wall due to gas moving behind the shock wave, causing the still air to be compressed. When the shock wave impinges on the wall, it must reflect. Then, a reflected shock wave moves out away from the wall, which further compresses the air in a manner that allows it to be at rest with the wall, as the reflected shock moves into the oncoming gas.

3.4 A Brief Introduction to Shock Tubes

A shock tube is a device that can be used to create a warm or hot, high-pressure, high-speed condition for a short period of time. During the period that this condition is generated, different types of physical phenomena can be studied. The advantage of a shock tube is these warm or hot, high-speed conditions can be developed relatively inexpensively compared to the cost of a heated flow generated with a large compression system or tank farm. One example of the use of shock tubes is in the study of chemical kinetics. As the moving shock wave passes across a region, an abrupt change in the temperature and pressure of the residual gas occurs. The reaction of the gas mixture to this abrupt change in conditions is useful in establishing the rate at which particular reactions occur at specific temperature and pressure conditions. Other phenomena that can be studied include compressible flow phenomena. On a small scale, a passing shock wave can be used to study the influence of a high-speed flow past an object. On a larger scale, large shock tubes have been used to create conditions for the transient testing of turbine sections in gas turbines. The Gas Turbine Laboratory at Ohio State University has a very large shock tube system, which has been operating there for over 20 years.

A shock tube is created by taking a pipe and separating a high-pressure region with a low-pressure region using a diaphragm, which is designed to rupture. A schematic sketch of a shock tube is shown in Figure 3.14. The high-side pressure and temperature are designated P_4 and T_4, respectively, while the low-side pressure and temperature are designated P_1 and T_1, respectively. As the diaphragm bursts, a normal shock wave is formed as the high-pressure gas expands into the low-pressure region. As the shock wave moves through the undisturbed gas, the gas is compressed across a moderately high-pressure ratio, and the gas is immediately accelerated to a high velocity and follows behind the wave. This velocity is similar to the gas velocity behind a moving shock wave, which was introduced in Section 3.3. The velocity of this compressed gas on the low-pressure side, V_3, is the same as the velocity on the high-pressure side V_2. However, one large difference is the gas on the low-pressure side is compressed by the expanding gas, while the gas from the high-pressure side moving to the right has been

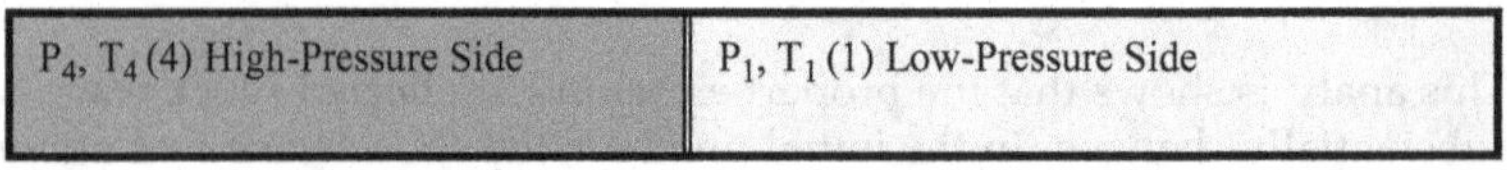

P$_4$, T$_4$ (4) High-Pressure Side	P$_1$, T$_1$ (1) Low-Pressure Side

FIGURE 3.14
Schematic of typical shock tube with high- and low-pressure sides divided by a diaphragm.

expanded to P_2. Due to the quasi-equilibrium between the expanded gas and the compressed gas, $V_2 = V_3$ and $P_2 = P_3$. The following figure, the shock tube sketch, gives a schematic depiction of a shock tube sometime after the diaphragm has been ruptured. The shock wave formed by the compression of the gas propagates into region (1) on the low-pressure side of the diaphragm. The shock wave leaves the compressed gas (2) moving behind it. When the diaphragm bursts, the high-pressure gas is expanded (3) and moves into the low-pressure region as the same speed as the compressed gas (2). In between the compressed low-pressure gas and the expanded high-pressure gas (3), an interface called the contact surface moves along at the velocity of the gas on either side. As the diaphragm ruptures, expansion waves begin to move into the high-pressure side at the speed of sound. The most-right-hand expansion wave will move into the high-pressure side at the speed of sound of the expanded gas from the high-pressure side, but starting from the gas moving at the velocity of V_3 and V_2. Consequently, the expansion wave quickly spreads out.

The phenomena of the moving shock wave created by the expanding gas can be analyzed using the approach outlined in Section 3.3. However, velocity V_2, and the pressure P_2 are unknowns. Additionally, the expansion of the high-pressure gas from P_4 to P_3, which produces velocity V_3, can be analyzed using the isentropic relationships similar to those introduced in Chapter 2, with the exception that some of the high-pressure gas does work on the expanding gas, creating a slightly higher velocity than the isentropic relationships would predict. P_3 and V_3 are unknown, but the high-pressure gas expansion drives the compression of the low-pressure gas. The condition that ties the expansion together with the compression shock is that both the velocity at the interface and the static pressure at the interface are equal. That is, $V_2 = V_3$ and $P_2 = P_3$.

The following figure depicts the transient flow phenomena in a shock tube a short time after the diaphragm has been ruptured. As the high-pressure gas expands into the region originally occupied by the low-pressure gas (P_1), the low-pressure gas is compressed to P_2, which is the same pressure (P_3) as the expanding gas from the high-pressure side. However, the temperature of the compressed gas, T_2, increases above T_1 due to the compression across the wave, while the temperature of the expanded high-pressure gas, T_3, drops below T_4 due to the expansion process. The change between the expanded high-pressure gas and the compressed low-pressure gas cannot be seen in the pressure distribution. However, the change in temperature is readily apparent across the contact surface.

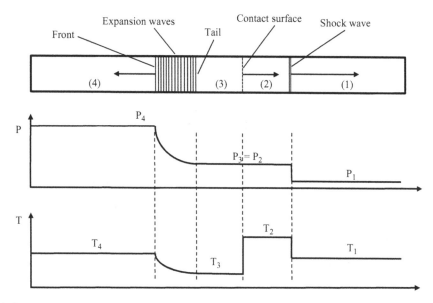

FIGURE 3.15
Schematic of flow phenomena in shock tube after diaphragm burst, including pressure and temperature distributions.

The expansion process between P_3 and P_4 is isentropic. However, in the expansion process, some of the upstream expanding gas does work on the downstream expanding gas. Consequently, noting P_4, P_3, and T_4, then T_3, the resulting V_3 determined using simple isentropic relationships is slightly lower than the actual V_3. Noting P_2, P_1, and T_1, the Mach number of the traveling shock wave can be determined, and T_2 and V_2 can be determined. At the interface, $P_2 = P_3$. Developments for V_3 and V_2 can be equated, and a relationship can be determined for P_4/P_1 based on P_2/P_1 and a_4 and a_1 as well as k_4 and k_1. However, the speed of sound and specific heat ratio are simply functions of temperature. The relationship is given next [1, 2]:

$$\frac{P_4}{P_1} = \frac{P_2}{P_1}\left[1 - \frac{(k_4-1)\left(\dfrac{a_1}{a_4}\right)\left(\dfrac{P_2}{P_1}-1\right)}{\sqrt{2k_1}\sqrt{2k_1+(k_1+1)\left(\dfrac{P_2}{P_1}-1\right)}}\right]^{\frac{-2k_1}{(k_1-1)}} \tag{3.17}$$

This relationship could be used to design a specific shock wave by choosing the pressure ratio P_2/P_1 and then calculating P_4/P_1. If the gases on both sides of the diaphragm are the same and $T_1 = T_4$, then the subscripts on the specific heat ratio, k, can be neglected and the ratio (a_1/a_4) just becomes 1. This relationship between P_2/P_1 and P_4/P_2 is tabulated in Appendix A.3.

Example 3.6: Shock Tube with Reflection

Given: A shock tube must be designed to operate at 650 K and 101 kPa for 2.5 msec in supersonic flow starting from a near ambient temperature. Assume that $T_1 = T_4 = 304$ K.

Find: Determine the distance from the contact surface that the object needs to be to produce a test duration of 2.5 msec. Also determine the minimum distance from object that the end of the tube must be to ensure that the reflected shock does not limit the duration of the test.

Analysis: The strength of the normal shock wave needed can be determined from the temperature ratio. $T_2 = 650$ K is the desired temperature of the test condition. $T_1 = 304$ K is the approximate temperature of the initial condition in the low-pressure section. $T_2/T_1 = 2.1382$
From the normal shock tables: $M_1 = 2.5$, $M_2 = 0.5130$, $P_2/P_1 = 7.125$, $T_2/T_1 = 2.1375$, and $\rho_2/\rho_1 = 3.3333$
From the shock tube table for $k = 1.4$: for $P_2/P_1 = 7.125$, $P_4/P_1 = 145.342$.
The low and high pressures in the shock tube can be determined from the shock tube ratios:

$$P_1 = \frac{P_2}{\dfrac{P_2}{P_1}} = \frac{101\,\text{kPa}}{7.125} = 14.175\,\text{kPa}$$

$$P_4 = P_1 * \frac{P_4}{P_1} = 14.175\,\text{kPa} \times 145.342 = 2060\,\text{kPa}$$

The velocity of the shock wave can be determined from the Mach number and speed of sound.

$$V_1 = M_1 a_1 = M_1\sqrt{kg_c R T_1} = 2.50\sqrt{1.4\,(1)\,287\,\frac{\text{J}}{\text{kg K}}\,304\,\text{K}} = 2.50 * 349.5\,\frac{\text{m}}{\text{s}}$$
$$= 873.7\,\text{m/s}$$

The gas velocity traveling behind the shock, $\vec{V_G}$, can be determined by the vector addition of the velocity of the moving shock wave, $\vec{V_S}$, plus the velocity of the gas leaving the shock, $\vec{V_2}$. This value is also equal to the absolute value of $|\vec{V_1}|$ less the absolute velocity of $|\vec{V_2}|$.

$$V_G = \left|\vec{V_S} + \vec{V_2}\right| = \left|\vec{V_1}\right| - \left|\vec{V_2}\right| = V_1\left(1 - \frac{\rho_1}{\rho_2}\right) = 873.7\,\frac{\text{m}}{\text{s}}\left(1 - \frac{1}{3.333}\right) = 611.6\,\frac{\text{m}}{\text{s}}$$

The distance to the object needed to develop the required time between the passing of the shock wave and the contact surface can be determined from the respective velocities and the required time.

$$t = \frac{L}{V_G} - \frac{L}{V_1}$$

$$L = \frac{t}{\left(\dfrac{1}{V_G} - \dfrac{1}{V_1}\right)} = \frac{0.0025\,\text{s}}{\left(\dfrac{1}{611.6\,\text{m}\,/\,\text{s}} - \dfrac{1}{873.7\,\text{m}\,/\,\text{s}}\right)} = 5.097\,\text{m}$$

The velocity of the reflected shock can be determined from the calculation of the velocity of the gas entering the shock.

$$V_{S/R} + V_G = V_G\left(\frac{k+1}{4}\right) + \sqrt{V_G^2\left(\frac{k+1}{4}\right)^2 + a_G^2}$$

The speed of sound in the compressed gas can be determined from the speed of sound formula.

$$a_G = \sqrt{kg_C R T_G} = \sqrt{1.4\,(1)\,287\,\frac{\text{J}}{\text{kg}\,\text{K}}\,650\,\text{K}} = 511.0\,\frac{\text{m}}{\text{s}}$$

Based on the gas velocity and the speed of sound of the gas, the reflected shock inflow velocity and absolute velocities can be determined.

$$V_{S/R} + V_G = 611.6\,\frac{\text{m}}{\text{s}}\left(\frac{1.4+1}{4}\right) + \sqrt{\left(611.6\,\frac{\text{m}}{\text{s}}\right)^2\left(\frac{1.4+1}{4}\right)^2 + \left(511.0\,\frac{\text{m}}{\text{s}}\right)^2}$$

$$= 996.1\,\frac{\text{m}}{\text{s}}$$

$$V_{S/R} = \left(V_{S/R} + V_G\right) - V_G = 996.1\,\frac{\text{m}}{\text{s}} - 611.6\,\frac{\text{m}}{\text{s}} = 384.5\,\frac{\text{m}}{\text{s}}$$

Based on the time required from the passing shock to the reflection and back, the end length needs to be calculated:

$$t = \frac{L_E}{V_S} + \frac{L_E}{V_{S/R}}$$

$$L_E = \frac{t}{\left(\dfrac{1}{V_S} + \dfrac{1}{V_{S/R}}\right)} = \frac{0.0025\,\text{s}}{\left(\dfrac{1}{873.7\,\text{m}\,/\,\text{s}} + \dfrac{1}{384.5\,\text{m}\,/\,\text{s}}\right)} = 0.667\,\text{m}$$

A simple schematic showing the movement of the normal shock then reflected shock is shown in Figure 3.16 as a function of time with the heavier solid line. The movement of the contact surface over time is shown with the dashed line. The leading and trailing expansion waves over times are shown with the lighter solid lines.

The shock tube is a practical application of moving and reflected shock waves with the complication of the expansion process. The shock tube problem was applied to develop specific conditions for a specific period of time in a short duration test. At higher pressure ratios, the contact surface moves at a large fraction of the shock wave velocity. In the present case, the contact

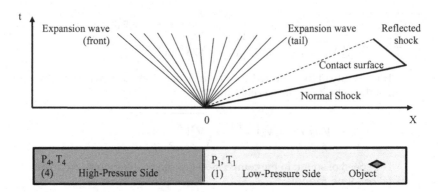

FIGURE 3.16
Schematic of shock tube and depiction of movement of flow phenomena after rupture.

surface moved at 70 percent of the speed of the shock requiring a long distance between the diaphragm location and test location or object to develop the required test duration. As the reflected shock moves much more slowly than the original shock wave, the end length of the tube can be much shorter than the distance to the location of interest from the diaphragm. Expansion waves initiate at the speed of sound at the rupture of the diaphragm moving back into the high-pressure side. The trailing expansion wave moves out from the contact surface relative to its velocity. As a result, when shock velocities are higher, the tail of the expansion waves actually moves forward.

Shock waves are an important phenomenon in many compressible flow situations. The equations for shock waves were developed from the momentum principle, conservation of mass, conservation of energy, and the ideal gas law. These equations were straightforward to develop, and the calculated values for M_2, P_2/P_1, T_2/T_1, ρ_2/ρ_1, and P_{T2}/P_{T1} versus M_1 are tabulated in Appendix A.2 for specific heat ratios, k of 1.4 and 1.3. Tables for shock tubes are also provided in Appendix A.3 for $k = 1.4$ and $k = 1.3$ assuming $a_1 = a_4$.

References

1. John, J.E.A., and T.G. Keith. 2006. *Gas Dynamics*, 3rd ed. Prentice Hall.
2. Saad, M.A. 1993. *Compressible Fluid Flow*, 2nd ed. Prentice Hall.

Chapter 3 Problems

1. The flow upstream from a normal shock wave in a wind tunnel has a Mach number of 2.3 and a static pressure of 35 kPa. The local static temperature has been estimated to be 255 K. Determine the Mach

number, static pressure, static temperature, and total temperature downstream from the normal shock wave. Also determine the total pressure loss.

2. A converging–diverging nozzle has an exit to throat area ratio of 1.45. The inlet total pressure is unknown, but the exit pressure is 100 kPa, and a normal shock sits at the exit of the nozzle. Determine the upstream total pressure as well as the Mach number before and after the shock at the nozzle exit plane.

3. The X-43 is a supersonic experimental aircraft that flies at up to Mach 9.6. Determine the total pressure and temperature that the leading edge of the aircraft experiences downstream from the very small bow shock wave, which occurs at the leading edge. You can assume that the initial portion of the bow shock at the leading edge is a normal shock wave. The free-stream conditions that the X-43 experiences is a static pressure of around 0.9 kPa and a local temperature of 230.5 K as it flies at approximately 32,000 m in altitude. The specific heat ratio k can be assumed to be 1.4.

4. A moving shock wave is generated by a high-speed train moving through a tunnel. The shock wave passes at 400 m/s. The ambient air that it moves into is 298 K and the ambient pressure is 100 kPa. Find the Mach number of the shock wave and estimate the speed of the train and the static pressure and temperature that is experienced by the train due to the compression of the shock wave.

5. A shock wave, assumed to be one dimensional, is generated by an explosion and moves through the still air at 690 m/s. The ambient temperature is 23°C and the ambient pressure is 100 kPa. As the shock wave passes a stationary observer, determine the static pressure and temperature that the observer would experience. Determine the velocity of the gas moving behind the shock wave. Calculate the total pressure and temperature that would be experienced by the stationary observer.

6. Air discharges from a large pipe at a velocity of 250 m/s. The air's temperature is 315 K and it discharges to an ambient pressure of 100 kPa. A valve in the line near the exit of the pipe suddenly closes and a reflected shock wave is formed, which moves into the pipe in the direction of the flow. Determine the speed of the reflected shock wave as well as the temperature and pressure of the air left behind the shock wave.

7. A moving shock wave traveling at a Mach number of 2 reflects from a large building. If the ambient temperature and pressure of the still air is 288 K and 101 kPa, determine the static pressure and temperature experienced by the wall of the building, as well as the speed at which the reflected shock wave initially travels away from the building.

8. A shock tube has an active length downstream from the diaphragm of 8.5 m. A target is placed 7.5 m downstream from the diaphragm. The pressure in the high-pressure side is 2.2 MPa, while the pressure in the low-pressure side is 15 kPa. The gas temperature on both sides of the diaphragm is 310 K. Determine the Mach number produced by the rupture of the diaphragm. Determine the pressure, temperature, and velocity of the compression wave generated by the rupture, as well as the velocity of the gas behind the shock wave. Determine the time that the compressed gas flow is experienced by the target before the contact surface passes over the target.

9. A shock tube is designed to generate a shock wave with a Mach number of 2.4. The design pressure to be generated is 150 kPa and the design temperature is 600 K. The target needs to experience the higher pressure and temperature conditions for 6 msec. Determine the required initial temperature and pressure to create the static temperature and pressure conditions. Determine the minimum length the target needs to be from the diagram to develop the condition for the required time. Determine how far the end of the diaphragm needs to be from the target to ensure the reflection does not limit the time of the test below the requirements.

4

Oblique Shock Waves

Supersonic flows cannot gradually adjust to the presence of an object in the flow because pressure waves travel at the speed of sound and are therefore unable to sense the presence of the object. One way supersonic flows can abruptly adjust to the presence of an object is through a normal shock wave. As we noted, this can result from the back pressure on a supersonic flow in a wind tunnel or compressible flow duct producing a stationary shock wave. The normal shock wave can also be produced by a moving projectile or a gas moving through a pipe or the response of a gas moving through a pipe to an abrupt valve closure. What type of situation can produce an oblique shock wave?

An oblique shock wave occurs when a supersonic flow is required to undergo an abrupt, but reasonably mild turn in flow. This turn can occur due to a sharp wedge encountered by the flow or a change in angle of a duct. Oblique shock waves are common facets of flow for supersonic aircraft. They also occur at supersonic inlets and supersonic flow in nozzles. They also typically occur at the leading edge and trailing edge of supersonic airfoils and the leading edge of supersonic aircraft. Oblique shocks are also common in the fans of turbofan engines, in compressors, and in supersonic turbines.

Oblique shock waves may initially seem challenging. However, the analysis of oblique shock waves is not greatly different from that of normal shock waves. In fact, an oblique shock wave is essentially a normal shock wave, but with a tangential flow component.

A weak oblique shock wave can occur due to a very small concave angle. However, the minimum angle of an oblique shock wave's angle, θ, to the flow will be the Mach angle or the arcsin($1/M$). The oblique shock wave equations will have two solutions for every turning angle or wedge angle, δ. However, in typical external flow situations, such as aircraft, the weak shock solution is always found, as there is no way to confine the back pressure to a level necessary to produce the strong shock solution. Additionally, there is a maximum deflection angle for oblique shocks. If this turning angle is exceeded, the shock wave will move off the location where the flow is required to turn and form a bow shock wave, which initially is similar to a normal shock wave. This chapter also addresses the subject of conical shock waves which have similarities and differences with oblique shock waves.

4.1 Oblique Shock Wave Equations

The development of oblique shock wave equations is very similar to the development of the normal shock equations. Similar to the development of the normal shock equations, the development of the oblique shock equations includes the use of conservation of mass, conservation of energy, the momentum principle, the ideal gas law, and the definition of the speed of sound. An oblique shock is very thin, so similar to the analysis for normal shock the area of an oblique shock is assumed to be constant. However, differences between the normal shock and oblique shock analyses include the reality that only the normal component of flow moves across the shock wave. Also the energy equation requires that the total temperature must stay the same across the shock. Only the normal velocity, not the tangential velocity, changes across an oblique shock.

In the following analysis, the flow moves from left to right, and the angle of the oblique shock, θ, is measured from a plane parallel to the oncoming flow and in a direction counterclockwise to the direction of this velocity vector. The deflection angle of the shock wave is also measured from this direction. In Figure 4.1, flow is parallel to a surface before an oblique shock wave turns the flow to the direction of the downstream surface. The deflection angle, δ, is the angle of the downstream surface to the upstream surface, and it is also the angle that the flow has been turned. The angle of the oblique shock, θ, is the angle that the oblique shock must be to change the direction of the approach flow by the deflection angle, δ.

4.1.1 Analysis of an Oblique Shock Wave

The analysis can begin by applying conservation of mass. However, our control volume surrounds the oblique shock wave, which sits slanted to the flow direction. Recalling the development of the control volume equations, only

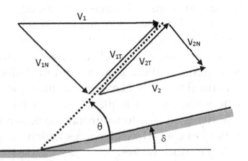

FIGURE 4.1
Schematic of velocity vectors for oblique shock of angle, θ, generated by concave angle, δ.

the normal component of velocity takes mass and other properties across the control volume, which in this instance is coincident with the shock. The integral around the test surface (Equation 1.20 for steady-state flow) can be interpreted as the outflow rate minus the inflow rate of mass from and to the control volume and written in terms of a one-dimensional uniform flow slanted at θ to the approach velocity.

$$\dot{P}_M = 0 = \oint_{CS} \rho V \cdot dA = \rho_2 V_{2N} A_2 - \rho_1 V_{1N} A_1 \tag{1.20}$$

Noting that the shock wave is very thin and $A_1 = A_2$, this simplifies to:

$$\rho_2 V_{2N} = \rho_1 V_{1N}$$

Again, this relationship will be used to integrate results from conservation of energy and the momentum principle. Applying the momentum principle (Equation 1.25 for steady-state flow) across an oblique shock wave, the vector nature of the equation suggests that it can be applied in the normal and tangential directions. The analysis can begin using the control volume form of the momentum principle with the steady-state assumption and with the flow in the normal direction to the oblique shock wave.

$$F_{S_N} + F_{B_N} + R_N = \oint_{CS} V_N \rho V \cdot dA / g_c \tag{1.25}$$

There are no physical attachments in the control volume, so reaction forces cannot be present. Also as this is a gas, body forces will be assumed to be negligible. These simplifications leave a balance between the change in pressure forces normal to the oblique shock wave equaling the outflow minus inflow of momentum, normal to the wave.

$$P_1 A_1 - P_2 A_2 = \frac{\rho_2 V_{2N}^2 A_2}{g_c} - \frac{\rho_1 V_{1N}^2 A_1}{g_c} \tag{4.1}$$

Area is then canceled, density is replaced with the ideal gas law and a relationship for pressure change is developed.

$$P_1 + \frac{P_1 V_{1N}^2}{R T_1 g_c} = P_2 + \frac{P_2 V_{2N}^2}{R T_2 g_c} \tag{4.2}$$

Factoring out pressure from each side:

$$P_1 \left(1 + \frac{V_{1N}^2}{R T_1 g_c}\right) = P_2 \left(1 + \frac{V_{2N}^2}{R T_2 g_c}\right) \tag{4.3}$$

Note that the denominator multiplied by the specific heat ratio k is equal to a^2. Multiplying the velocity terms on each side by (k/k), the relationship for pressure change across an oblique shock wave can be written in terms of the normal Mach number and the specific heat ratio.

$$P_1\left(1+k\,M_{1N}^2\right)=P_2\left(1+k\,M_{2N}^2\right) \tag{4.4}$$

This relationship is more often given in terms of the static pressure ratio across a shock wave.

$$\frac{P_2}{P_1}=\frac{\left(1+k\,M_{1N}^2\right)}{\left(1+k\,M_{2N}^2\right)} \tag{4.5}$$

The momentum equation normal to an oblique shock wave was essentially the same as the momentum equation across a normal shock wave. However, in the present case, only the velocities normal to the oblique shock wave were used. What does momentum tell us about the tangential velocities?

$$F_{S_T}+F_{B_T}+R_T=\oint_{CS}V_T\,\rho\,V\cdot dA\,/\,g_c$$

Again, similar to normal momentum, reaction and body forces can be ignored. Pressure forces need to be evaluated. Although pressure changes across an oblique shock wave in the normal direction, there is no change in the tangential direction. The pressure forces can be evaluated above or below the oblique shock wave, and there is no change in pressure in the tangential direction to produce a change in tangential velocity. Therefore, assuming no area change, tangential momentum can be written.

$$0=\frac{\rho_2V_{2N}V_{2T}}{g_c}-\frac{\rho_1V_{1N}V_{1T}}{g_c} \tag{4.6}$$

Applying the result from continuity, this leads us to the conclusion that $V_{2T}=V_{1T}$.

Conservation of energy essentially indicates that the total enthalpy stays constant across a shock wave. In terms of an oblique shock wave, the principle can be written.

$$h_T=h_2+\frac{V_2^2}{2g_c}=h_1+\frac{V_1^2}{2g_c}=h_2+\frac{V_{2N}^2+V_{2T}^2}{2g_c}=h_1+\frac{V_{1N}^2+V_{1T}^2}{2g_c}$$

Previously, from tangential momentum, $V_{1T}=V_{2T}$, which means the tangential terms can be eliminated.

$$h_2+\frac{V_{2N}^2}{2g_c}=h_1+\frac{V_{1N}^2}{2g_c} \tag{4.7}$$

Assuming a constant C_P and dividing through by $C_P=kR/(k-1)$ and factoring T.

$$T_2\left(1+\frac{(k-1)}{2}\frac{V_{2N}^2}{k\,R_g c\,T_2}\right) = T_1\left(1+\frac{(k-1)}{2}\frac{V_{1N}^2}{k\,R_g c\,T_1}\right) \tag{4.8}$$

The relationship of $k\,g_c\,R\,T$ is identified as the square of the speed of sound, and our new relationship becomes a function of the normal Mach number on each side of the shock:

$$T_2\left(1+\frac{(k-1)}{2}M_{2N}^2\right) = T_1\left(1+\frac{(k-1)}{2}M_{1N}^2\right) \tag{4.9}$$

The results for conservation of mass, static pressure ratio, and static temperature ratio across an oblique shock wave are identical to the relationships that were developed for normal shock waves, with the exception that all the current relationships have been put in terms of the Mach number or velocity normal to the oblique shock wave. Consequently, substituting these temperature and pressure ratios into continuity using the ideal gas law and the definition of the speed of sound it can be shown that:

$$M_{2N} = \sqrt{\frac{M_{1N}^2 + \dfrac{2}{k-1}}{M_{1N}^2\dfrac{2k}{k-1} - 1}} \tag{4.10}$$

Clearly, an oblique shock wave is essentially a normal shock wave across the oblique shock surface with a tangential component of velocity. The other necessary constraint is that $V_{2T} = V_{1T}$.

4.2 Supersonic Flow Over an Abrupt Wedge

When a supersonic flow encounters a wedge in its flow path it must turn abruptly. The upstream flow cannot sense the wedge, as the flow is moving faster than pressure waves. In this section, the equations developed for oblique shocks in the previous section will be applied in an example.

Example 4.1: Oblique Shock Example

Given: A supersonic flow has a Mach number of 2.0 and it is moving parallel to a flat plate. At the point of interest, the flow encounters a concave corner that produces an oblique shock wave that has an angle of 41° as shown in Figure 4.2.

Find: Determine the strength of the oblique shock wave in terms of M_{1N}, M_{2N}, P_2/P_1, T_2/T_1, M_{1T}, M_{2T}, M_2, and P_{T2}/P_{T1}, and calculate the angle of the turn required to produce this oblique shock.

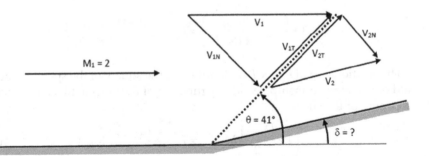

FIGURE 4.2
Schematic of oblique shock of angle, θ, generated by concave angle, δ.

Analysis: This problem can be most easily analyzed if put in terms of Mach number. The analysis can begin by determining the Mach number of the flow normal to the oblique shock wave. Finding M_{1N} and M_{1T}.

$$M_{1N} = M_1 \sin(\theta) = 2\sin(41°) = 1.3121 \tag{4.11}$$

$$M_{1T} = M_1 \cos(\theta) = 2\cos(41°) = 1.5094 \tag{4.12}$$

Next, finding M_{2N} will allow finding static conditions across the oblique shock wave.

$$M_{2N} = \sqrt{\frac{M_{1N}^2 + \frac{2}{k-1}}{M_{1N}^2 \frac{2k}{k-1} - 1}} = \sqrt{\frac{1.3121^2 + \frac{2}{1.4-1}}{1.3121^2 \frac{2*1.4}{1.4-1} - 1}} = 0.7799$$

The static pressure and temperature ratios can be computed from this section's relationships.

$$\frac{P_2}{P_1} = \frac{\left(1 + k M_{1N}^2\right)}{\left(1 + k M_{2N}^2\right)} = \frac{1 + 1.4 \times 1.3121^2}{1 + 1.4 \times 0.7799^2} = 1.8419$$

$$\frac{T_2}{T_1} = \frac{\left(1 + (k-1)/2\, M_{1N}^2\right)}{\left(1 + (k-1)/2\, M_{2N}^2\right)} = \frac{\left(1 + 0.4/2 \times 1.3121^2\right)}{\left(1 + 0.4/2 \times 0.7799^2\right)} = 1.1985$$

The tangential Mach number can be determined from the knowledge that the tangential velocity remains constant across the oblique shock wave.

$$V_{2T} = V_{1T} = M_{2T}a_2 = M_{1T}a_1 \tag{4.13}$$

$$M_{2T} = M_{1T}\frac{a_1}{a_2} = M_{1T}\frac{\sqrt{k\, g_c\, R\, T_1}}{\sqrt{k\, g_c\, R\, T_2}} = M_{1T} / \sqrt{\frac{T_2}{T_1}} = \frac{1.5094}{\sqrt{1.1985}} = 1.3787 \tag{4.14}$$

At this point, noting that the normal and tangential components of Mach number can be combined using the square root of the sum of the squares, the Mach number downstream of the oblique shock wave can be determined.

$$M_2 = \sqrt{M_{2N}^2 + M_{2T}^2} = \sqrt{.7799^2 + 1.3787^2} = 1.584$$

The total pressure ratio across the shock wave can be determined starting with the static pressure ratio. Later, the static to total pressure ratios before and after the shock wave can be multiplied and divided, respectively.

$$\frac{P_{T2}}{P_{T1}} = \frac{P_2}{P_1} \frac{P_1}{P_{T1}} \frac{P_{T2}}{P_2} = 1.8419 \times \frac{0.1278}{0.2409} = 0.9772$$

However, this total pressure ratio can simply be taken from the normal shock table using M_{1N}. The angle of the wedge, δ, can be determined using the arcsin of M_{2N}/M_2.

$$\theta - \delta = \operatorname{asin}\left(\frac{M_{2N}}{M_2}\right) = \operatorname{asin}\left(\frac{0.7799}{1.584}\right) = 29.496° \tag{4.15}$$

$$\delta = \theta - (\theta - \delta) = 41° - 29.496° = 11.504° \tag{4.16}$$

The analysis indicated that the change in angle of the flow was 11.504° due to a concave angle. This analysis shows a practical approach to solving oblique shock wave problems if the angle of the oblique shock wave with respect to the flow direction is known. The procedure involves breaking the incident flow down into normal and tangential components based on the angle of the oblique shock. The normal component of the incident Mach number upstream of the oblique shock was used to find the downstream normal Mach number. With M_{1N} and M_{2N}, the static pressure and temperature ratios were determined based on the relationships for the ratio of these properties across an oblique shock. These equations are identical to those used for normal shocks, which means that the normal shock tables can also be used in the development of an oblique shock solution. Noting that the tangential velocity remains constant across the oblique shock, the M_{2T} can be assessed from M_{1T} divided by the square root of T_2/T_1. Finally, M_2 can be determined from its orthogonal components and the angle of the oblique shock with respect to the direction of M_2 can be found. This essentially determines the deflection angle.

The angle of the oblique shock is often unknown, and the oblique shock angle then needs to be found based on the deflection angle. There are oblique shock charts (see Appendix A.4) that provide information on the shock angle given the deflection. However, these charts are often difficult to use accurately. Developing a set of oblique shock equations in a spreadsheet allows a quick method to iterate to an acceptably close solution. Additionally, Table A.4 provides the oblique shock angle for a given Mach number and deflection angle. However, due to the wide range of information in this Mach number, angle space it has been developed at Mach number increments of 0.1 and deflection angles increments of 1°. Charts are also provided in some books, which provide the downstream Mach number, given the inlet Mach number and deflection angle. However, calculations are much more accurate, and they are the suggested procedure.

4.3 Supersonic Inlets, Nozzles, and Airfoils

Supersonic inlets are designed to diffuse supersonic flows, while nozzles are designed to accelerate flows. However, each system can involve flows with oblique shock waves. Supersonic airfoils are designed to create lift and are also a location where oblique shocks form. Any type of shock produces a substantial increase in pressure. In the creation of lift, an increase in the pressure over the lower portion of the wing is critical. However, at supersonic speeds, pressure drag or "wave drag" can produce a substantial force, which impedes flight. A supersonic diffuser is designed to result in a pressure rise. A shock wave can produce a substantial pressure rise. However, a normal shock wave often results in a substantial total pressure loss, which must be made up by a jet engine's fan in supersonic flight. Consequently, keeping the total pressure loss at a minimum is very important in supersonic diffusers for aircraft. One approach to diffusing a supersonic flow with low losses is the development of an oblique shock diffuser. An oblique shock diffuser, a nozzle with an oblique shock at the exit, and a supersonic airfoil will be analyzed or discussed in part in this section.

Example 4.2: Oblique Shock Diffuser

A supersonic aircraft is designed to fly at a Mach number of 1.8 as shown in Figure 4.3. The engine manufacturer has plans to design a single oblique shock diffuser as part of the process to diffuse the supersonic flow. They have proposed a diffuser with a wedge angle of 8.5°. Compare the total pressure loss for this supersonic diffuser, which includes the oblique shock and then normal shock with a normal shock diffuser in terms of the total pressure loss of the system.

Analysis: The analysis can begin by estimating the shock angle provided in Table A.4. At 8°, $\theta = 41.6734°$ and at 9°, $\theta = 42.8385°$. This averages to an oblique shock angle of 42.2559°. The previous procedure can now be used to determine the system's total pressure loss and the Mach number and pressure, which are downstream from the diffuser.

Finding M_{1N} and M_{1T}.

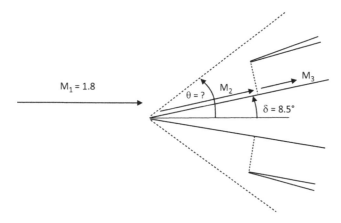

FIGURE 4.3
Schematic of one oblique shock diffuser.

$$M_{1N} = M_1 \sin(\theta) = 1.8 \sin(42.2559°) = 1.2104$$

$$M_{1T} = M_1 \cos(\theta) = 1.8 \cos(42.2559°) = 1.3323$$

Next, finding M_{2N} will allow finding static conditions across the oblique shock wave. These values can be found using the normal shock tables.

Normal shock table: $M_{1N} = 1.21$, $M_{2N} = 0.8360$, $P_2/P_1 = 1.5415$, $T_2/T_1 = 1.1343$, $P_{T2}/P_{T1} = .9918$

The tangential Mach number can be determined from the knowledge that the tangential velocity remains constant across the oblique shock wave.

$$V_{2T} = V_{1T} = M_{2T}a_2 = M_{1T}a_1$$

$$M_{2T} = M_{1T} / \sqrt{\frac{T_2}{T_1}} = \frac{1.3323}{\sqrt{1.1343}} = 1.2506$$

At this point, the normal and tangential components of Mach number can be combined using the square root of the sum of the squares and the Mach number downstream of the oblique shock wave can be determined.

$$M_2 = \sqrt{M_{2N}^2 + M_{2T}^2} = \sqrt{.8360^2 + 1.2506^2} = 1.5046$$

The normal shock table can be used again to find M_3 and the total pressure loss of the normal shock. Interpolating between $M = 1.50$ and $M = 1.51$, the following values are used from the table.

Normal shock table: $M_2 = 1.505$, $M_3 = 0.6993$, $P_3/P_2 = 2.4759$, $T_3/T_2 = 1.3236$, $P_{T3}/P_{T2} = 0.9282$.

The total pressure loss across the oblique diffuser shock system can be characterized by the total pressure ratio.

$$\frac{P_{T3}}{P_{T1}} = \frac{P_{T2}}{P_{T1}} \frac{P_{T3}}{P_{T2}} = 0.9918 * 0.9282 = 0.9206$$

The corresponding static pressure rise at $M_3 = 0.70$ is

$$\frac{P_3}{P_1} = \frac{P_2}{P_1}\frac{P_3}{P_2} = 1.5415 * 2.4759 = 3.8166$$

The normal shock has a total pressure loss, $P_{T2}/P_{T1} = 0.8127$, $M_2 = 0.6165$, and $P_2/P_1 = 3.6133$. The comparison shows that the oblique shock diffuser results in a 13 percent increase in the total pressure.

Checking the angle of the wedge, δ, is determined using the arcsin of the M_{2N}/M_2.

$$\theta - \delta = \text{asin}\left(\frac{M_{2N}}{M_2}\right) = \text{asin}\left(\frac{0.836}{1.5046}\right) = 33.754°$$

$$\delta = \theta - (\theta - \delta) = 42.256° - 33.754° = 8.502°$$

Our procedure with our new table, Table A.4.1, gives good accuracy.

Example 4.3: Overexpanded Nozzle

Oblique shocks can often occur at the exits of nozzles, particularly during startup when the inlet total pressure of the system has not reached the design value. In this case, the converging diverging nozzle is considered overexpanded, and the flow accommodates the increase in pressure at the nozzle exit with an oblique shock. In the following example, the oblique shock that accommodates the back pressure at the exit of the nozzle is determined.

Given: A converging diverging nozzle with exit to throat area ratio of 2.0 is driven by a total pressure of 545 kPa, and it expands to a back pressure of 100 kPa as shown in Figure 4.4. You can assume that the inlet total air temperature is 500 K.

Find: Determine the exit plane pressure and exit Mach number. Also determine the strength of the oblique shock, the oblique shock angle, the deflection angle, and the Mach number after the shock.

Sketch:

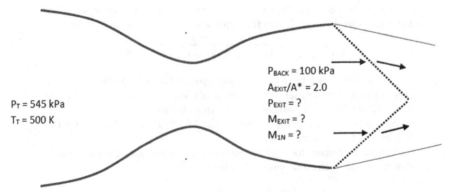

$P_{BACK} = 100$ kPa
$A_{EXIT}/A^* = 2.0$
$P_{EXIT} = ?$
$M_{EXIT} = ?$
$M_{1N} = ?$

$P_T = 545$ kPa
$T_T = 500$ K

FIGURE 4.4
Schematic of overexpanded converging diverging (C/D) nozzle with oblique shocks at exit.

Analysis: This example shows an isentropic flow to the exit of the nozzle where the flow adjusts abruptly to the backplane pressure through an oblique shock. The area ratio of the nozzle, $A_{EXIT}/A^* = 2.0$. As the flow is isentropic up until the oblique shock, the isentropic Mach number tables can be used to find the flow condition at the exit plane of the nozzle.

$$M_1 = 2.20, P/P_T = 0.093522, T/T_T = 0.50813, A/A_* = 2.0050$$

Based on the exit to throat area ratio, the design exit Mach number is very close to 2.20. The exit static pressure will be:

$$P_1 = P_{T1} * \frac{P_1}{P_{T1}} = 545 \text{ kPa} * 0.093522 = 50.97 \text{ kPa}$$

At the exit plane, the flow must adjust to the backplane pressure. As the flow is supersonic, the adjustment is abrupt through an oblique shock. The pressure ratio of the oblique shock will be:

$$\frac{P_2}{P_1} = \frac{100 \text{ kPa}}{50.97 \text{ kPa}} = 1.9620$$

The strength of the resulting oblique shock can be found through the normal shock tables. At a Mach number of 1.35, the following values are found:

Normal shock table: $M_1 = 1.35$, $M_2 = 0.7618$, $P_2/P_1 = 1.9596$, $T_2/T_1 = 1.2226$, $P_{T2}/P_{T1} = 0.9697$.

The pressure ratio produced by $M_{1N} = 1.35$ and $M_{2N} = 0.7618$ is a very close match with the pressure ratio resulting from the overexpansion. As a result, the oblique shock angle can be calculated.

$$\theta = \text{asin}\left(\frac{M_{1N}}{M_1}\right) = \text{asin}\left(\frac{1.35}{2.20}\right) = 37.85°$$

The tangential Mach number can be determined now that the oblique shock angle has been found.

$$M_{1T} = M_1 \cos(37.85°) = 1.7372$$

The tangential Mach number on the downstream side of the shock can be found with the help of the shock temperature ratio, T_2/T_1, based on the idea that the tangential velocity is constant.

$$M_{2T} = \frac{M_{1T}}{\sqrt{T_2/T_1}} = \frac{1.7372}{\sqrt{1.2226}} = 1.5711$$

The Mach number downstream from the oblique shock can be determined from the normal and tangential components of the downstream Mach number.

$$M_2 = \sqrt{M_{2N}^2 + M_{2T}^2} = \sqrt{0.7618^2 + 1.5711^2} = 1.746$$

At this point, the deflection angle can be easily determined:

$$\theta - \delta = \mathrm{asin}\left(\frac{M_{2N}}{M_2}\right) = \mathrm{asin}\left(\frac{0.7618}{1.5711}\right) = 25.87°$$

$$\delta = \theta - (\theta - \delta) = 37.85° - 25.87° = 11.98°$$

Consequently, the deflection angle will be 12.0°, and the flow will move toward the center line.

Discussion: The flow in the present problem expanded over a 5.45 to 1 pressure ratio, which is sufficient to induce supersonic flow. However, the area ratio of the nozzle proscribed the exit Mach number to be $M_1 = 2.20$. A quick check of the isentropic tables indicated that the exit flow would expand to a pressure well below the backplane pressure requiring an oblique shock to adjust the flow to the backplane pressure. This oblique shock turned the flow inward toward the center line. What will happen along the center line? The flow has no place to go along the center line, so the flow must turn back to a direction parallel to the midplane. How does this adjustment occur? This issue will be dealt with in the next section.

4.3.1 Oblique Shocks on Airfoils

Flow over an airfoil can also result in an oblique shock or shocks. Figure 4.5 shows a sketch of a diamond-shaped airfoil moving through the air at a supersonic speed. At the bottom of the airfoil, the flow initially goes through an oblique shock to accommodate the air's concave turn downward. Here, the pressure rises on the bottom creating lift. Over the downstream section on the bottom, the flow must expand around a convex corner. This change in direction occurs a bit more gradually through a series of expansion fans.

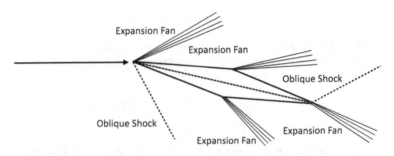

FIGURE 4.5
Schematic of diamond shaped supersonic airfoil showing oblique shocks and expansions.

The pressure drops on the bottom of this airfoil with the expansion. However, the pressure is still above the atmospheric pressure, and the flow at the bottom makes a final expansion to turn more closely toward the horizontal direction. The flow over the top of the airfoil undergoes an initial expansion and then a second expansion lowering the pressure on the top surface of the airfoil, creating an additional lift over the airfoil surface. However, close to the trailing edge on the top surface of the airfoil, the flow must turn up toward the direction of the inlet flow, and this change in direction is accomplished with an oblique shock wave. The oblique shock waves that are analyzed for this airfoil shape can be evaluated based on the change in angle in the concave direction. The analysis of a convex turn requires a series of expansion waves to expand around the corner and Prandtl–Meyer expansion fans will be addressed in another chapter.

4.4 Oblique Shock Reflections

Oblique shock reflections occur when after an initial oblique shock a second oblique shock must form to turn the flow back to the original direction. Reflected oblique shocks sometimes arise due to issues, such as symmetry conditions, after an oblique shock at the exit of a nozzle forces the flow inward or once an oblique shock forms due to a concave wall, it must turn back to the original direction due to the wall on the opposite side of a duct. Both situations can be examined.

Example 4.4: Oblique Shock Reflection

Given: A supersonic flow initially has a Mach number of 2.0, and it is moving parallel to a wall of a duct. The flow initially encounters a corner that produces an oblique shock wave that has an angle of 41°. The resulting oblique shock wave turns the flow 11.5°, which is the direction of the corner. The resulting Mach number, M_2, is 1.584. This flow is shown in Figure 4.6.

Find: Determine the strength of the reflected oblique shock wave in terms of M_{2N}, M_{3N}, P_3/P_2, T_3/T_2, M_{2T}, M_{3T}, M_3, and P_{T3}/P_{T2}, and calculate the angle of this oblique shock required to produce this turn.

Analysis: Based on our oblique shock wave table at $M = 1.6$, the oblique shock angle, $\theta = 52.8839°$ and $54.889°$ at deflection angles, $\delta = 11°$ and $12°$, respectively. Also at $M = 1.5$, $\theta = 59.4651°$ and $64.3588°$ at $\delta = 11°$ and $12°$, respectively. Unfortunately, for $M = 1.5$, the location is close to the maximum deflection angle, δ_{MAX} of $12.11267°$ at $\theta = 66.59°$ and our interpolated value for $\theta = 55.17°$, which yields a deflection angle of $11.70°$. Writing down the oblique shock equations and iterating on θ ($\theta = 54.737°$) until $\delta = 11.5°$

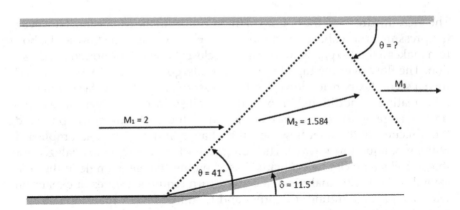

FIGURE 4.6
Schematic of oblique shock and oblique shock reflection caused by a concave corner.

goes very quickly on a spreadsheet. The equations with answers are shown below starting with M_{2N} and M_{2T}.

$$M_{2N} = M_2 \sin(\theta) = 1.584 \sin(54.737°) = 1.2934$$

$$M_{2T} = M_2 \cos(\theta) = 1.584 \cos(54.737°) = 0.9145$$

Next, finding M_{3N} will allow finding static conditions across the reflected oblique shock wave.

$$M_{3N} = \sqrt{\dfrac{M_{2N}^2 + \dfrac{2}{k-1}}{M_{2N}^2 \dfrac{2k}{k-1} - 1}} = \sqrt{\dfrac{1.2934^2 + \dfrac{2}{k-1}}{1.2934^2 \dfrac{2k}{k-1} - 1}} = 0.7894$$

The static pressure and temperature ratios can be computed from this section's relationships.

$$\frac{P_3}{P_2} = \frac{\left(1 + k\, M_{2N}^2\right)}{\left(1 + k\, M_{3N}^2\right)} = \frac{1 + 1.4 \times 1.2934^2}{1 + 1.4 \times 0.7894^2} = 1.7849$$

$$\frac{T_3}{T_2} = \frac{\left(1 + (k-1)/2\, M_{2N}^2\right)}{\left(1 + (k-1)/2\, M_{3N}^2\right)} = \frac{\left(1 + 0.4/2 \times 1.2934^2\right)}{\left(1 + 0.4/2 \times 0.7894^2\right)} = 1.1867$$

The tangential Mach number can be determined from the knowledge that the tangential velocity remains constant across the reflected oblique shock wave.

$$M_{3T} = M_{2T} / \sqrt{\frac{T_3}{T_2}} = \frac{0.9145}{\sqrt{1.1867}} = 0.8395$$

At this point, the normal and tangential components of Mach number are combined to determine the downstream Mach number of the reflected oblique shock wave.

$$M_3 = \sqrt{M_{3N}^2 + M_{3T}^2} = \sqrt{.7894^2 + 0.8395^2} = 1.1523$$

The total pressure loss across the reflected oblique shock wave can be interpolated from the normal shock tables at $M_{2N} = 1.2934$ producing $P_{T3}/P_{T2} = 0.9805$. If static conditions are needed downstream, they can be determined from the pressure and temperature ratios.

Example 4.5: Overexpanded Nozzle with Oblique Shock Reflection

The nozzle example used in Section 4.3 had an overexpanded condition, which adjusted to the higher backplane pressure through an oblique shock wave as shown in Figure 4.7. The oblique shock wave turned the flow inward toward the center line on both the top and bottom of the two-dimensional nozzle. Clearly, both streams cannot be diverted toward each other without some type of adjustment. Also because the flow is still supersonic, this adjustment cannot occur gradually; it is abrupt. This abrupt response must occur at the intersection of the initial oblique shock waves originating at the upper and lower exits of the two-dimensional nozzle. The flow must adjust with the upper and lower flows abruptly turning parallel to one another. These turns will clearly result in concave corners, resulting in a second oblique shock wave reflected from the original at the center plane of the nozzle. As the original turn inward was 12°, the deflection angle on the reflected shock will also be 12°.

The original oblique shock wave was the mechanism that the overexpanded flow leaving the nozzle used to adjust to the higher backplane pressure. This second oblique shock wave is the means that the flow adjusts to the center line condition. However, as the flow undergoes a second oblique shock, the static pressure of the flow downstream from the second oblique shock wave will increase. The resulting local static pressure downstream from the second oblique shock wave will now be above the local backplane pressure. This will require the flow to expand back down to the local backplane pressure at the edge of the flow plane.

Given: Based on the analysis in Example 4.3, the Mach number downstream from the oblique shock, $M_2 = 1.746$, the static pressure has adjusted to the backplane pressure at $P_2 = 100$ kPa, and the oblique shocks have turned the flow inward at a deflection angle, $\delta = 12°$.

Wanted: Noting the flow must adjust at the center line and turn back parallel to the discharge direction, determine the oblique shock angle, the downstream Mach number, and the pressure after the reflected oblique shock.

Sketch:

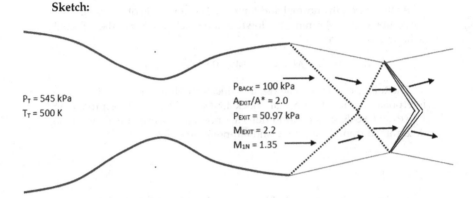

$P_T = 545$ kPa
$T_T = 500$ K

$P_{BACK} = 100$ kPa
$A_{EXIT}/A^* = 2.0$
$P_{EXIT} = 50.97$ kPa
$M_{EXIT} = 2.2$
$M_{1N} = 1.35$

FIGURE 4.7
Schematic of overexpanded C/D nozzle with oblique shocks and expansions.

Analysis: The oblique shock table gives an oblique shock angle (θ) of 50.168° at a Mach number of 1.7 for a deflection angle of 12° and an oblique shock angle (θ) of 46.686° at a Mach number of 1.8 for a deflection angle of 12°. Interpolating at a Mach number of 1.746 estimates the oblique shock angle to be 48.567°. This gives an adequately close deflection angle (δ) of 12.07°. A deflection angle of 12° from a reflected shock produces an oblique shock angle of 48.46°. The resulting downstream Mach number and pressure can be calculated using our oblique shock equations.

$$M_{2N} = M_2 \sin(\theta) = 1.746 \sin(48.46°) = 1.3069$$

$$M_{2T} = M_2 \cos(\theta) = 1.746 \cos(48.46°) = 1.1578$$

Next, finding M_{3N} will allow finding static conditions across the reflected oblique shock wave.

$$M_{3N} = \sqrt{\frac{M_{2N}^2 + \dfrac{2}{k-1}}{M_{2N}^2 \dfrac{2k}{k-1} - 1}} = \sqrt{\frac{1.3069^2 + \dfrac{2}{k-1}}{1.3069^2 \dfrac{2k}{k-1} - 1}} = 0.7825$$

The static pressure and temperature ratios can be computed from this section's relationships.

$$\frac{P_3}{P_2} = \frac{\left(1 + k\,M_{2N}^2\right)}{\left(1 + k\,M_{3N}^2\right)} = \frac{1 + 1.4 * 1.3069^2}{1 + 1.4 * 0.7825^2} = 1.8259$$

$$\frac{T_3}{T_2} = \frac{\left(1 + (k-1)/2\,M_{2N}^2\right)}{\left(1 + (k-1)/2\,M_{3N}^2\right)} = \frac{\left(1 + 0.4/2 \times 1.3069^2\right)}{\left(1 + 0.4/2 \times 0.7825^2\right)} = 1.1952$$

The tangential Mach number can be determined from the knowledge that the tangential velocity remains constant across the reflected oblique shock wave.

$$M_{3T} = M_{2T} / \sqrt{\frac{T_3}{T_2}} = \frac{1.1578}{\sqrt{1.1952}} = 1.0591$$

At this point, the normal and tangential components of Mach number are combined to determine the downstream Mach number of the reflected oblique shock wave.

$$M_3 = \sqrt{M_{3N}^2 + M_{3T}^2} = \sqrt{.7894^2 + 1.0591^2} = 1.3168$$

Summary: The reflected oblique shock will produce a Mach number of 1.317 and a static pressure ratio of 1.8259. The pressure behind the oblique shock will rise to 182.6 kPa, which is well above the back-plane pressure. At the edge of the shock pattern, the pressure must match the backplane pressure. This requires that the flow expands outwardly to the backplane pressure of 100 kPa. This expansion cannot occur in a discrete oblique shock, as was noted in the development of the normal shock section. The expansion will occur through a process known as Prandtl–Meyer expansion fans. This topic will be addressed in the next chapter.

4.5 Conical Shock Waves

The leading edge region of many missiles and high speed projectiles is in the shape of a cone. A shock wave which is attached to a cone proceeding directly into the flow, with no angle of attack will have a shock wave which is conical in shape and concentric to the body. The pressure recovery on the surface of the cone will be related to the Mach number and the semivertex angle of the cone, sigma (σ). Similar to oblique shock waves, conical shock waves have both a weak and a strong shock solution. However, similar to isolated wedges only the weak shock solution is believed to occur for an isolated cone. A cone with a relatively small semivertex angle is known to produce a relatively weak shock and a relatively mild pressure recovery. Consequently, a cone with a relatively small semivertex angle will have a reduced level of pressure drag due to the pressure on the conical surface.

Conical shock waves differ from oblique shock waves in that the pressure behind the conical shock is not constant but has an additional rise after the initial shock wave toward the surface of the cone. This pressure rise is concentric and increases on conical surfaces on intermediate angles or rays between the shock wave with angle theta (θ) to the flow and surface with angle sigma (σ) to the flow. Also unlike oblique shock waves, the angle of the flow changes as the flow begins to move through the shock wave and downstream in the general direction of flow. The flow across the shock is governed by oblique shock relationships while the following flow is additionally compressed isentropically. The solution for conical shock

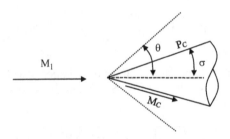

FIGURE 4.8
Conical shock wave at the leading edge of a cone.

waves has been compiled and published in NACA Report 1135 [3] and the solutions have been plotted in terms of shock angle theta (θ), surface pressure coefficient, and surface Mach number as a function of the semivertex angle sigma (σ) of the cone and free stream Mach number in a series of charts presented in Appendix A.4.3. A schematic of the flow is shown below in Figure 4.8.

Example 4.6: High Speed Projectile with a Conical Leading Edge

A ship-based projectile with a conical leading edge is designed to travel at a Mach number of 5. The leading edge cone has a semivertex angle of 12°. Assume that the local pressure is atmospheric (101,325 Pa). Find the angle of the conical shock wave, the surface pressure, and the Mach number over the cone. Compare these values to similar values but for a two-dimensional wedge with a half-wedge angle of 12°.

Analysis: This analysis can begin by determining the angle of the conical shock wave originating on the vertex point of the cone. Based on Figure A.4.3A the shock angle can be interpolated at a Mach number of 5 to be 17.4°. The pressure on the cone, P_C, is needed to estimate the drag on the projectile. The pressure can be determined using the pressure coefficient provided from Figure A.4.4A at the Mach and semivertex angle. Here the pressure coefficient is given as 0.1025 and can be interpreted as:

$$C_P = \frac{P_C - P_\infty}{q_\infty}$$

The dynamic pressure, q_∞ can be determined as follows:

$$q_\infty = \rho_\infty \frac{U_\infty^2}{2} = \frac{k}{2} M_\infty^2 \, P_\infty = \frac{1.4}{2} 5^2 \, 101325 \, Pa = 1{,}773{,}188 \, Pa$$

Consequently, the pressure on the cone can be calculated to be:

$$P_C = P_\infty + C_P \, q_\infty = 101325 \, Pa + 0.1025 * 1{,}773{,}188 \, Pa = 283{,}077 \, Pa$$

The oblique shock of a two-dimensional wedge has some similarities to the conical leading edge. An analysis of a 12° half angle for a wedge under similar conditions can be undertaken based on the approach in Section 4.2. The oblique shock angle resulting from the wedge at an approach Mach number of 5 is tabulated as 21.2845° in Table A.4.1 in the Appendix. The normal and tangential inlet Mach numbers can be determined with the sine and cosine of the oblique shock angle.

$$M_{1N} = M_1 \sin(\theta) = 5\sin(21.2845°) = 1.8150$$

$$M_{1T} = M_1 \cos(\theta) = 5\cos(21.2845°) = 4.6589$$

The downstream normal Mach number and static pressure ratio can be calculated using the equations developed in this chapter or interpolated from the normal shock table, Table A.2.1.

$$M_{2N} = 0.61318, P_2 / P_1 = 3.6766, T_2 / T_1 = 1.5428$$

The resulting pressure downstream from the shock can be determined from the pressure ratio.

$$P_2 = P_1 \frac{P_2}{P_1} = 101325\,\text{Pa} \times 3.6766 = 372,529\,\text{Pa}$$

The downstream Mach number, M_2, can be determined from the normal and tangential components. Noting the tangential velocity stays constant.

$$M_{2T} = \frac{M_{1T}}{\sqrt{\frac{T_2}{T_1}}} = \frac{4.6589}{\sqrt{1.5428}} = 3.7508$$

$$M_2 = \sqrt{M_{2T}^2 + M_{2N}^2} = \sqrt{3.7508^2 + 0.61318^2} = 3.8006$$

This comparison between the conical shock wave on a cone and the oblique shock wave on a wedge under similar conditions shows that the wedge has a stronger shock due to a steeper angle resulting in a higher pressure rise and a lower downstream Mach number. Also the flow field downstream from an oblique shock is simpler than a conical shock flow field having both a uniform static pressure field and downstream Mach number and flow direction.

References

[1] John, J.E.A., and T.G. Keith. 2006. *Gas Dynamics*, 3rd ed. Prentice Hall.

[2] Saad, M.A. 1993. *Compressible Fluid Flow*, 2nd ed. Prentice Hall.

[3] Ames Research Staff, 1953, *Equations, Tables, and Charts for Compressible Flow*, NACA Report 1135.

Chapter 4 Problems

1. An ambient airflow generates a 44° oblique shock wave as it travels over a 10° wedge. Determine the value of the Mach number.

2. An airflow traveling at Mach 2.2 over a wedge of an unknown angle generates a 32° oblique shock wave. The air temperature is 250 K and the air pressure is 35 kPa. Determine the normal and tangential Mach number of the upstream flow with respect to the oblique shock wave. Determine the normal and tangential components of the downstream flow. Determine the static temperature rise and static pressure rise across the shock and the downstream Mach number. Also determine the angle of the wedge.

3. An airflow with a Mach number of 1.9 approaches a 9° wedge. The ambient temperature is 240 K and the ambient pressure is 30 kPa. Determine the oblique shock angle for the weak shock solution. Determine the downstream Mach number, the static pressure, and temperature. Also determine the total pressure loss from the oblique shock wave.

4. A single oblique shock diffuser is designed for an engine inlet. The design inlet Mach number is 2.1 and the wedge angle is 11°. After the oblique shock, the flow takes a normal shock into the inlet diffuser of the engine. Determine the downstream Mach numbers, M_2 and M_3, as well as the total pressure loss of the system and compare it to a normal shock diffuser. Assume that the inlet static pressure is 10 kPa and the inlet static temperature is 230 K.

5. A double oblique shock diffuser is designed for the same engine. Again, the inlet Mach number is 2.1 but the flow is designed to undergo two consecutive 8° wedges. After two oblique shocks, the flow takes a normal shock into the inlet diffuser of the engine. Determine the downstream Mach numbers, M_2, M_3, and M_4, as well as the total pressure loss of the system and compare it with a normal shock diffuser. Again, assume that the inlet static pressure is 10 kPa and the inlet static temperature is 230 K.

6. A converging diverging nozzle has an inlet total pressure of 390 kPa and has an exit to throat area of 1.45. The back pressure to this nozzle is 100 kPa. Determine the design Mach number and the static pressure at the exit plane. Determine the angle of the oblique shock that is required to adjust the pressure back to ambient as well as the resulting angle that the flow must turn inward after this oblique shock. Determine the Mach number downstream from the oblique shock.

7. Airflow with a Mach number of 2.4 is generated by a converging diverging nozzle. In a downstream test section, a 10° wedge

is introduced and an oblique shock forms and deflects the flow upward. At the top of the duct, the flow then reflects to turn back at a 10° angle to flow parallel to the upper wall. If the initial static pressure is 45 kPa, determine the downstream Mach number and static pressure after the initial oblique shock, and calculate the Mach number and pressure of the flow after the reflected oblique shock.

8. Take the results from problem 6 noting that, after the oblique shock, the flow turns inwardly. However, along the center line of the nozzle, the flow must turn back parallel to the center line access through a second oblique shock. Calculate the angle of this oblique shock required to turn the flow back to the parallel direction. Calculate the pressure and Mach number downstream from the reflected oblique shock. What needs to happen to the flow next?

9. A projectile with a conical leading edge moves through air at a Mach number of 4.0. The semivertex angle is 11° and the ambient pressure is 100,000 Pa. Find the angle of the conical shock as well as the static pressure on the cone as well as the Mach number just off the conical surface.

5

An Introduction to Prandtl–Meyer Flow

In Chapter 3, normal shock equations were developed from the principles of conservation of mass, conservation of energy, and the momentum principle in combination with the ideal gas law and equation for the speed of sound. The analysis demonstrated that a normal shock wave results in a total pressure loss in the flow stream. In the discussion of normal shocks, it was apparent that compression waves could reinforce one another, while expansion waves could not. Moreover, a discrete expansion shock would violate the second law of thermodynamics, as it would result in an increase in the total pressure. When sonic or supersonic flow encounters a sudden change in direction through a convex corner, the change in direction is accomplished through a series of expansion fans called Prandtl–Meyer flow. In this chapter, the flow equations for Prandtl–Meyer expansion fans will be developed. Subsequently, analysis using Prandtl–Meyer expansions will be studied. Later, the next portion of the chapter will deal with Prandtl–Meyer reflections. Finally, the maximum possible turning angle for a Prandtl–Meyer expansion will be examined.

5.1 Prandtl–Meyer Expansion Fans

When supersonic flow encounters an abrupt change in direction due to the presence of a change in geometry such as a wedge or a change in the direction of the flow surface, it must respond abruptly as supersonic flow cannot communicate this disturbance to the upstream flow. In the case of a wedge shape, which turns the flow in a concave direction, an oblique shock wave forms. Based on Chapter 4, an oblique shock wave has similar features to a normal shock wave, with the addition of a tangential component of velocity that does not change. An oblique shock wave is very thin similar to a normal shock wave and produces an abrupt increase in static pressure and temperature and a loss in total pressure. If the Mach number of the flow component normal to the wave is not too much larger than 1, then the compression is weak, and this abrupt change is close to isentropic. By developing a surface with a series of very small concave angles, the compression near to the surface can be almost isentropic in nature. However, the rays of these weak shocks will converge together, away from the wall to form a discrete oblique shock wave, which is stronger.

When supersonic flow encounters an abrupt change in direction but through a convex turn or a sudden expansion due to an abrupt reduction in pressure, expansion waves are produced. As noted in Chapter 3, expansion waves cannot reinforce one another to form a discrete expansion shock. First, as the flow expands, the trailing waves move into cooler air and their speed must be slower than the preceding expansion waves. Moreover, the formation of a discrete expansion wave would cause an increase in the total pressure, which is a violation of the second law of thermodynamics, and consequently is impossible. Instead of forming a discrete expansion wave, the flow expands incrementally through a series of weak expansion waves, which project out at the Mach angle, μ, relative to the local flow angle. As the flow negotiates a convex corner or an abrupt change in pressure, these Mach waves fan out from the corner forming a fan starting with the Mach angle relative to the initial flow angle and ending with the corresponding Mach angle relative to the final flow angle. An example of this would be flow over a thin diamond-shaped airfoil. Prandtl–Meyer expansion fans could form as flow moves around the convex corners at the maximum thickness of the airfoil. These expansion fans could also form if the angle of attack of the flow was large enough to form a convex turn of the initial flow over the top of the airfoil. Similar expansion fans can form at the exit of an underexpanded nozzle or if an expansion is needed to equalize pressure or effect a change in the flow direction. In the next section, the equations of motion needed to address the analysis of these expansions are addressed.

5.2 Prandtl–Meyer Flow Equations

The analysis for oblique shocks looked at flow in a concave corner. Continuity, energy, and normal and tangential momentum were applied, and the resulting equations for the normal component of continuity, momentum, and energy looked identical to the normal shock equations. Prandtl–Meyer flow occurs over a convex corner through an expansion. Our discussion about how expansion waves behave as flow negotiates a convex corner, as shown in Figure 5.1, is similar. Yet, expansion waves cannot coalesce into a discrete wave similar to a shock. Here the flow is isentropic.

As the flow expands over differential angle $d\nu$ similar to an oblique shock wave, the tangential velocity is expected not to change, as there is no pressure change in the tangential direction.

$$V\cos(\mu) = (V + dV)\cos(\mu + d\nu) \tag{5.1}$$

Using the identity, $\cos(a+b) = \cos(a)\cos(b) - \sin(a)\sin(b)$, and noting $\cos(d\nu) = 1$ and $\sin(d\nu) = d\nu$.

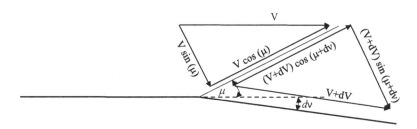

FIGURE 5.1
Expansion around a convex corner.

$$V\cos(\mu) = (V+dV)\big[\cos(\mu)\cos(dv)-\sin(\mu)\sin(dv)\big]$$
$$= (V+dV)\big[\cos(\mu)-\sin(\mu)dv\big]$$

Multiplying out terms, canceling, and dropping products of second-order terms:

$$\cancel{V\cos(\mu)} = \cancel{V\cos(\mu)} + dV\cos(\mu)-V\sin(\mu)dv$$

$$dV\cos(\mu) = V\sin(\mu)dv \tag{5.2}$$

$$\frac{dV}{V} = dv\tan(\mu)$$

In this case, the angle μ is the Mach angle and is equal to $\sin^{-1}(1/M)$. Therefore, $\sin(\mu) = 1/M$ and the $\cos(\mu) = \sqrt{(M^2-1)/M^2}$.

$$\frac{dV}{V} = \frac{dv}{\sqrt{M^2-1}} \tag{5.3}$$

Velocity can be written in terms of Mach number and the speed of sound.

$$V = M\sqrt{k\,g_C\,RT}$$

Differentiating V using the chain rule:

$$dV = dM\sqrt{k\,g_C\,RT} + M\sqrt{k\,g_C\,R}\,\frac{dT}{2\sqrt{T}} \tag{5.4}$$

Consequently dV/V can be written as:

$$\frac{dV}{V} = \frac{dM}{M} + \frac{1}{2}\frac{dT}{T} \tag{5.5}$$

The ratio, dV/V can now be eliminated, but dT/T should also be developed in terms of the Mach number.

$$T = \frac{T_T}{1+\dfrac{k-1}{2}M^2} \tag{2.12}$$

$$dT = -\frac{T_T(k-1)MdM}{\left(1+\dfrac{k-1}{2}M^2\right)^2} \tag{5.6}$$

Consequently, dT/T can be written in terms of the Mach number and the specific heat ratio, k.

$$\frac{dT}{T} = -\frac{(k-1)MdM}{1+\dfrac{k-1}{2}M^2} \tag{5.7}$$

Resulting in the following relationship for dV/V.

$$\frac{dV}{V} = \frac{dM}{M} - \frac{\dfrac{(k-1)}{2}MdM}{1+\dfrac{k-1}{2}M^2} \tag{5.8}$$

If a common denominator for the preceding equation is developed, then:

$$\frac{dV}{V} = \frac{dM\left(1+\dfrac{k-1}{2}M^2\right)}{M\left(1+\dfrac{k-1}{2}M^2\right)} - \frac{\dfrac{(k-1)}{2}M^2dM}{\left(1+\dfrac{k-1}{2}M^2\right)M} = \frac{dM}{M\left(1+\dfrac{k-1}{2}M^2\right)} \tag{5.9}$$

Combining the two equations for dV/V yields the following relationship for $d\nu$.

$$d\nu = \frac{dM}{M}\frac{\sqrt{M^2-1}}{1+\dfrac{k-1}{2}M^2} \tag{5.10}$$

This equation can be integrated from a reference Mach number to the Mach number of interest. The logical reference Mach number is 1. By inspection, the reference kernel is 0 at a Mach number of 1. Flow must be supersonic for the possibility of Prandtl–Meyer expansion waves to occur. By assuming that $\nu_{REF} = 0°$, values of ν versus Mach number can be tabulated. The integral of the right-hand side from a reference value of 1 can be calculated using the result shown next [1, 2]:

$$\nu(M) = \sqrt{\frac{k+1}{k-1}}\tan^{-1}\sqrt{\frac{k-1}{k+1}(M^2-1)} - \tan^{-1}\sqrt{M^2-1} \tag{5.11}$$

The preceding equation is called the Prandtl–Meyer function, and it is tabulated in Appendix A.5 for specific heat ratios of 1.4 and 1.3. The Prandtl–Meyer function can be related to an initial Mach number and a change in the flow angle $\Delta\nu$.

5.3 Prandtl–Meyer Expansions

Consider supersonic flow over a surface. As the flow moves along the surface, the surface takes a sudden change in angle over a convex corner. As the flow negotiates the corner, a series of Prandtl–Meyer expansion fans extend out in the shape of a fan as the flow expands around the corner. The Prandtl–Meyer function can be used to determine the final Mach number, as well as the range in the fan angle around the corner.

> **Example 5.1: Prandtl–Meyer Expansion Around a Convex Corner: Shown in Figure 5.2**
>
> **Given:** A flow with a free-stream Mach number of 1.9 encounters an abrupt convex angle of 6°, $k = 1.4$.
>
> **Wanted:** Determine the resulting Mach number as well as the angle of the expansion fan.
>
> **Sketch:**

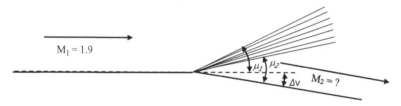

FIGURE 5.2
Supersonic flow expansion around a convex corner.

Analysis: The supersonic flow has been expanded through an angle of 6°. The approach Mach number is given at 1.9. Based on the Prandtl–Meyer function in Appendix A.5 for a specific heat ratio of 1.4, the function value is 23.5861°. The final value for the Prandtl–Meyer function can be determined by adding the expansion angle, Δv, to the initial Prandtl–Meyer function, giving an angle of 29.5861°. By inspection of Table A.5 for a $k = 1.4$, M_2 is between 2.11 and 2.12. Using linear interpolation, the downstream Mach number, M_2, is equal to 2.118. The fan angle can be determined using the Mach angles at the two Mach numbers. At $M_1 = 1.9$, the Mach angle $\mu_1 = 31.76°$. At $M_2 = 2.118$, the Mach angle $\mu_2 = 28.17°$. The resulting fan angle, ϕ, will be:

$$\phi = \mu_1 - \mu_2 + \Delta v = 31.76° - 28.17° + 6.0° = 9.59°$$

The expansion around the convex corner produces a series of expansion fans that were present over an angle of 9.59° and resulted in a downstream Mach number of 2.118. Expansion fans are an isentropic process, so isentropic properties can be used to determine the resulting pressure ratio. These relationships can be developed with our isentropic equations or using the isentropic tables, noting that for this isentropic process, the total pressure remains constant.

$$\frac{P_2}{P_1} = \frac{P_T}{P_1}\frac{P_2}{P_T} = \left[1+\frac{(k-1)}{2}M_1^2\right]^{k/_{k-1}} \Big/ \left[1+\frac{(k-1)}{2}M_2^2\right]^{k/_{k-1}} = \frac{6.7006}{9.410} = 0.7121$$

Consequently, the 6° expansion took the flow from a Mach number of 1.9 to 2.118 with a corresponding isentropic static pressure drops through a ratio of 0.7121. The static temperature ratio can be determined in a similar manner.

Example 5.2: Underexpanded Nozzle

An underexpanded nozzle will also rapidly adjust to the exit pressure through a Prandtl–Meyer expansion fan. The flow will expand outward from the nozzle to adjust to the back pressure on the nozzle. The following problem provides an example of an underexpanded nozzle and the resulting expansion and flow turning.

Given: A converging–diverging nozzle with an inlet total pressure of 1000 kPa is expanded to a back pressure of 100 kPa, as shown in Figure 5.3. The nozzle is designed to an exit to throat area of 1.6875 for a design Mach number of 2.0.

Find: Determine the Mach number after the expansion as well and the flow turning angle and the fan angle.

Sketch:

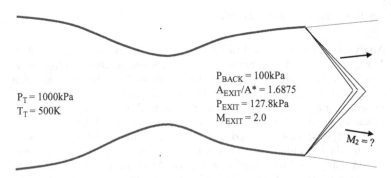

FIGURE 5.3
Underexpanded converging–diverging nozzle with exit expansion fan.

Analysis: The Mach number after the expansion can be found based on isentropic relationship and the expansion ratio. The turning angle can be determined based on the initial and final Prandtl–Meyer function values.

$$M_2 = \left\{ \left[\left(\frac{P_T}{P} \right)^{k - \frac{1}{k}} - 1 \right] \frac{2}{k-1} \right\}^{\frac{1}{2}} = \left\{ \left[\left(\frac{1000\,\text{kPa}}{100\,\text{kPa}} \right)^{0.4/_{1.4}} - 1 \right] \frac{2}{0.4} \right\}^{\frac{1}{2}} = 2.157$$

Based on the Prandtl–Meyer function, $v(M_1) = 26.38°$ and $v(M_2) = 30.61°$, which gives an outward turning angle of 4.23°. The Mach angle of the exit flow at $M_1 = 2.0$ is 30.0°, while the Mach angle at $M_2 = 2.157$ is 27.62°. The resulting fan angle can be determined as follows:

$$\phi = \mu_1 - \mu_2 + \Delta v = 30.0° - 27.62° + 4.23° = 6.61°$$

At this exit pressure, the nozzle is only mildly underexpanded. Consequently, the outward turning is only about 4.23°. The expansion fan is calculated to be about 6.61°. However, what happens to the flow along the centerline of the nozzle? The flow cannot continue outward at the centerline. What happens in this region? At the centerline, the flow must expand again to turn parallel to the nozzle centerline. This process is called a Prandtl–Meyer reflection. The subject of the next section will be Prandtl–Meyer reflections.

On occasion, supersonic inlets are designed with very gradual concave curvature. In this case, supersonic flow is taken through what could be viewed as a series of very weak compressions. These compression waves will converge off the surface and form an oblique shock wave. However, near the surface where these compression waves are gradual, the compression is almost isentropic and the Prandtl–Meyer function using a negative Δv could be used to estimate the pressure rise from this compression. However, the flow directly downstream from the oblique shock above where the compression waves collect would experience a total pressure loss.

5.4 Prandtl–Meyer Reflections

Prandtl–Meyer reflections can occur at the exit of an underexpanded nozzle. Initially, the flow must expand outwardly to match the back pressure at the exit. However, at the symmetry plane, it is impossible for the flow both above and below the plane to move away from the centerline. Consequently, the flow that moves out from the nozzle must expand again at the centerline to turn the flow parallel.

A Prandtl–Meyer reflection can also occur in a flow in a duct when a convex corner turns the flow outward, causing the flow to need to expand again at the far wall to adjust the flow to move parallel to the far wall. In addition to reflections in an underexpanded nozzle, Prandtl–Meyer reflections can also occur downstream from an overexpanded nozzle that originally adjusts to the exit pressure through an oblique shock, which turns it inwardly. The symmetry at the centerline causes the flow to turn parallel through a second oblique shock that drives the pressure above atmospheric. The flow then expands through a Prandtl–Meyer expansion to match the pressure, but turning outward.

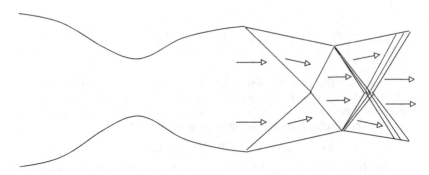

FIGURE 5.4
Overexpanded nozzle showing oblique shocks followed by expansion fans.

The flow then expands through a Prandtl–Meyer reflection to expand the flow back to a parallel direction to the nozzle flow direction. This series of oblique shock waves and expansion waves is shown in Figure 5.4.

> **Example 5.3: Prandtl–Meyer Reflection from an Expansion Around a Convex Corner shown in Figure 5.5**
>
> **Given:** A flow with a free-stream Mach number of 1.5 encounters an abrupt convex angle of 8°, $k = 1.4$.
>
> **Wanted:** Determine the resulting Mach number as well as the angle of the expansion fan.
>
> **Sketch:**

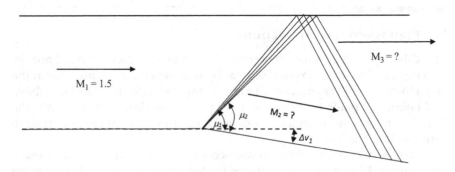

FIGURE 5.5
Supersonic flow around convex corner with expansion fan and reflection.

> **Analysis:** The supersonic flow has been initially expanded through an angle of 8°. The approach Mach number is given at 1.5. Based on the Prandtl–Meyer function in Appendix A.5 for a specific heat ratio of 1.4,

the function value is 11.905°. The final value for the Prandtl–Meyer function can be determined by adding the expansion angle, Δv, to the initial Prandtl–Meyer function, giving an angle of 19.905°. By inspection of Table A.5 for $k = 1.4$, M_2 is between 1.77 and 1.78. Using linear interpolation, the downstream Mach number, M_2, is equal to 1.7717. The fan angle can be determined using the Mach angles at the two Mach numbers. At $M_1 = 1.5$, the Mach angle $\mu_1 = 41.81°$. At $M_2 = 1.7717$, the Mach angle $\mu_2 = 34.36°$. The resulting fan angle, ϕ, will be:

$$\phi = \mu_1 - \mu_2 + \Delta v = 41.81° - 34.36° + 8.0° = 15.45°$$

The expansion around the convex corner produces a series of expansion fans that were present over an angle of 15.45° and resulted in a downstream Mach number of 1.7717. Expansion fans are an isentropic process, so isentropic properties can be used to determine the resulting pressure ratio. These relationships can be developed with our isentropic equations or using the isentropic tables noting that for this isentropic process, the total pressure remains constant.

$$\frac{P_2}{P_1} = \frac{P_T}{P_1}\frac{P_2}{P_T} = \left[1 + \frac{(k-1)}{2}M_1^2\right]^{k/_{k-1}} \bigg/ \left[1 + \frac{(k-1)}{2}M_2^2\right]^{k/_{k-1}} = \frac{3.671}{5.5029} = 0.6671$$

Consequently, the 8° expansion took the flow from a Mach number of 1.5 to 1.7717 with a corresponding isentropic static pressure drop through a ratio of 0.6671. The nature of the resulting reflection is more complicated as the fan has now spread and the reflection is no longer discrete. However, if the second expansion can be looked at from an overall perspective, a method similar to the original expansion can be used to approximate the reflection. Considering that the flow downstream from the original expansion fan is now deflected away from the far wall at an angle of 8°, the flow will need to expand again by 8° to adjust to the wall. A similar procedure can be used to estimate the Prandtl–Meyer function. Adding 8° to the Prandtl–Meyer function at $M_2 = 1.7717$, the Prandtl–Meyer function now becomes 27.905° and M_3 is determined to be 2.0557. The total pressure remains constant, so the pressure ratio can be estimated from either P_1 and M_1 or P_2 and M_2.

$$\frac{P_3}{P_2} = \frac{P_T}{P_2}\frac{P_3}{P_T} = \left[1 + \frac{(k-1)}{2}M_2^2\right]^{k/_{k-1}} \bigg/ \left[1 + \frac{(k-1)}{2}M_3^2\right]^{k/_{k-1}} = \frac{5.5029}{8.5350} = 0.6447$$

Consequently, the 8° reflection took the flow from a Mach number of 1.7717 to about 2.0557 with an estimated isentropic static pressure drop through a ratio of 0.6447.

Prandtl–Meyer expansion fans spread across the expansion angle, so reflections of expansion fans are not discrete and can occur over a relatively broad region. This causes their analysis to be more complicated than oblique shock reflections. However, reflections of Prandtl–Meyer expansion fans are also common features of both underexpanded nozzles and overexpanded nozzles after the initial system of oblique shocks.

5.5 Maximum Turning Angle for Prandtl–Meyer Flow

The maximum turning angle of flow might be related to a situation where say a rocket nozzle was expanding to the pressure of space. This phenomenon might be an issue, as the exhaust of the nozzle could impinge on the one of the surrounding surfaces of the aircraft. Then, depending on the sensitivity of the surface and the composition and temperature of the exhaust, the impingement of the jet might cause a problem with the surface or a sensor. The maximum deflection angle can be estimated from the Prandtl–Meyer function. The function has an M^2 term within the arctangent of the square root sign. As the pressure ratio gets very high as it would at the edge of space, the resulting Mach number would get very high and the arctangent would return an angle close to 90°. Consequently, for flow, the maximum turning angle in radians becomes

$$v(\infty) = \sqrt{\frac{k+1}{k-1}}\frac{\pi}{2} - \frac{\pi}{2} = \left(\sqrt{\frac{k+1}{k-1}} - 1\right)\frac{\pi}{2} \tag{5.12}$$

The maximum turning angle in degrees becomes

$$v(\infty) = \left(\sqrt{\frac{k+1}{k-1}} - 1\right)90°$$

For a gas with a specific heat ratio, $k = 1.4$, the maximum turning angle is determined to be 130.45°.

> **Example 5.4: Mach 3 Helium Jet Exhausting to Vacuum**
>
> **Given:** A very small rocket nozzle is used to make tiny adjustments to a spacecraft's attitude. The propulsion gas is helium and the rocket nozzle has a design exit Mach number of 3.
>
> **Wanted:** Find the maximum turning angle for helium and find the turning angle expected from a Mach = 3 nozzle.
>
> **Analysis:** The maximum turning angle for helium can be determined from the Prandtl–Meyer function, noting that the specific heat ratio of helium, a noble gas, is 5/3. The Prandtl–Meyer function for a Mach number of 3 also needs to be determined.

$$v(\infty) = \left(\sqrt{\frac{k+1}{k-1}} - 1 \right) 90° = \left(\sqrt{\frac{\frac{5}{3}+1}{\frac{5}{3}-1}} - 1 \right) 90° = 90°$$

For helium at a Mach number of 3, the Prandtl–Meyer function is

$$v(3) = \sqrt{\frac{\frac{5}{3}+1}{\frac{5}{3}-1}} \tan^{-1} \sqrt{\frac{\frac{5}{3}+1}{\frac{5}{3}-1}(3^2-1)} - \tan^{-1} \sqrt{3^2-1} = 38.94°$$

The resulting flow exiting the nozzle at a Mach number of 3 could be expected to turn:

$$90° - 38.94° = 51.06°$$

Consequently, depending on the location of the jet and the design of the spacecraft, it is unlikely that the helium jet would impinge on any surrounding surfaces.

References

1. John, J.E.A., and T.G. Keith. 2006. *Gas Dynamics*, 3rd ed. Prentice Hall.
2. Saad, M.A. 1993. *Compressible Fluid Flow*, 2nd ed. Prentice Hall.

Chapter 5 Problems

1. Flow moves at a supersonic Mach number of 1.6 when suddenly the side of the duct expands outward with a convex corner of 12°. If the initial pressure in the duct is 100 kPa, determine the downstream pressure, Mach number, and the fan angle.

2. Flow discharges from the choked converging nozzle. The inlet total pressure of the flow to the nozzle is 250 kPa and the nozzle discharges to a backplane pressure of 100 kPa. Find the exit pressure of the nozzle and determine the angle of the flow downstream from the expansion fans exiting from this two-dimensional nozzle. Determine the Mach number of the flow downstream from the expansion fans.

3. Flow discharges from a converging–diverging nozzle with an exit to throat area ratio of 2.4. The upstream total pressure driving the nozzle is 2 MPa, while the backplane pressure is 100 kPa. Determine

the pressure and Mach number of the flow at the exit plane. What happens to the flow just downstream of the exit plane. What angle does the flow turn? What is the flow's Mach number after the turn?

4. A very thin supersonic airfoil can be simulated as a thin flat plate. If the incoming Mach number is 2.0 and the airfoil has an angle of attack of 3°, determine the lift and drag on this wing. Assume the chord length of the wing is 1 m and the local static pressure is 20 kPa. What is the static pressure at the bottom of the wing and the top of the wing? What is the Mach number above and below the airfoil? Describe what will happen downstream from the airfoil. Repeat the analysis for 6°.

5. A supersonic airfoil is designed based on a wedge ½ angle of 4°. The front half of the symmetric airfoil expands at a total angle of 8°, while the back half contracts similarly. The airfoil encounters supersonic airflow with a Mach number of 2.3 and a static pressure of 10 kPa. The airfoil has an angle of attack of 5°. Determine the static pressure experienced by the four sections of the wing as well as the local Mach number. Note that the lift coefficient can be defined as:

$$C_L = \frac{F_L}{\rho U_\infty^2 / 2 \times Chord} = \frac{F_L}{P_\infty \frac{kM^2}{2} \times Chord}$$

Find the lift and the drag coefficient for the airfoil.

6. Take the situation of the converging–diverging nozzle in problem 3 and determine what happens to the flow along the centerline downstream from the expansion fans. How does the flow turn back to the parallel direction? Calculate the approximate pressure and Mach number downstream from the reflected expansion fan. What would the pressure be in that region? What would happen next?

7. A rocket nozzle has been designed for an exit Mach number of 3.5. However, while it is used as an upper stage booster, it will also see service in space where the pressure ratio is essentially infinite. Determine the turning angle of the flow in space if the specific heat ratio can be estimated at $k = 1.4$.

6

Applications

Chapter 6 of *An Introduction to Compressible Flow* reviews the subject matter of the previous five chapters through a series of applications. These applications include the startup of a supersonic test section, single and double oblique shock diffusers, supersonic airfoils, and supersonic nozzles. The conceptual material for each example is briefly reviewed and the application is addressed. Initially, as the problematic situation is explained, the subject matter is reviewed and the equations used to solve the problem are reintroduced and used in the problem development. Next, a similar example problem is solved taking advantage of the tables and previous concepts to develop the answer.

6.1 Supersonic Wind Tunnel Startup

Supersonic wind tunnels are useful in testing for flow over supersonic aircraft and projectiles. Supersonic flow over objects cannot be sensed upstream so flow must adjust abruptly to the presence of an aircraft or projectile. However, depending on the geometry of the model, certain portions of the flow can act as though it is supersonic and certain portions of the flow can act as though it is subsonic. The flow can also be significantly influenced by the angle of attack or the yaw angle of the flow. Supersonic testing can provide information on lift and drag as well as moments about the aircraft in various configurations and with changes to control surfaces. Most of the design work on aircraft is done using computational fluid dynamics these days but verification testing of aircraft is often considered useful if not critical at some point in the development process.

The conceptual construction and startup of a supersonic wind tunnel is useful in understanding flow through converging–diverging nozzles and converging–diverging diffusers. An initial and critical component for a supersonic wind tunnel will be a converging–diverging nozzle which is needed to accelerate the flow from the subsonic to the supersonic regime. Next a test section is needed to provide the test object with the flow environment for testing. This test section will typically be constant area in cross-section as changes in cross-section are expected to affect the Mach number. Downstream from the supersonic test section a converging–diverging diffuser will be needed to recover as much pressure as possible to minimize the steady-state power demands of the tunnel. Finally, some type of compressor or blower will be needed to provide both pressure rise and flow during start up and during steady-state operation.

Upon startup the converging–diverging nozzle will operate as a subsonic flow accelerating in the converging section and decelerating in the diverging section before the test section. Subsonic flow will be established throughout the test section and the converging–diverging diffuser. As the flow rate is increased the initial converging–diverging nozzle will choke but initially subsonic flow will remain through the converging–diverging nozzle, test section, and diffuser. As the downstream pressure is reduced, supersonic flow will begin to be established in the diverging section of the converging–diverging nozzle. However, the downstream pressure initially will not be low enough to maintain supersonic flow and the flow in the diverging section of the nozzle will undergo a normal shock back to subsonic flow to match the downstream back pressure conditions. Flow downstream of the shock will be subsonic and will remain subsonic throughout the test section. As the back pressure is lowered further, the normal shock in the nozzle will move progressively downstream to the exit of the converging–diverging nozzle. If flow around a test object and its accompanying losses are not considered, then this will be the most demanding flow situation in terms of the system power requirements and this condition would be expected to produce the largest level of total pressure losses. One requirement of this situation will be the size of the downstream nozzle. It will need to be large enough to swallow the flow downstream from a normal shock in the test section. Since the test section is designed for a particular Mach number due to the geometry of the converging–diverging nozzle, the normal shock in the test section can be expected to be driven by the design Mach number. As indicated in chapter three, a normal shock will be accompanied with an increase in the area of the virtual throat. This increase in $A*$ means the downstream nozzle must have a larger throat to be able to accommodate the flow after the normal shock. This requirement can be shown by equating the flow rates of the two nozzles at the upstream and downstream throats and writing these flow rates in terms of their total conditions recognizing that each nozzle will be choked and the total temperature between the upstream nozzle and downstream diffuser will not change.

$$\dot{m}_1 = \dot{m}_2$$

$$\frac{P_{T1}}{RT_T} A_1^* \sqrt{k\, g_C\, R\, T_T} \left(1 + \frac{k-1}{2}\right)^{\frac{-(k+1)}{2(k-1)}} = \frac{P_{T2}}{RT_T} A_2^* \sqrt{k\, g_C\, R\, T_T} \left(1 + \frac{k-1}{2}\right)^{\frac{-(k+1)}{2(k-1)}} \quad (2.24)$$

Eliminating similar terms this relationship becomes

$$P_{T1}\, A_1^* = P_{T2}\, A_2^*$$

This relationship requires that the increase in the throat area of the downstream converging–diverging diffuser will be inversely proportional to the drop in the total pressure across the shock to allow for the downstream flow. Actual test sections will include boundary layer losses and friction and total pressure losses due to the testing of the model and they will require an even larger downstream throat during initial tunnel startup.

Example 6.1: Analysis of the Flow and Pressure Drop Requirements for a Supersonic wind Tunnel

A small supersonic test section is being designed for an educational experiment in compressible flow. The inlet total temperature is set at 365 K and the inlet total pressure is set at 40,000 Pa. The test section design Mach number is to be set at a value of 2.2. The test section cross-sectional area is set at 0.0075 m² as shown schematically in Figure 6.1.

a) Show a pressure distribution through the converging–diverging nozzle, test section, and the converging–diverging diffuser section for subsonic flow.

b) During start up find the test section pressure and volumetric flow rate required to choke the wind tunnel.

c) For the design Mach number in the test section find the test section pressure and volumetric flow rate assuming that a normal shock exists in the test section just upstream from the diffuser. Also determine the minimum size of the throat for the diffuser to accept the flow downstream of the shock.

d) Find the required power for the wind tunnel based on a shock in the test section.

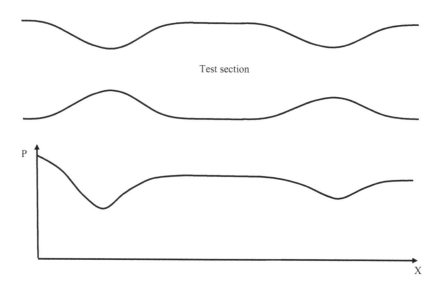

FIGURE 6.1
Schematic of supersonic wind tunnel nozzle, test section, and diffuser with pressure distribution for subsonic flow.

a) Subsonic flow through the test section, even with the initial nozzle choking will accelerate the flow into the converging section of the nozzle with a resulting static pressure drop until the throat of the nozzle as shown in Figure 6.1. At the throat as the area begins to increase the velocity of this compressible flow will decrease and the pressure rises. Inside the test section, the pressure will be below the total pressure but above the pressure anywhere downstream from the throat. As the flow enters the supersonic diffuser with its converging–diverging geometry, it initially speeds up with the resulting pressure drop then slows down in the diverging section with an accompanying increase in pressure.

b) The test section will be designed to produce a Mach number of 2.2. Consequently, the design for the nozzle will have a throat to exit area ratio which will produce a Mach number of 2.2. However, the initial choking in the upstream nozzle occurs at a back pressure that corresponds to the pressure ratio for the subsonic solution for the same area ratio. Recall the A/A^* versus Mach number figure from Chapter 2 which shows a subsonic and supersonic solution for every A/A^* other than 1. The required information is in the isentropic Mach number tables.

Isentropic Mach number table: $M = 2.2$, $A/A^* = 2.005$
Isentropic Mach number table: $M = 0.30$, $A/A^* = 2.035065$
Isentropic Mach number table: $M = 0.31$, $A/A^* = 1.976507$
Based on the isentropic Mach number tables the subsonic Mach number for the same area ratio is between 0.30 and 0.31. Interpolation suggests a value of $M = 0.305$.
The resulting static pressure will be

$$\frac{P}{P_T} = \frac{1}{\left(1+\frac{k-1}{2}M^2\right)^{k/(k-1)}} = \frac{1}{\left(1+\frac{1.4-1}{2}0.305^2\right)^{1.4/(1.4-1)}} = 0.93752 \qquad (2.14)$$

$$P = P_T \times \frac{P}{P_T} = 40,000\,\text{Pa} \times 0.93752 = 37501\,\text{Pa}$$

The volumetric flow rate can be determined from the local Mach number and the speed of sound. Noting the Mach number the static temperature can be determined (2.13).

$$\frac{T}{T_T} = \frac{1}{\left(1+\frac{k-1}{2}M^2\right)} = \frac{1}{\left(1+\frac{1.4-1}{2}0.305^2\right)} = 0.98174$$

$$T = T_T \times \frac{T}{T_T} = 365\,\text{K} \times 0.98174 = 358.3\,\text{K}$$

The speed of sound and the velocity in the test section can now be easily determined.

$$a = \sqrt{k\,g_C\,R\,T} = \sqrt{1.4\,(1)\,287\,\frac{\text{J}}{\text{kg}\,\text{K}}\,358.3\,\text{K}} = 379.4\,\text{m}/\text{s}$$

$$V = M\,a = 0.305 \times 379.4\,\frac{\text{m}}{\text{s}} = 115.7\,\text{m}/\text{s}$$

The volumetric flow rate will be equal to the velocity times the area.

$$\dot{V} = V\,A = 115.7\,\frac{\text{m}}{\text{s}} \times 0.0075\,\text{m}^2 = 0.868\,\frac{\text{m}^3}{\text{s}}$$

Consequently, the volumetric flow requirements and back pressure to total pressure ratio to operate at the choked subsonic design point for the converging–diverging nozzle is not greatly challenging for this isentropic flow situation.

c) During wind tunnel start up as the back pressure to total pressure ratio is decreased and the volumetric flow rate is increased a normal shock is expected to form just downstream of the throat. This normal shock wave is expected to move out from its initial location close to the throat to the test section and eventually move to a position just upstream from the converging–diverging diffuser. This condition with a normal shock in the test section is expected to be the most challenging condition for tunnel operation. Analyzing this problem involves finding the supersonic isentropic condition and then the resulting volumetric flow rate.

Isentropic Mach number table: $M = 2.2$, $A/A^* = 2.005$, $P/P_T = 0.093522$, $T/T_T = 0.50813$

Normal Shock table: $M_1 = 2.2$, $M_2 = 0.547056$, $P_2/P_1 = 5.48$, $T_2/T_1 = 1.8569$, $P_{T2}/P_{T1} = 0.62814$

The converging–diverging nozzle is designed to produce a Mach number of 2.2 in the test section. Based on our isentropic Mach number tables at a Mach number of 2.2 the exit to throat area will be 2.005 resulting in a throat area of 0.00374 m². In the region of supersonic flow in the test section, the static pressure will be reduced to a value of less than 10 percent of the total pressure and the static temperature will be roughly half of the total temperature. The static pressure and temperature downstream from the shock can be determined using the static to total pressure and temperature ratios from the isentropic tables together with the static pressure and temperature ratios across the normal shock.

$$P_2 = P_T \frac{P}{P_T} \frac{P}{P_1} = 40,000\,\text{Pa} \times 0.093522 \times 5.48 = 20,500\,\text{Pa}$$

$$T_2 = T_T \frac{T_1}{T_T} \frac{T_2}{T_1} = 365\,\text{K} \times 0.50813 \times 1.8569 = 344.4\,\text{K}$$

The downstream static temperature will allow determination of the downstream speed of sound and with the downstream Mach number, the velocity after the shock (Equation 2.11).

$$a_2 = \sqrt{k\,g_C\,R\,T_2} = \sqrt{1.4\,(1)\,287\,\frac{\text{J}}{\text{kg K}}\,344.4\,\text{K}} = 372.0\,\frac{\text{m}}{\text{s}}$$

$$V_2 = M_2\,a_2 = 0.547 \times 372.0\,\frac{\text{m}}{\text{s}} = 203.5\,\frac{\text{m}}{\text{s}}$$

The required volumetric flow rate for the system becomes

$$\dot{V} = V_2\,A = 203.5\,\frac{\text{m}}{\text{s}} \times 0.0075\,\text{m}^2 = 1.526\,\frac{\text{m}^3}{\text{s}}$$

The size of the downstream throat area will be significantly larger than the upstream throat area based the relationships between throats and total pressure ratios.

$$P_{T1}\,A_1^* = P_{T2}\,A_2^*\ \text{or}\ A_2^* = A_1^* \times \frac{P_{T1}}{P_{T2}} = \frac{0.00374\,\text{m}^2}{0.62814} = 0.005955\,\text{m}^2$$

d) The last aspect of this problem is to estimate the power requirements for the compressor in the current scenario. The downstream static pressure is assumed to be the compressor inlet pressure and the inlet total pressure is assumed to be the compressor discharge pressure. However, if the downstream static temperature is assumed to be the compressor inlet temperature and an isentropic compression is assumed to occur then the calculated inlet total temperature will exceed the inlet total temperature in the problem. Also this would not account for the isentropic efficiency of the compressor. This means that in a steady-state situation, energy must be removed from the air prior to the compressor. The temperature needed to produce the inlet total pressure can be determined from a combination of the isentropic relationship between temperature and pressure and the isentropic compressor efficiency.

$$\frac{T_2}{T_1} = \left(\frac{P_2}{P_1}\right)^{k-1/k} \tag{1.50}$$

$$\eta_C = \frac{W_{Ideal}}{W_{Actual}} = \frac{h_{2s} - h_1}{h_2 - h_1} = \frac{T_{2s} - T_1}{T_2 - T_1} \tag{6.1}$$

$$T_2 = T_1 \left\{ 1 + \left[\left(\frac{P_2}{P_1} \right)^{k-\frac{1}{k}} - 1 \right] / \eta_C \right\} \tag{6.2}$$

Solving for T_1 based on the appropriate pressures:

$$T_1 = \frac{T_2}{\left\{ 1 + \left[\left(\frac{P_2}{P_1} \right)^{k-\frac{1}{k}} - 1 \right] / \eta_C \right\}} = \frac{365\,\mathrm{K}}{\left\{ 1 + \left[\left(\frac{40,000\,\mathrm{Pa}}{20,500\,\mathrm{Pa}} \right)^{0.4/\!1.4} - 1 \right] / 0.8 \right\}} = 289.0\,\mathrm{K}$$

The resulting required inlet temperature which achieves an inlet total temperature of 365 K is about 289 K assuming an isentropic compressor efficiency of around 80 percent. A first law balance around the compressor allows an estimate of the compressor power requirements to run this small supersonic wind tunnel.

$$\dot{W} = \dot{m}\,C_P\,(T_2 - T_1) \tag{6.3}$$

At this point the mass flow rate, $\dot{m}$, is unknown. However, the mass flow rate is simply the density times the volumetric flow rate which is known. The density can be calculated from the ideal gas law (Equation 1.31).

$$\rho = \frac{P}{RT} = \frac{20,500\,\mathrm{Pa}}{287\,\dfrac{\mathrm{J}}{\mathrm{kg\,K}} * 344.4\,\mathrm{K}} = 0.2074\,\frac{\mathrm{kg}}{\mathrm{m}^3}$$

$$\dot{m} = \rho\,\dot{V} = 0.2074\,\frac{\mathrm{kg}}{\mathrm{m}^3} \times 1.526\,\frac{\mathrm{m}^3}{\mathrm{s}} = 0.3166\,\frac{\mathrm{kg}}{\mathrm{s}}$$

Now calculating the compressor power:

$$\dot{W} = \dot{m}\,C_P\,(T_2 - T_1) = 0.3166\,\frac{\mathrm{kg}}{\mathrm{s}} \times 1.005\,\frac{\mathrm{kJ}}{\mathrm{kg}} \times (365\,\mathrm{K} - 289\,\mathrm{K}) = 26.9\,\mathrm{kW}$$

So the power input to the compressor is about 27 kW or 36 HP and the required cooling rate upstream of the compressor would ideally be the same to drive this flow continuously through a compressible flow facility. A figure showing the tunnel pressure distribution for this condition is shown below in Figure 6.2.

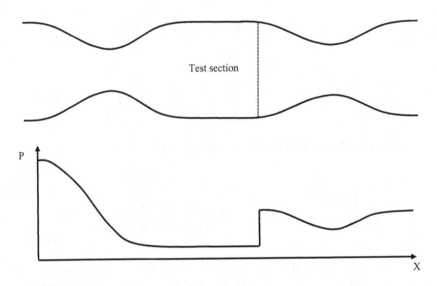

FIGURE 6.2
Schematic of supersonic wind tunnel nozzle, test section, and diffuser with pressure distribution
for supersonic flow and normal shock at test section exit.

Discussion: Based on the isentropic flow equations the minimum compressor compression ratio and volumetric flow rate needed to drive the
tunnel to choked flow is modest. However, this result assumes isentropic flow which does not account for friction nor the difficulty normally
encountered to diffuse flow from a sonic condition to a low subsonic
condition. The most difficult start-up condition occurs when a normal
shock wave sits inside the test section of the wind tunnel. The strong
normal shock produces a substantial loss in total pressure that has the
practical design requirement that the throat of the downstream converging–diverging diffuser must be larger to allow the same flow through
as the upstream nozzle. The design of a supersonic test section is actually quite challenging as changes in surface angle can produce oblique
shocks or Prandtl–Meyer expansion fans. Also most supersonic diffusers downstream from test sections are designed with oblique shocks
to reduce the total pressure loss encountered in the pressure recovery
process.

6.2 Oblique Shock Diffusers

One challenging aspect of the design of supersonic air breathing propulsion
engines is the recovery of pressure from supersonic flow. Aircraft which fly
at Mach number not too much past a value of 2.0 typically integrate single

or double oblique shock diffusers into their engine inlets. Pressure recovery from supersonic to subsonic flow is much better if flow undergoes one or more oblique shocks prior to the final pressure recovery through a normal shock. One of the practical difficulties of designing a single or double oblique shock diffuser is the potential of the boundary layer to separate if the pressure rise taken is too large. Consequently, single oblique shock diffusers are more common than double oblique shock diffusers. The previous section also suggested oblique shock diffusers were common in the diffuser section of supersonic wind tunnels. This approach would certainly reduce total pressure losses. However, as long as a normal shock must exist in the test section during start up the exit diffuser would need to be designed to swallow the flow downstream of this shock.

Example 6.2: Analysis of a Single and Double Oblique Shock Diffuser

Oblique shock diffusers are designed help recover pressure in a supersonic inlet efficiently and to slow the supersonic flow down prior to taking a normal shock somewhere inside the inlet of the engine. Using one-dimensional flow relationships, it would be possible to "optimize" the pressure recovery with either a single or double oblique shock diffuser. However, an additional consideration is the potential of the flow to separate after an oblique shock. This separation could cause a substantial blockage of flow into the inlet to an engine and consequently cause some substantial performance issues in the engine. Consequently, avoiding flow separation in an oblique shock diffuser prior to an engine would be an important design goal. Analyzing the potential of a boundary layer to separate due to an oblique shock is a complicated problem. Current computational fluid dynamics methods are likely to provide useful engineering guidance. A simpler approach might be to rely on experimental data on separation deflection angles for oblique shocks. Figure 6.3 was taken from NACA Research Memorandum E51L26. It provides a figure showing the results of an experimental investigation of the relationship between deflection angle and separation for two-dimensional flow over a Mach number range between about 1.33 and 3.0. Choosing a deflection angle comfortably below the separation line in this figure is likely a good approach in developing a single oblique shock diffuser with a robust operational range.

   In our initial oblique shock diffuser analysis, a Mach number of 1.8 will be assumed with a single wedge angle of 10° as presented in Figure 6.4. This angle is 3° below the separation angle of 13° at a Mach number of 1.8 and it should provide a reasonably robust operational design. In this analysis the pressure recovery for this single oblique shock diffuser will be compared to the pressure recovery for a normal shock diffuser.

   The sketch shows the inlet flow Mach number of 1.8 followed by an oblique shock as the flow abruptly adjusts to a deflection angle of 10°. This deflection of the supersonic flow results in the formation of an oblique shock which not only changes the flow direction but also slows the flow normal to the oblique shock direction also resulting in rise in the

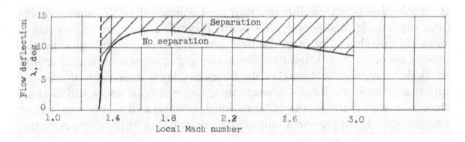

FIGURE 6.3
Effect of Mach number and flow deflection angle (λ, °) required for separation in two dimensional flow, $k = 1.4$, taken from NACA RM E51L26.

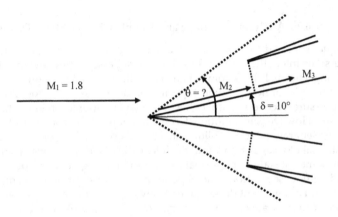

FIGURE 6.4
Sketch of single oblique shock diffuser showing flow downstream from oblique shock as well as flow downstream from a normal shock in the inlet to the engine.

static press. Downstream from the oblique shock a normal shock sits at the entrance to the engine inlet.

Analysis: Typically, determining the angle of an oblique shock wave of a supersonic flow responding to a deflection angle is an iterative process. However, using the oblique shock tables provided in Appendix A.4, the corresponding shock angle due to a deflection angle can be interpolated. In our current case the free stream Mach number is 1.8 and the deflection angle is 10° corresponding to an oblique shock angle of 44.0567°. The normal component of the flow across the oblique shock wave undergoes a pressure rise while the tangential velocity stays constant.

$$M_{1N} = M_1 \sin\theta = 1.8 \sin 44.0567° = 1.2517$$

$$M_{1T} = M_1 \cos\theta = 1.8 \cos 44.0567° = 1.2936$$

The normal component essentially undergoes a normal shock and the downstream component, M_{2N}, can be calculated using Equation (4.10):

$$M_{2N} = \sqrt{\frac{M_{1N}^2 + \dfrac{2}{k-1}}{M_{1N}^2 \dfrac{2k}{k-1} - 1}} = \sqrt{\frac{1,2517^2 + \dfrac{2}{1.4-1}}{1.2517^2 \dfrac{2*1.4}{1.4-1} - 1}} = 0.8117$$

Since there is no pressure variation in the tangential direction the tangential velocity remains constant and the downstream tangential Mach number, M_{2T}, can be calculated using Equation (4.14). However, Equation (4.14) requires the temperature ratio across the oblique shock which can be determined using Equation (4.9).

$$\frac{T_2}{T_1} = \frac{\left(1+(k-1)/2\,M_{1N}^2\right)}{\left(1+(k-1)/2\,M_{2N}^2\right)} = \frac{\left(1+0.4/2*1.25171^2\right)}{\left(1+0.4/2*0.8117^2\right)} = 1.1604$$

The static pressure ratio can be determined similarly using Equation (4.5)

$$\frac{P_2}{P_1} = \frac{\left(1+k\,M_{1N}^2\right)}{\left(1+k\,M_{2N}^2\right)} = \frac{\left(1+1.4*1.25171^2\right)}{\left(1+1.4*0.8117^2\right)} = 1.6611$$

With the temperature ratio across the shock the tangential Mach number is simple to determine.

$$M_{2T} = M_{1T}/\sqrt{\frac{T_2}{T_1}} = \frac{1.2936}{\sqrt{1.1604}} = 1.2008$$

The downstream Mach number can be determined using the sum of the squares of the downstream components.

$$M_2 = \sqrt{M_{2N}^2 + M_{2T}^2} = \sqrt{.8117^2 + 1.2008^2} = 1.4494$$

As a check on the calculation, the deflection angle can be determined by subtracting the angle of the oblique shock with respect to the downstream flow from the upstream flow. The angle of the oblique shock with respect to the downstream flow can easily be determined.

$$\theta - \delta = \mathrm{asin}\left(\frac{M_{2N}}{M_2}\right) = \mathrm{asin}\left(\frac{0.8117}{1.4494}\right) = 34.0567°$$

The deflection angle is equal to the oblique shock angle, θ, less the downstream shock angle, $\theta - \delta$.

$$\delta = \theta - (\theta - \delta) = 44.0567° - 34.0567° = 10.000°$$

Obviously, the calculated resolution of the angles is higher than what we could expect to achieve in any real flow. At this point we would like to determine the total pressure loss across the oblique shock. The most efficient approach is to look up the total pressure loss using the normal component of the incident flow to the shock and the normal shock table. Based on the normal shock table for an inlet Mach number of 1.25 the downstream to upstream total pressure ratio P_{T2}/P_{T1} is 0.9871. However, at this point the static pressure ratio, P_2/P_1, is known across the shock and so is the upstream and downstream Mach numbers. These upstream and downstream Mach numbers can be used to determine P/P_T before and after the shock. The total pressure ratio can be determined stating from the static pressure ratio and then using the static to total pressure ratios appropriately.

$$\frac{P_{T2}}{P_{T1}} = \frac{P_2}{P_1} \frac{P_{T2}}{P_2} \frac{P_1}{P_{T1}}$$

The static to total pressure ratios across the shock can actually be determined using the calculated static temperature ratio across the shock raised to the $k/(k-1)$.

$$\frac{P_{T2}}{P_{T1}} = \frac{P_2}{P_1} / \left(\frac{T_2}{T_1}\right)^{k/k-1} = \frac{1.6611}{1.1604^{1.4/0.4}} = 0.9868$$

The small difference between the value determined above and the value from the shock table is due to the small difference between the calculated normal shock, $M_{1N} = 1.2517$, and the value used with the table, $M = 1.25$.

The last portion of this problem is to determine the total pressure loss across the shock that is assumed to sit at the entrance of the engine inlet. Based on this problem the Mach number just upstream of this normal shock is 1.45. The total pressure ratio across this shock can be determined from the normal shock table for a specific heat ratio, $k = 1.4$. Based on the normal shock table, $P_{T3}/P_{T2} = 0.9448$. The total loss across this supersonic diffuser can be determined from the product of the two total pressure ratios.

$$\frac{P_{T3}}{P_{T1}} = \frac{P_{T2}}{P_{T1}} \frac{P_{T3}}{P_{T2}} = \frac{P_{T3}}{P_{T1}} = 0.9868 \times 0.9448 = 0.9324$$

Consequently, the total pressure loss across this single shock diffuser is about 6.76 percent. How does this compare to a normal shock diffuser? The answer to this question can immediately be determined from the normal shock table. At an inlet Mach number of 1.8, the total pressure ratio across the shock is 0.8127. This means the total pressure loss for a normal shock diffuser would be 18.73 percent which is nearly three times the loss for the single oblique shock diffuser.

The analysis for a double oblique shock diffuser is very similar to a single oblique shock diffuser except two oblique shocks occur before the inlet to the engine where a normal shock is assumed to sit. In this second analysis the incident Mach number will be assumed to be 2.2 and two 8° wedge angles produce the two oblique shocks prior to the inlet. Based on Figure 6.3 the first wedge angle is well below the separation limit which is just under 12°. The second wedge angle should be shallow enough not to cause a separation as long as sufficient space is provided for the turbulent boundary layer to recover. However, these practical issues cannot be addressed by the current flow analysis.

Based on the oblique shock table $k = 1.4$ (Appendix A.4), a Mach number of 2.2 will produce an oblique shock angle of 32.8269° when encountering an 8° deflection angle. The normal and tangential components of Mach number with respect to the shock can easily be determined.

$$M_{1N} = M_1 \sin\theta = 2.2\sin 32.8269° = 1.2247$$

$$M_{1T} = M_1 \cos\theta = 2.2\cos 32.8269° = 1.8276$$

The downstream normal Mach number and the pressure and temperature ratios across the shock can easily be interpolated from tables or determined from the normal shock relationships.

Normal shock: $M_{2N} = 0.8272$, $P_2/P_1 = 1.5832$, $T_2/T_1 = 1.1435$, $P_{T2}/P_{T1} = 0.9902$

The downstream tangential Mach number, M_{2T} can be calculated from the upstream value and the temperature ratio.

$$M_{2T} = M_{1T} / \sqrt{\frac{T_2}{T_1}} = \frac{1.8276}{\sqrt{1.1435}} = 1.7091$$

The Mach number downstream from the initial oblique shock can be determined from the downstream components.

$$M_2 = \sqrt{M_{2N}^2 + M_{2T}^2} = \sqrt{.8272^2 + 1.7091^2} = 1.8987$$

The downstream Mach number is close enough to $M_2 = 1.9$ that the oblique shock tables can be used again to find the oblique shock angle, $\theta = 39.2722°$ for a deflection angle of 8°. Now the previous calculations can be repeated.

$$M_{2N} = M_2 \sin\theta = 1.9\sin 39.2722° = 1.2027$$

$$M_{2T} = M_2 \cos\theta = 1.9\cos 39.2722° = 1.4709$$

Again, the downstream normal Mach number and the pressure and temperature ratios across the shock can easily be interpolated from tables or determined from the normal shock relationships.

Normal shock: $M_{3N} = 0.8405$, $P_3/P_2 = 1.5209$, $T_3/T_2 = 1.1297$, $P_{T3}/P_{T2} = 0.9925$

The downstream tangential Mach number after the second oblique shock, M_{3T} can be calculated from the upstream value and the temperature ratio.

$$M_{3T} = M_{2T} / \sqrt{\frac{T_3}{T_2}} = \frac{1.4709}{\sqrt{1.1297}} = 1.3839$$

The Mach number downstream from the second oblique shock can be determined from the downstream components.

$$M_3 = \sqrt{M_{3N}^2 + M_{3T}^2} = \sqrt{0.8405^2 + 1.3839^2} = 1.6191$$

The final portion of this analysis is to determine the overall total pressure loss. The total pressure ratio in the final normal shock, noting $M_3 \cong 1.62$, is $P_{T3}/P_{T4} = 0.888$. The overall total pressure ratio across the three shocks is simply the product of these ratios.

$$\frac{P_{T4}}{P_{T1}} = \frac{P_{T2}}{P_{T1}} \frac{P_{T3}}{P_{T2}} \frac{P_{T4}}{P_{T3}} = \frac{P_{T4}}{P_{T1}} = 0.9902 * 0.9925 * 0.888 = 0.8727$$

This equates to a total pressure loss of around 12.73 percent. If a normal shock was used to diffuse the air the total pressure ratio at $M_1 = 2.2$ is 0.6281 which suggests a 37.18 percent loss. Consequently, oblique shock diffusers are very effective mechanisms in helping supersonic flows recover pressure. However, care must be taken to avoid boundary layer separation which could cause major blockage to the engine inlet.

6.3 Supersonic Airfoils

Supersonic flows which encounter airfoils must abruptly adjust to the presence of the airfoils as pressure waves cannot move upstream faster than the speed of sound. Consequently, supersonic airfoils tend to be thin with relatively sharp leading edge regions. Often supersonic airfoils are approximated as flat plates or thin diamond shapes for the purpose conceptual analysis. In this section a flat plate airfoil and a thin diamond airfoil will be analyzed using the shock expansion method which uses the analysis of oblique shocks and Prandtl–Meyer expansion fans to determine the pressure fields above and below the airfoil. Since this section analyzes airfoils, the analyses will be extended to determine the lift and drag coefficients of the airfoils.

Example 6.3: Flat Plate Supersonic Airfoil

In this example a flat plate supersonic airfoil encounters a supersonic flow at a Mach number of 2.3 as sketched in Figure 6.5. The angle of attack is 4°. Determine the flow above and below the airfoil as well as the pressure ratio above and below the airfoil and also the lift and drag coefficients.

Sketch: The state of the flow and the pressure below the airfoil can be determined from an oblique shock analysis for a 4° deflection angle for an incident Mach number of 2.3 with θ= 28.9057°, Table A.4.1.

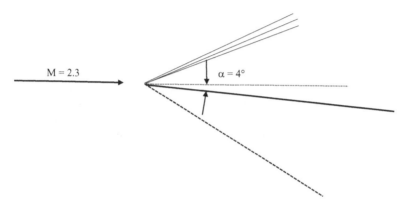

FIGURE 6.5
Sketch of flat plate supersonic airfoil showing oblique shock attached to leading edge below the airfoil and an expansion fan over the upper surface.

$$M_{1N} = M_1 \sin\theta = 2.3 \sin 28.9057° = 1.1118$$

$$M_{1T} = M_1 \cos\theta = 2.3 \cos 28.9057° = 2.0135$$

The normal Mach number downstream of the oblique shock and the pressure and temperature ratios across the shock can either be interpolated from tables or calculated from the normal shock relationships.

Normal shock: $M_{2N} = 0.9028, P_2/P_1 = 1.2753, T_2/T_1 = 1.0724,$
$$P_{T2}/P_{T1} = 0.9985$$

This part of the analysis shows that the total pressure loss is very low for this small deflection angle, suggesting that for these small angles the flow is almost isentropic. With the temperature ratio and the upstream tangential Mach number the downstream tangential Mach number can be determined. Next, the square root of the squared and summed downstream components can be used to determine the Mach number.

$$M_{2T} = M_{1T} / \sqrt{\frac{T_2}{T_1}} = \frac{2.0135}{\sqrt{1.0742}} = 1.9443$$

$$M_2 = \sqrt{M_{2N}^2 + M_{2T}^2} = \sqrt{.9028^2 + 1.9443^2} = 2.1437$$

Over the top of the airfoil, an expansion fan is needed to expand the flow around the 4° convex corner the flow must negotiate. Recall that the Prandtl–Meyer function was used to develop a functional relationship between Mach number and the Prandtl–Meyer function angle, given by Equation (5.11).

$$v(M) = \sqrt{\frac{k+1}{k-1}} \tan^{-1} \sqrt{\frac{k-1}{k+1}(M^2-1)} - \tan^{-1}\sqrt{M^2-1} \qquad (5.11)$$

This Prandtl–Meyer function is tabulated for $k = 1.4$ in Appendix A.5.1 yielding the angle v for the approach flow as 35.2828°. The downstream flow on the upper surface will have a Prandtl–Meyer function of 39.2828° or 4° larger than the approach flow. This change in flow angle is around the convex corner is calculated to produce a downstream Mach number of 2.464. Either the isentropic tables or the isentropic relationships can be used to determine the pressure ratio across the expansion fan noting the total pressure remains constant across an expansion fan.

$$\frac{P_{2u}}{P_1} = \frac{P_{2u}}{P_T}\frac{P_T}{P_1} = \frac{\left(1+\frac{k-1}{2}M_1^2\right)^{k/k-1}}{\left(1+\frac{k-1}{2}M_{2u}^2\right)^{k/k-1}} = \frac{\left(1+\frac{0.4}{2}2.3^2\right)^{1.4/0.4}}{\left(1+\frac{0.4}{2}2.464^2\right)^{1.4/0.4}} = 0.7738$$

At this point by referencing the upstream static pressure, P_1 or P_∞ the lift force and drag force on the wing of chord length, C, in Figure 6.5 could be calculated. The lift and drag coefficients could also be determined. It should be noted that the lift force is considered the force per unit length on the airfoil perpendicular to the free stream velocity and the drag force is considered the force per unit length on the airfoil in the direction opposite to the free stream velocity. An airfoil is considered to be a two-dimensional section of a wing. The lift coefficient is the lift force normalized by $\rho U_\infty^2/2 * C$ and the drag coefficient is the drag force normalized by the same relationship. This normalized force can be rewritten in terms of the Mach number and free stream static pressure, P_1 or P_∞.

$$\rho U_\infty^2 / 2g_c \times C = \frac{P_\infty U_\infty^2}{RT 2g_c}C = \frac{P_\infty k M_\infty^2}{2}C$$

The lift force on the airfoil can be determined in terms of the pressure on the upper and lower surfaces and the projection of the airfoil chord length onto a plane parallel to the free stream velocity.

$$L = (P_{2L} - P_{2U}) \times C \cos\alpha$$

Taking the ratio of lift over the dynamic pressure times the chord, cancelling the chord length.

$$C_L = \frac{\left(\dfrac{P_{2L}}{P_\infty} - \dfrac{P_{2U}}{P_\infty}\right)\cos\alpha}{k\,M_\infty^2\Big/2} = \frac{(1.2753 - 0.7738)\cos 4°}{1.4*2.3^2\Big/2} = 0.1351$$

The determination of the drag coefficient for this simple flat plate airfoil is very similar to the lift coefficient except that in the case of drag, the flat plate airfoil is projected onto a plane normal to the streamwise velocity. Consequently, drag will be calculated as:

$$D = (P_{2L} - P_{2U}) \times C \sin\alpha$$

And the drag coefficient becomes.

$$C_D = \frac{\left(\dfrac{P_{2L}}{P_\infty} - \dfrac{P_{2U}}{P_\infty}\right)\sin\alpha}{k\,M_\infty^2\Big/2} = \frac{(1.2753 - 0.7738)\sin 4°}{1.4*2.3^2\Big/2} = 0.00945$$

So in this simplified case the lift to drag ratio is equal to 14.3 to 1. However, note that the present analysis was based on isentropic flow and no skin friction losses were accounted for. Also recognize the calculated drag in the current analysis was wave drag which is a consequence of supersonic flow.

Example 6.4: Diamond-shaped Supersonic Airfoil

The analysis of supersonic symmetrical diamond shaped airfoil is somewhat more complicated than the analysis of a flat plate supersonic airfoil (Figure 6.6). In the present sketch, the angle of attack, α, is greater than the half wedge angle of the supersonic airfoil which means the flow over the upper surface of the airfoil encounters a convex turn causing a Prandtl–Meyer expansion fan. On the bottom of the airfoil the flow undergoes a concave angle producing an oblique shock wave. On the upper and lower aft portions of the airfoil the flow expands around a convex corner as the flow moves from the front to the aft half of the airfoil. In the present problem, the free stream Mach number is given at 2.4 and similar to the previous problem the angle of attack, α will be set at 4°. The half wedge angle of the current airfoil is set at 3°.

Sketch:

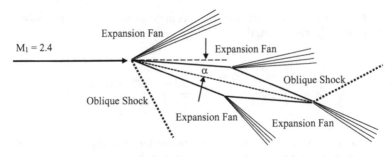

FIGURE 6.6
Sketch of a symmetrical diamond shaped airfoil in a supersonic flow.

Analysis: The analysis will determine the pressure ratio and downstream Mach number as the flow moves over each of the segments of this symmetrical diamond shaped airfoil. On the lower portion of the airfoil the flow initially is deflected through a concave angle of 7° before expansion around a convex angle of 6°. Consequently, the analysis on the bottom surface begins with an oblique shock analysis where the initial normal and tangential Mach number components are found. The oblique shock table gives a shock angle of 30.2507° at a Mach number of 2.4 with a deflection angle of 7°.

$$M_{1N} = M_1 \sin\theta = 2.4 \sin 30.2507° = 1.2091$$

$$M_{1T} = M_1 \cos\theta = 2.4 \cos 30.2507° = 2.0732$$

With the normal component of the oblique shock, the properties across the normal shock can be determined from either the normal shock table by interpolation or using the normal shock equations.

Normal shock: $M_{2N} = 0.8366$, $P_2/P_1 = 1.5389$, $T_2/T_1 = 1.1337$, $P_{T2}/P_{T1} = 0.9919$

The tangential Mach number and the static temperature ratio across the shock can be used to determine the downstream tangential Mach number and then the downstream Mach number.

$$M_{2T} = M_{1T} / \sqrt{\frac{T_2}{T_1}} = \frac{2.0732}{\sqrt{1.1337}} = 1.9471$$

$$M_2 = \sqrt{M_{2N}^2 + M_{2T}^2} = \sqrt{.8366^2 + 1.9471^2} = 2.1192$$

The key components here are the pressure ratio, P_{2L}/P_1, and the downstream Mach number $M_{2L} = 2.1192$. These values can be used to find the Prandtl–Meyer function value and later the pressure ratio, P_{3L}/P_1. The downstream Mach number can be rounded to a value of 2.12. Based on the Prandtl–Meyer function table at $M = 2.12$, $\nu = 29.6308°$. As the turning angle is 6° the new Prandtl–Meyer function value is 35.6308° which corresponds to a Mach number by interpolation of 2.3543. The downstream Mach number allows the calculation of the pressure ratio across the expansion fan.

$$\frac{P_{3L}}{P_{2L}} = \frac{P_{3L}}{P_T}\frac{P_T}{P_{2L}} = \frac{\left(1+\frac{k-1}{2}M_{2L}^2\right)^{k/k-1}}{\left(1+\frac{k-1}{2}M_{3L}^2\right)^{k/k-1}} = \frac{\left(1+\frac{0.4}{2}2.12^2\right)^{1.4/0.4}}{\left(1+\frac{0.4}{2}2.3543^2\right)^{1.4/0.4}} = 0.6931$$

However, this ratio references to the pressure under section 2L and the upstream ratio needs to be used to reference to the free stream pressure.

$$\frac{P_{3L}}{P_1} = \frac{P_{3L}}{P_{2L}}\frac{P_{2L}}{P_1} = 0.6931 \times 1.5389 = 1.0666$$

The analysis for the upper surfaces is similar to the last section. First a 1° expansion occurs and then another 6° expansion occurs. Starting with the inlet Mach number at $M = 2.4$, $\nu = 36.7465°$. Adding 1° to the Prandtl–Meyer function results in a Mach number, $M_{2u} = 2.4416$. The resulting pressure ratio becomes

$$\frac{P_{2u}}{P_1} = \frac{P_{2u}}{P_T}\frac{P_T}{P_1} = \frac{\left(1+\frac{k-1}{2}M_1^2\right)^{k/k-1}}{\left(1+\frac{k-1}{2}M_{2u}^2\right)^{k/k-1}} = \frac{\left(1+\frac{0.4}{2}2.4^2\right)^{1.4/0.4}}{\left(1+\frac{0.4}{2}2.4416^2\right)^{1.4/0.4}} = 0.9371$$

The section expansion results in a Prandtl–Meyer function of 43.7465° which interpolates to a downstream section Mach number of 2.7058. The ratio of P_{3u}/P_1 can be calculated directly since both changes in angle were expansions and therefore isentropic.

$$\frac{P_{3u}}{P_1} = \frac{P_{3u}}{P_T}\frac{P_T}{P_1} = \frac{\left(1+\frac{k-1}{2}M_1^2\right)^{k/k-1}}{\left(1+\frac{k-1}{2}M_{3u}^2\right)^{k/k-1}} = \frac{\left(1+\frac{0.4}{2}2.4^2\right)^{1.4/0.4}}{\left(1+\frac{0.4}{2}2.7058^2\right)^{1.4/0.4}} = 0.6224$$

Now that all the local static to free stream static pressure ratios have been calculated around the airfoil, the lift and drag coefficients can be determined. At this point calculating the horizontal and vertical distances for the segments with respect to the free stream Mach number is needed.

$$L_{2u} = L_{3L} = \frac{C\cos(\alpha - 3°)}{2\cos 3°} = \frac{C\,0.99985}{2\times 0.99863} = 0.5006 \times C$$

$$L_{2L} = L_{3U} = \frac{C\cos(\alpha + 3°)}{2\cos 3°} = \frac{C\,0.99255}{2\times 0.99863} = 0.49695 \times C$$

The vertical components of the airfoil can be determined similarly.

$$H_{2U} = H_{3L} = \frac{C\sin(\alpha - 3°)}{2\cos 3°} = \frac{C\,0.01745}{2\times 0.99863} = 0.00874 \times C$$

$$H_{2L} = H_{3U} = \frac{C\sin(\alpha + 3°)}{2\cos 3°} = \frac{C\,012187}{2\times 0.99863} = 0.06102 \times C$$

The lift force due to pressure can be calculated using the pressure and segments.

$$L = P_{2L} \times 0.49695\,C + P_{3L} \times 0.5006\,C - P_{2U} \times 0.5006\,C - P_{3U} \times 0.49695\,C$$

The lift coefficient is simply the lift divided by $P_\infty \times k\,M_\infty^2 / 2 \times C$. This results in the following value after cancelling C, the chord length in the denominator and numerator.

$$C_L = \left[\left(\frac{P_{2L}}{P_\infty} - \frac{P_{3U}}{P_\infty} \right) \times 0.49695 + \left(\frac{P_{3L}}{P_\infty} - \frac{P_{2U}}{P_\infty} \right) \times 0.5006 \right] / \left(k\,M_\infty^2 / 2 \right)$$

Adding in the pressure ratios:

$$C_L = \left[\left(1.5389 - 0.6224 \right) \times 0.49695 + \left(1.0666 - 0.9371 \right) \times 0.5006 \right] / \left(1.4 \times 2.4^2 / 2 \right) = 0.129$$

The drag force due to pressure can be calculated using the pressure and height segments.

$$D = P_{2L} \times 0.06102\,C + P_{3L} \times 0.00874\,C - P_{2U} \times 0.00874\,C - P_{3U} \times 0.06102\,C$$

The drag coefficient is the drag force divided by $P_\infty \times k\,M_\infty^2 / 2 \times C$. This results in the following value after cancelling C in the denominator and numerator.

$$C_D = \left[\left(\frac{P_{2L}}{P_\infty} - \frac{P_{3U}}{P_\infty} \right) \times 0.06102 + \left(\frac{P_{3L}}{P_\infty} - \frac{P_{2U}}{P_\infty} \right) \times 0.00874 \right] / \left(k\,M_\infty^2 / 2 \right)$$

Adding in the pressure ratios:

$$C_D = \left[\left(1.539 - 0.6224 \right) \times 0.06102 + \left(1.0666 - 0.937 \right) \times 0.00874 \right] / \left(1.4 \times 2.4^2 / 2 \right) = 0.0142$$

When the lift coefficient in this example is compared to the one in Example 6.3, the value is just a bit lower, largely due to the slightly higher Mach number in the analysis. However, the drag coefficient due to pressure is substantially higher than the value in Example 6.3. This increased drag is due to the thickness of the airfoil compared with the flat plate airfoil which was assumed to have no thickness. Consequently, in supersonic flow over an airfoil lift is typically closely proportional to the angle of attack and drag is a function of the angle of attack, the thickness ratio, and the camber. In this example the airfoil was assumed to have no camber.

6.4 Overexpanded and Underexpanded Supersonic Nozzles

Supersonic nozzles are required for supersonic wind tunnels, supersonic aircraft engines and rockets. During startup supersonic nozzles can often operate in an overexpanded condition as the pressure that feeds the nozzle increases. Rocket engines often go through a range of pressure ratios due to the change in back pressure as they rise through the atmosphere. This variation on back pressure can cause rocket nozzles to experience operations ranging from overexpansion to underexpansion. Both an overexpanded and underexpanded condition will be examined in the present section.

Example 6.5: Overexpanded Supersonic Nozzle

In this example a two-dimensional converging–diverging nozzle is designed for an exit Mach number of 2.5. However, in this start up condition, the flow is overexpanded and needs to adjust to the back pressure on the nozzle through an oblique shock as shown in Figure 6.7 below. The resulting downstream flow will be analyzed. In this example the ambient pressure is assumed to be 101.325 kPa while the nozzle exit pressure is 50 kPa during this start up condition. The initial oblique shock allows the exit plane pressure to adjust to the back pressure. However, in making this adjustment the flow turns inward. At the symmetry plan of this two-dimensional nozzle the flow must turn parallel and it does so with a second oblique shock. While the flow is now parallel, the pressure is now above the ambient or back plane pressure and the flow must expand out through a Prandtl–Meyer expansion fan to match it. However, now the flow is moving outward and this cannot endure at the symmetry plane so the flow must expand again to flow parallel. After the second Prandtl–Meyer expansion, the pressure is now lower than the ambient and another series of oblique shocks will develop.

Analyze the downstream flow from the nozzle.

Sketch:

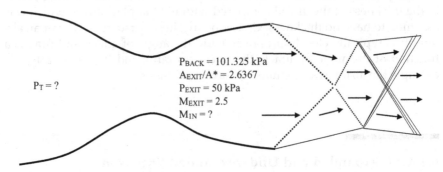

$P_T = ?$

$P_{BACK} = 101.325$ kPa
$A_{EXIT}/A^* = 2.6367$
$P_{EXIT} = 50$ kPa
$M_{EXIT} = 2.5$
$M_{1N} = ?$

FIGURE 6.7
Sketch of an overexpanded nozzle showing series of oblique shocks and expansions.

Analysis: $P_{EX} = 50$ kPa, $P_{AMB} = 101.325$ kPa, $P_{AMB}/P_{EX} = 2.0265$
N/S Table, $M_1 = 1.37$, $M_2 = 0.7527$, $P_2/P_1 = 2.02305$, $T_2/T_1 = 1.2354$
The oblique shock angle can be determined from the required normal shock for the pressure rise:

$$\theta = \sin^{-1}\frac{M_{1N}}{M_1} = \sin^{-1}\frac{1.37}{2.5} = 33.23°$$

With the oblique shock angle the tangential Mach number component can be determined first before the shock and then afterwards.

$$M_{1T} = M_1 \cos\theta = 2.5\cos 33.23° = 2.0912$$

$$M_{2T} = M_{1T} / \sqrt{\frac{T_2}{T_1}} = \frac{2.0912}{\sqrt{1.2354}} = 1.8814$$

With the downstream normal and tangential Mach number components, M_2 can be found.

$$M_2 = \sqrt{M_{2N}^2 + M_{2T}^2} = \sqrt{.7527^2 + 1.8814^2} = 2.0264$$

Now the deflection angle can be determined by first finding $\theta - \delta$.

$$\theta - \delta = \sin^{-1}\frac{M_{2N}}{M_2} = \sin^{-1}\frac{0.7527}{2.0264} = 21.805°$$

$$\delta = \theta - (\theta - \delta) = 33.23° - 21.805° = 11.425°$$

At this point, the flow is pointing in and it must undergo a second oblique shock to turn parallel to its current direction. Interpolating from the oblique shock table, $\theta = 40.39°$, and the normal and tangential components of the Mach numbers can be determined.

$$M_{2N} = M_2 \sin\theta = 2.0264 \sin 40.39° = 1.3131$$

$$M_{2T} = M_2 \cos\theta = 2.0264 \cos 40.39° = 1.5434$$

Rounding the normal component to 1.31 the normal shock table can be used:

N/S Table, $M_{2N} = 1.31$, $M_{3N} = 0.7809$, $P_3/P_2 = 1.8354$, $T_3/T_2 = 1.1972$

Next the downstream tangential components and Mach number can be determined.

$$M_{3T} = M_{2T} / \sqrt{\frac{T_3}{T_2}} = \frac{1.5434}{\sqrt{1.1972}} = 1.4106$$

With the downstream normal and tangential Mach number components, M_2 can be found.

$$M_3 = \sqrt{M_{3N}^2 + M_{3T}^2} = \sqrt{.7809^2 + 1.4106^2} = 1.6123$$

The deflection angle, δ, can be checked by finding $\theta - \delta$.

$$\theta - \delta = \sin^{-1}\frac{M_{3N}}{M_3} = \sin^{-1}\frac{0.7809}{1.6123} = 28.97°$$

$$\delta = \theta - (\theta - \delta) = 40.39° - 28.97° = 11.42°$$

The issue at this point is the pressure has risen by a ratio of 1.8354 through the second oblique shock and the static pressure after the shock is now over 83 percent higher than the ambient pressure. Consequently, the flow must adjust to this lower ambient pressure and due to the supersonic flow it must adjust abruptly through an expansion fan. This Prandtl–Meyer expansion is an isentropic process so either our isentropic relationships or the isentropic Mach number tables can be used. Rounding the downstream Mach number to the nearest one-hundredth, $M = 1.61$.

Isentropic tables: $M_3 = 1.61$, $P_3/P_T = 0.2318$
Dividing through by the pressure ratio, $P_3/P_2 = 1.8354$, noting $P_2 = P_4 = P_{AMB}$:

$$\frac{P_4}{P_T} = \frac{P_3}{P_T} \times \frac{P_4}{P_3} = \frac{0.2318}{1.8354} = 0.1263$$

Isentropic tables: $M_4 = 2.01$, $P_4/P_T = 0.1258$
So the resulting expansion required to match the back pressure produces a Mach number, $M_4 = 2.01$. The Prandtl–Meyer function can be

used to determine the resulting turning angle which will turn the flow outward.

Prandtl–Meyer function Table: $M_3 = 1.61$, $\nu_3 = 15.1615°$, $M_4 = 2.01$, $\nu_4 = 26.655°$

Based on these two values the turning angle of the flow will be an outward angle of:

$$\nu_4 - \nu_3 = 26.655° - 15.1615° = 11.494°$$

The resulting fan angle will be

$$fan\ angle = \nu_4 - \nu_3 + \mu_3 - \mu_4 = 26.655° - 15.1615° + 38.398° - 29.836° = 20.055°$$

The pressure of the flow now matches the ambient pressure. However, the flow is turned out from the centerline. Consequently, the flow much expand one more time to turn parallel to centerline. Since the fan angle is spread out at the centerline, the resulting flow angle and Mach number can only be approximated. The resulting Prandtl–Meyer function can be estimated to be

$$\nu_5 = \nu_4 + \nabla\nu = 26.655° + 11.494° = 38.148°$$

Based on the Prandtl–Meyer function table this corresponds to a Mach number of 2.4585. Rounding to a Mach number of 2.46 and using the isentropic Mach number tables.

Isentropic tables: $M_5 = 2.46$, $P_5/P_T = 0.06229$

The resulting local pressure downstream from this expansion fan can be determined from the static to total pressure ratios since P_4 is ambient and the flow is isentropic meaning total pressure stays constant.

$$P_5 = P_4 \times \frac{P_T}{P_4} \times \frac{P_5}{P_T} = 101.325\ \text{kPa} \times \frac{1}{0.1258} \times 0.06229 = 50.17\ \text{kPa}$$

Some small inaccuracies due to approximation have produced a pressure which is slightly lower than it would actually be. However, after undergoing the series of oblique shocks and expansion fans, this analysis suggests that this flow is very similar to the original flow and another series of oblique shocks and expansions could be expected.

Example 6.6: Underexpanded Supersonic Nozzle

An under expanded nozzle occurs when the pressure ratio across the nozzle is higher than the nozzle design allows for. An example of this might be a rocket nozzle as it flies up through the atmosphere into lower and lower pressures. Since the pressure within the nozzle is higher than the surrounding or back pressure, the flow must expand. Figure 6.8 presents a sketch of an underexpanded nozzle with a series of expansion fans and then an oblique shock. In our example $M_{EXIT} = 2.2$, $P_{EXIT} = 100$ kPa, and $P_{BACK} = 50$ kPa.

Analyze the downstream flow from the nozzle.

Sketch:

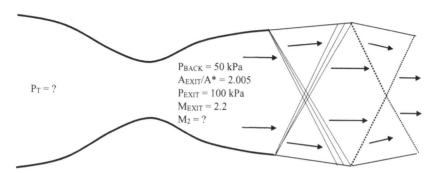

FIGURE 6.8
Sketch of underexpanded nozzle showing series of expansions and oblique shocks.

Analysis: $M_{EX} = 2.2$, Isentropic Mach number Tables: $P/P_T = 0.093522$
  Since the back pressure is lower than the exit plane pressure and the exit flow is supersonic, the nozzle is underexpanded and must expand outward to match the back plane pressure. This expansion occurs gradually through a series of expansion fans. This expansion is isentropic so the total pressure is going to stay constant. The total pressure can be determined from the exit pressure and the static to total pressure ratio.

$$P_T = \frac{P_{EX}}{\frac{P}{P_T}} = \frac{100\ kPa}{0.093522} = 1069.3\ kPa$$

The Mach number just downstream from the expansion can be determined from the resulting static to total pressure ratio after the expansion.

$$\frac{P_{BACK}}{P_T} = \frac{P_{EX}}{P_T} * \frac{P_{BACK}}{P_{EX}} = 0.093522 \times \frac{50\ kPa}{100\ kPa} = 0.046761$$

The exit Mach number can be interpolated from the isentropic Mach number tables.

Isentropic Tables: $M = 2.64$, $P/P_T = 0.04711$, $M = 2.65$, $P/P_T = 0.046389$
  Interpolating between the two, $M_2 = 2.64482$ and rounding $M_2 = 2.645$. The flow at the exit of the nozzle expands from a Mach number of 2.2 to approximately 2.645. The expansion angle can be determined using the Prandtl–Meyer flow table. The Prandtl–Meyer function is 31.7325° at a Mach number of 2.2 and 42.4181° at a Mach number of 2.645. Consequently, the flow is determined to expand outwardly from this two-dimensional nozzle at an angle of 10.69°.
  Along the centerline, the flow cannot continue flowing out above and below the symmetry plane of the two-dimensional nozzle at an angle of

10.69° so the flow is forced once again to expand outwardly at an angle of 10.69°. However, after the first expansion, the expansion is distributed over the fan angle so the analysis becomes approximate. However, if the flow were to expand discretely across an angle of 10.69° from a Mach number of 2.645, the Prandtl–Meyer function would be expected to increase from a value of 42.42° to an angle of 53.11° resulting in a Mach number of 3.18 after the second expansion. Based on our isentropic Mach number relationships P/P_T at this Mach number is 0.020832. Consequently, the second expansion produces the following static pressure.

$$P_3 = P_{EX} \times \frac{P_T}{P_{EX}} \times \frac{P_3}{P_T} = 100\,\text{kPa} \times \frac{1}{0.093522} \times 0.020832 = 22.275\,\text{kPa}$$

This analysis shows that the pressure downstream from the second expansion is now below the back pressure of 50 kPa. Consequently, an oblique shock is required to increase the pressure to the level of the back pressure. The pressure ratio across the oblique shock will be.

$$\frac{P_{BACK}}{P_3} = \frac{50\,\text{kPa}}{22.275\,\text{kPa}} = 2.245$$

Based on the normal shock tables:

Normal Shock: $M_{3N} = 1.44$, $M_{4N} = 0.72345$, $P_4/P_3 = 2.2525$, $T_4/T_3 = 1.2807$
    The angle of the oblique shock can be found from the arcsine of the ratio of M_{3N}/M_3.

$$\theta = \sin^{-1}\frac{M_{3N}}{M_3} = \sin^{-1}\frac{1.44}{3.18} = 26.92°$$

The tangential component can be determined from the oblique shock angle. The downstream tangential component can be determined from the upstream component and the static temperature ratio across the oblique shock.

$$M_{3T} = M_3 \cos 26.92° = 2.8353$$

$$M_{4T} = \frac{M_{3T}}{\sqrt{T_4/T_3}} = \frac{2.8353}{\sqrt{1.2807}} = 2.5054$$

The downstream Mach number can be determined from the root sum square of the components.

$$M_4 = \sqrt{M_{4T}^2 + M_{4N}^2} = \sqrt{2.5054^2 + 0.7234^2} = 2.6078$$

The turning angle can be determined from the angle between the downstream components and the downstream flow.

$$\theta - \delta = \sin^{-1}\frac{M_{4N}}{M_4} = \sin^{-1}\frac{0.7235}{2.6078} = 16.11°$$

The resulting turning angle can be determined by subtracting the downstream oblique shock angle from the upstream oblique shock angle.

$$\delta = \theta - (\theta - \delta) = 26.92° - 16.11° = 10.81°$$

Since the flow is now turned inward, another oblique shock is needed to turn the flow back parallel to the centerline of the flow. This second oblique shock will result in a static pressure above the back pressure and require an expansion to meet the back pressure boundary condition. Of course this analysis is also only approximate due to the spreading out of the expansion waves.

References

1. John, J.E.A. and Keith, T.G., 2006, *Gas Dynamics*, 3rd ed, Prentice Hall.
2. Saad, M.A., 1993, *Compressible Fluid Flow*, 2nd ed., Prentice Hall.

Chapter 6 Problems

1. A supersonic wind tunnel is designed to operate at a test section Mach number of 2.4. The inlet total pressure is 100 kPa and the inlet total temperature is 370 K. The nozzle has a throat area of 0.004 m². (a) Find the area ratio and the static to total pressure ratio at the nozzle exit required to achieve this Mach number. (b) Determine the minimum size that the downstream throat can be to accept the flow during start up when a normal shock appears in the test section. (c) Determine the total pressure loss due to a shock in the test section. (d) Calculate tunnel flow rate and the power required to operate the tunnel when a shock exists in the test section. You may assume that the air is isentropically compressed from the downstream total pressure to the upstream total pressure.

2. Calculate the ratio of the downstream total pressure to the inlet total pressure for a single oblique shock diffuser with a 9° wedge angle operating at an inlet Mach number of 2.0. Compare this total pressure ratio to the ratio of normal shock diffuser. A normal shock can be assumed to sit at the inlet to the engine after the initial oblique shock.

3. Calculate the ratio of the downstream total pressure to the inlet total pressure for a double oblique shock diffuser with two 9° wedge angles operating at an inlet Mach number of 2.4. Compare this total

pressure ratio to the ratio of normal shock diffuser. A normal shock can be assumed to sit at the inlet to the engine after the two oblique shocks.

4. Using the shock-expansion method determine the pressure ratio above and below a flat plate airfoil at angles of attack of 1°, 3°, and 5° encountering a free-stream Mach number of 2.2. Also, determine the lift and drag coefficients and the lift to drag ratio. Show all your work for one angle and results for the other angles.

5. Using the shock-expansion method determine the pressure ratio above and below a symmetrical diamond shaped airfoil with a half wedge angle of 3° at angles of attack of 3°, and 5° encountering a free-stream Mach number of 2.2. Also, determine the lift and drag coefficients and the lift to drag ratio. Compare these results to problem 4 if it has also been assigned.

6. A converging–diverging nozzle is designed to operate at an exit Mach number of 2.1. The exit pressure is 50 kPa, while the back pressure is 80 kPa. Describe and calculate the process that the flow uses to adjust to the back pressure. Note that the flow is not parallel to the symmetry plane. What must happen next? Determine the resulting flow process which resolves the flow at the midplane. Determine the resulting pressure due to this flow reflection? Describe how the flow continues to react. Is the nozzle considered underexpanded or overexpanded?

7. A converging–diverging nozzle is designed to operate at an exit Mach number of 2.25. The exit pressure is 100 kPa and the back pressure is 70 kPa. Describe and analyze the process that the flow uses to adjust to the difference between the exit and back pressure. After the flow adjustment the flow will no longer to be parallel to the symmetry plane. What will happen next? Describe and calculate the flow process which occurs to adjust the flow back to be parallel to the symmetry plane. Describe what happens next. Is the nozzle described as underexpanded or overexpanded?

7

Linearized Flow

Linearized flow is an approximate method which allows a quick and reasonably accurate estimate of the impact of a small change in flow angle on local pressure. This method can be used to estimate the pressure coefficient due to small changes in flow angle and it can be applied to supersonic airfoils to estimate lift and drag coefficients. The basis of linearized flow is similar to Prandtl–Meyer expansion fans in that it assumes that flow is isentropic. However, while Prandtl–Meyer flow is based on the variation Mach number due to changes in angle with convex flow, linearized flow estimates variation in pressure coefficient with either mild convex or concave angles. Linearized flow can also be applied to the estimate of lift and drag on simple airfoil shapes. In this chapter, linearized flow equations are developed for application to changes in pressure coefficient in supersonic flows. Linearized flow equations are also developed to estimate lift and drag for flat plate and other thin airfoil shapes in supersonic flow. A comparison between linearized flow and the shock expansion method is given for flat plate and thin diamond-shaped airfoils.

7.1 Introduction to Linearized Flow

Supersonic flow over airfoils is substantially different than subsonic flow. Subsonic airfoils are sensed by the flow due to pressure waves allowing the flow to adjust to the airfoil's presence. As a result, streamlining is possible. Subsonic airfoils typically have negligible pressure drag at low angles of attack as the flow largely stays attached. As the flow accelerates from the stagnation point, which will typically be slightly below the camber line near the leading edge, up over the top of the airfoil, a low pressure results at the front of the airfoil. This "suction" condition results in negligible pressure drag at low angles of attack. Supersonic flows are unable to "sense" the presence of an airfoil and as a result must abruptly adjust to the airfoils presence. This difference results in substantial differences between subsonic and supersonic airfoils.

Subsonic airfoils typically have a moderate maximum thickness. One well-known airfoil series, the NACA 24XX series airfoil, was found to be able to produce a maximum lift coefficient when the thickness to chord ratio was

12 percent. This "optimum" thickness is due to a trade-off between a thicker leading edge which reduces the acceleration around the leading edge and as a result reduces the airfoil's susceptibility to separate downstream from this region and separation in the aft region of the airfoil due to the flow diffusing around the airfoil's thickness. Supersonic airfoils tend to be thin with sharp leading and trailing edge regions. Rounded leading edges which are too blunt could result in an initial bow shock wave which sits off the leading edge and would result in a substantial pressure loss. Additionally, supersonic airfoils are subject to wave drag. The abrupt nature of the response of a supersonic flow to an airfoil results in a higher pressure force on surfaces facing the free-stream direction and lower pressures on surfaces opposite to the free-stream direction. The leading edge suction which is present on subsonic airfoils is not possible on supersonic airfoils. Since the response of the supersonic flow to the airfoil results on higher pressure on surfaces facing the free-stream direction and lower pressures on surfaces away from the free-steam direction, pressure drag, increases with both angle of attack and thickness. Consequently, supersonic airfoils are typically thin and have sharp leading and trailing edges. This issue will be explored in this chapter as supersonic airfoils are analyzed using both linearized flow and the shock expansion method.

In the previous chapter, the shock expansion method was used to determine drag on a thin diamond-shaped airfoil with a low angle of attack. The adjustment of the flow to the bottom surface resulted in a weak oblique shock. In the analysis of the oblique shock, the total pressure ratio, P_{T2}/P_{T1}, was 0.9919, which indicates that the total pressure loss was almost negligible. The resulting Mach number can be predicted closely using the Prandtl–Meyer function table using a negative turning angle, $\Delta\nu$. The resulting static pressure ratio, P_2/P_1, can then be determined using isentropic relationships. This nearly isentropic nature of weak concave turning angles was noticed early on and a simple method to estimate the pressure coefficient over an airfoil was developed. The shock expansion method is an accurate approach to determine pressures over a simple flat plate or wedge-shaped airfoil and the resulting lift and drag. However, this method would be very involved if used to determine the lift and drag or moment over a curved airfoil.

7.2 Development of the Linearized Pressure Coefficient

The development of the Prandtl–Meyer function involved looking at flow over a differential convex angle. The convex angle requires the flow to expand around the corner. This results in an expansion normal to the Mach wave. However, the flow tangent to the Mach wave remains unchanged as there is no pressure variation in the direction tangent to the Mach wave.

This relationship results in the equality (Equation 5.1) between tangential velocity upstream of the Mach wave (V *cosine* μ) and the expanded flow ($V + dV$) times the cosine of the Mach angle plus the differential angle ($\mu + d\nu$). Through a trigonometric identity the differential velocity normalized by the velocity (dV/V) can be equated to the differential angle ($d\nu$) times the tangent of the Mach angle ($\tan(\mu)$) in Equation (5.2). Noting the sine of the Mach angle (μ) is equal to the speed of sound over the velocity ($a/V = 1/M$) then the cosine of the Mach angle can be equated to ($\sqrt{(M^2-1)}/M$). Substituting for the tangent with the sine and cosine of the Mach angle results in a relationship between the normalized differential velocity and the differential angle normalized on ($\sqrt{(M^2-1)}$) or Equation (5.3).

$$\frac{dV}{V} = \frac{d\nu}{\sqrt{M^2 - 1}} \tag{5.3}$$

If Equation (5.3) is integrated over a small increment it can be approximated as:

$$\frac{V - V_\infty}{V_\infty} = \frac{\nu}{\sqrt{M_\infty^2 - 1}} \tag{7.1}$$

Conventionally, the Greek symbol, θ, is used for the turning angle instead of ν but is defined as the direction opposite to ν.

$$\frac{V_\infty - V}{V_\infty} = \frac{\theta}{\sqrt{M_\infty^2 - 1}} \tag{7.2}$$

This relationship based on Equation (5.3), can be equated to pressure change using the equation between differential pressure and differential velocity developed in Chapter 2 (Equation 2.19).

$$-dP = \frac{\rho V dV}{g_c} \tag{2.19}$$

If Equation (2.19) is integrated across a small incremental velocity the result is

$$-\Delta P = \frac{\rho\left[(V + \Delta V)^2 - V^2\right]}{2g_c} = \frac{\rho\left[2V\Delta V + \Delta V^2\right]}{2g_c} \tag{7.3}$$

If the differential velocity ΔV is small then squared term can be dropped and the equation can be linearized similar to Equations (7.1) and (7.2).

$$P_\infty - P = \frac{\rho V_\infty (V - V_\infty)}{g_c} \tag{7.4}$$

Equation (7.4) can be developed into a pressure coefficient equation if the pressure change is divided by $-\rho V_\infty^2 / 2g_c$ resulting in the following result.

$$C_P = \frac{P - P_\infty}{\rho V_\infty^2 / 2g_C} = \frac{2\left(V_\infty - V\right)}{V_\infty} \qquad (7.5)$$

Equating Equation (7.5) to Equation (7.2) the following linearized relationship for pressure coefficient versus turning angle is developed for small angles in supersonic flow.

$$C_P = \frac{P - P_\infty}{\rho V_\infty^2 / 2g_C} = \frac{2\theta}{\sqrt{M_\infty^2 - 1}} \qquad (7.6)$$

7.3 Linearized Flow Over Airfoils

Linearized flow over airfoils is restricted to supersonic flows over thin airfoils with modest angles of attack. At higher angles to the free-stream flow the pressure coefficient estimates can increasingly vary from oblique shock or Prandtl–Meyer solutions. Supersonic flow at lower Mach numbers is especially sensitive to increasing angles. A typical situation is pictured in Figure 7.1 below.

Figure 7.1 shows a thin symmetrical airfoil in supersonic flow at a modest angle of attack, α. When the surface of the airfoil creates a positive or concave angle with the free-stream flow then the pressure coefficient, C_P, is positive and when the surface creates a negative or convex angle with the free-stream flow the C_P is negative. Generally, when the angle of attack, α, is positive the airfoil surface will create a positive lift. Also note that based on the figure that all surfaces which create a positive C_P have a surface normal which projects in a direction which partially opposes on free-stream velocity vector while all the airfoil surfaces which have a negative C_P have a surface normal which projects in a direction which is partially in the direction of the free-stream velocity vector. These observations indicate that this airfoil will have notable pressure or wave drag which increases with thickness and angle of attack.

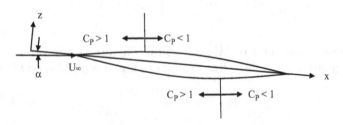

FIGURE 7.1
Coordinate system and features of supersonic flow past a thin airfoil.

The pressure coefficient is shown in Equation (7.6) to be a function of the surface angle, θ. The surface angle on our thin airfoil with respect to the x axis will be $\theta = \sin^{-1} dz/dx$. However, for a thin airfoil within a close approximation, $\theta \approx dz/dx$. If the angle of attack, α, is added in then the upper surface angle $\theta_U \approx dz/dx - \alpha$ and the lower surface angle, θ_L, is approximately equal to $\theta_L \approx -dz/dx + \alpha$. This relationship is shown schematically in Figure 7.2 below.

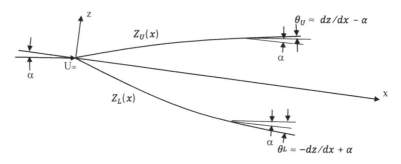

FIGURE 7.2
Schematic of thin airfoil showing approximate surface angle.

The differential lift along the surface of the airfoil is related to the pressure on the bottom of the airfoil less the pressure on the top of the airfoil. The differential surface force on the lower surface will be related to $dl_L = P_L \, ds_L \cos \theta_L$ while the differential surface force on the upper surface can be related to $dl_U = P_U \, ds_U \cos \theta_U$. For modest angles of attack, the distance along the airfoil can be approximated by $dx \approx ds_L \cos \theta_L \approx ds_U \cos \theta_U$ and the differential lift can be estimated as:

$$dl \approx (P_L - P_U) dx \qquad (7.7)$$

The lift force can be determined by integrating the differential lift from the leading edge to the trailing edge of the airfoil and the lift coefficient of an airfoil can be found by dividing through by the dynamic pressure, $\rho U_\infty^2 / 2g_C$, and the chord length, c. The pressure coefficient along the lower surface is equal to the local lower surface pressure less the free-stream pressure divided by the dynamic pressure. The pressure coefficient for the upper surface is similar except the local pressure on the upper surface is used. If the upper pressure coefficient at a given x location is subtracted from the lower pressure coefficient, the free-stream pressure is eliminated so the differential lift coefficient can be determined from the difference in the lower less upper pressure coefficients divided by the chord length, c.

$$dC_l \approx (C_{PL} - C_{PU}) d\left(\frac{x}{c}\right) \qquad (7.8)$$

If θ_L and θ_U are substituted into Equation (7.6) for substitution into C_{PL} and C_{PU} in Equation (7.8), the following equation for the differential lift is generated.

$$dC_l \approx \frac{2}{\sqrt{M_\infty^2 - 1}} \left(2\alpha + \frac{dz_L}{dx} - \frac{dz_U}{dx} \right) d\left(\frac{x}{c} \right) \tag{7.9}$$

If dz/dx is integrated from 0 to c on the top or the bottom, it simply results in a value of zero as dz/dx is integrated from where $z = 0$ at the origin back to the trailing edge where $z = 0$. Consequently, the lift coefficient for a thin airfoil in a supersonic flow at a modest angle of attack is

$$C_l = \frac{4\alpha}{\sqrt{M_\infty^2 - 1}} \tag{7.10}$$

The drag force on the airfoil is related to the pressure force on the airfoil normal to a plane perpendicular to the free-stream velocity. On the top of the airfoil this means as the differential force against the airfoil is positive when the surface is moving upward perpendicular from the free-stream velocity and negative when the surface is moving downward perpendicular to the free-steam velocity.

$$dd_U = P_U \sin\theta_U \, ds \approx P_U \left(-\alpha + \frac{dz_U}{dx} \right) dx \tag{7.11}$$

The differential drag force on the lower surface of the airfoil is similar except that as the surface is going downward relative to the free-stream velocity vector the differential force due to pressure is added and it is subtracted assuming the direction of the bottom surface relative to the free-stream velocity vector turns upwards.

$$dd_L = -P_L \sin\theta_U \, ds \approx P_L \left(\alpha - \frac{dz_L}{dx} \right) dx \tag{7.12}$$

Similar to the lift coefficient, the drag coefficient can be determined by integrating the differential drag force on the upper and lower surfaces of the airfoil from the leading edge to the trailing edge of the airfoil and then dividing by the dynamic pressure and the chord length. Adding Equations (7.11) and (7.12) together the differential drag force on the top and bottom surface can be approximated by

$$dd \approx \left[(P_L - P_U) * \alpha - P_L \frac{dz_L}{dx} + P_U \frac{dz_U}{dx} \right] dx \tag{7.13}$$

Comparison of Equation (7.13) with Equation (7.7) shows that by inspection, the first component of differential drag is related to differential lift times the angle of attack, α. This term can easily be put in terms of a pressure coefficient. For clarity it is appropriate to break Equation (7.13) into two equations.

One could be accurately described as drag due to lift and the other as drag due to thickness. The differential drag coefficient due to lift and be written in terms of the local pressure coefficients as shown below

$$dC_{d-l} \approx \left(C_{PL} - C_{PU}\right)\alpha\, d\left(\frac{x}{c}\right) \tag{7.14}$$

If θ_L and θ_U are substituted into Equation (7.6) for substitution into C_{PL} and C_{PU} in Equation (7.14), the following equation for the differential drag due to lift is generated.

$$dC_{d-l} \approx \frac{2\alpha}{\sqrt{M_\infty^2 - 1}}\left(2\alpha + -\frac{dz_L}{dx} - \frac{dz_U}{dx}\right)d\left(\frac{x}{c}\right) \tag{7.15}$$

The second and third term in Equation (7.13) can be put in terms of a pressure coefficient if $P_\infty dz_L/dx$ and $P_\infty dz_U/dx$ are added and subtracted. This new equation represents drag due to thickness. Upon integration across the airfoil, the last two terms in Equation (7.16) become zero.

$$dC_{d-t} \approx \left(-C_{PL}\frac{dz_L}{dx} + C_{PU}\frac{dz_U}{dx} - \frac{2}{kM_\infty^2}\frac{dz_L}{dx} + \frac{2}{kM_\infty^2}\frac{dz_U}{dx}\right)d\left(\frac{x}{c}\right) \tag{7.16}$$

If θ_L and θ_U are substituted into Equation (7.6) for substitution into C_{PL} and C_{PU} in Equation (7.16) the following Equation (7.17) for the differential drag due to thickness is generated. Upon integration it should be noted that all the terms with a single gradient in (dz/dx) go to zero noting z starts at the z-origin and ends at the z-origin.

$$dC_{d-t} \approx \left\{\frac{2}{\sqrt{M_\infty^2 - 1}}\left[-2\alpha\left(\frac{dz_L}{dx} + \frac{dz_U}{dx}\right) + \left(\frac{dz_L}{dx}\right)^2 + \left(\frac{dz_U}{dx}\right)^2\right] \right.$$
$$\left. -\frac{2}{kM_\infty^2}\left[\left(\frac{dz_L}{dx} - \frac{dz_U}{dx}\right)\right]\right\}d\left(\frac{x}{c}\right) \tag{7.17}$$

Upon integration Equation (7.15) can be interpreted as the drag coefficient due to lift and can be written as

$$C_{d-l} \approx \frac{4\alpha^2}{\sqrt{M_\infty^2 - 1}} = \alpha\, C_l \tag{7.18}$$

Looking at Equation (7.17) only the squared terms remain upon integration. One simplification which can be made is to report the integrals of the gradient squared as the mean squared values of the slope.

$$\overline{\sigma_U^2} = \frac{1}{c}\int_0^c\left(\frac{dz_U}{dx}\right)^2 dx \tag{7.19a}$$

$$\overline{\sigma_L^2} = \frac{1}{c} \int_0^c \left(\frac{dz_L}{dx} \right)^2 dx \qquad (7.19b)$$

Upon integration of Equation (7.16) the drag coefficient due to thickness in supersonic flow over a thin airfoil at a modest angle of attack can be approximated as:

$$C_{d-t} \approx \frac{2}{\sqrt{M_\infty^2 - 1}} \left(\overline{\sigma_u^2} + \overline{\sigma_L^2} \right) \qquad (7.20)$$

Generally speaking drag due to lift and drag due to thickness in supersonic flow are considered as wave drag. However, in addition to wave drag, drag due to skin friction can also be important. In general the drag coefficient for a supersonic airfoil can be estimated using the three components of drag:

$$C_{d-total} = C_{d-lift} + C_{d-thickness} + C_{d-skin\ friction} \qquad (7.21)$$

Note that skin friction drag will not be addressed in this course. Another assumption which should be noted in this chapter is that an airfoil is a two-dimensional section of a wing. A wing on the other hand has a finite length and it is subject to issues which arise at the tips and at the junction and due to the fuselage in supersonic flow.

7.4 Comparisons with the Shock Expansion Method

The general method for approximating the pressure coefficient for linearized flow was developed in Section 7.2. This linearized flow method was analytically applied to flow over airfoil sections in Section 7.3. The present section uses the relationships developed in Section 7.3 to estimate the lift and drag coefficients of supersonic airfoils and compares them to results using the shock expansion method. This technique is restricted to thin airfoils at modest angles of attack. As angles of attack get higher and/or airfoils become too thick, the increasing non-linear effects on these flows and the resulting losses begin to reduce the accuracy of these linearized approximations. This section will begin with an analysis of a flat plate airfoil and results for two Mach numbers and a range of angles of attack will be presented. Later, a thin diamond-shaped airfoil will be analyzed and results will compared with the shock expansion method over a range of angles of attack and two Mach numbers. Finally, an airfoil with a continuously varying surface curvature will be analyzed using the linearized flow method to show its usefulness.

Example 7.1: Analysis of Flat Plate Airfoil Comparing Linearized Flow with Shock Expansion

In this example a flat plate airfoil similar to the one shown in Figure 6.5 is analyzed for lift and drag coefficients. The free-stream Mach number is set at 1.7 and the angle of attack is set at 5°. Equation (7.10) will be applied to determine the lift coefficient. Note that the angle used in the equation must be in radians.

$$C_l = \frac{4\alpha}{\sqrt{M_\infty^2 - 1}} = \frac{4 * 5° * \dfrac{\pi}{180°}}{\sqrt{1.7^2 - 1}} = 0.2539$$

The lift coefficient is simply a function of the angle of attack and the free-stream Mach number. Determining the drag coefficient is similarly easy. Equation (7.18) can be applied to determine the drag coefficient noting that our notional flat plate airfoil has no thickness. If the airfoil had a thickness then Equation (7.20) could be added to Equation (7.18) to add drag due to thickness.

$$C_{d-l} \approx \frac{4\alpha^2}{\sqrt{M_\infty^2 - 1}} = \frac{4 * \left(5° * \dfrac{\pi}{180°} \right)^2}{\sqrt{1.7^2 - 1}} = 0.02216$$

The lift to drag ratio will simply be $1/\alpha$ for this flat plate airfoil as can be seen from the equations for the lift and the drag coefficients resulting in an estimated lift to drag ratio of 11.46.

The shock expansion method is now used to determine the lift and drag for this ideal airfoil. Similar to Figure 6.5, the flow over the bottom of the airfoil goes through a concave angle downward at 5° indicating an oblique shock will occur. The flow over the top surface results in a convex turn indicating an expansion fan must occur. The analysis can be started with the oblique shock analysis on the bottom surface of the airfoil. The angle of attack is 5° while the free-stream Mach number is 1.7. Based on the oblique shock table in Appendix A.4, at these conditions the oblique shock makes an angle of 41.029° with the free-stream. With the shock angle and the Mach number, the normal and tangential components of the free-stream Mach number can be determined relative to the oblique shock.

$$M_{1N} = M_1 \sin\theta = 1.7 \sin 41.029° = 1.1159$$

$$M_{1T} = M_1 \cos\theta = 1.7 \cos 41.029° = 1.2824$$

The normal component of the Mach number downstream from the oblique shock can be interpolated from the normal shock tables or calculated from the normal shock relationships.

Normal shock: $M_{2N} = 0.8996, P_2/P_1 = 1.2862, T_2/T_1 = 1.0751, P_{T2}/P_{T1} = 0.9985$

The low total pressure loss shows that this oblique shock is almost isentropic. Since the pressure ratio has been determined, the remainder of the oblique shock analysis is not required. The flow over the top of the airfoil moves through a $5°$ convex turn requiring an expansion. The Prandtl–Meyer function, ν, for a Mach number of 1.7 is $17.81°$. Since the turning angle $\Delta \nu = 5°$, $\nu_2 = 22.71°$ which corresponds to a Mach number of 1.8726. Based on our isentropic relationship for pressure noting the total pressure, P_T, is constant:

$$\frac{P_{2u}}{P_1} = \frac{P_{2u}}{P_T}\frac{P_T}{P_1} = \frac{\left(1 + \frac{k-1}{2}M_1^2\right)^{k/k-1}}{\left(1 + \frac{k-1}{2}M_{2u}^2\right)^{k/k-1}} = \frac{\left(1 + \frac{0.4}{2}1.7^2\right)^{1.4/0.4}}{\left(1 + \frac{0.4}{2}1.8726^2\right)^{1.4/0.4}} = 0.7684$$

Based on Example 6.3, the lift coefficient for a flat plate airfoil can be determined from the local static to free-stream pressure ratio on the lower surface less the local static to free-stream pressure ratio on the upper surface times the cosine of the angle of attack divided by the dynamic pressure.

$$C_L = \frac{\left(\frac{P_{2L}}{P_\infty} - \frac{P_{2u}}{P_\infty}\right)\cos\alpha}{k\,M_\infty^2/2} = \frac{(1.2862 - 0.7684)\cos 5°}{1.4 * 1.7^2/2} = 0.2550$$

This shock expansion analysis gives a lift coefficient which is only about 0.43 percent higher than the linear analysis result which is quite acceptable at this lower angle of attack. The drag coefficient due to lift is similar to the lift calculation. However, the sine of the angle of attack is used rather than the cosine.

$$C_D = \frac{\left(\frac{P_{2L}}{P_\infty} - \frac{P_{2u}}{P_\infty}\right)\sin\alpha}{k\,M_\infty^2/2} = \frac{(1.2862 - 0.7684)\sin 5°}{1.4 * 1.7^2/2} = 0.02231$$

This drag coefficient compares well with the linearized flow model which estimates a drag coefficient of 0.02216 which is within 0.7 percent of the value determined from the shock expansion method. A comparison between lift and drag coefficients determined using the shock expansion method compared to the linear method is presented in Table 7.1 at Mach numbers of 1.7 and 2.1 for angles of attack between $1°$ and $10°$. Over this range the comparison is very consistent but showing increased second-order effects for higher angles and lower Mach numbers.

TABLE 7.1

Comparison of Flat Plate Airfoil Lift and Drag Using Shock Expansion and Linear Method

α (°)	1	2	3	4	5	6	7	8	9	10
M_1	1.7	1.7	1.7	1.7	1.7	1.7	1.7	1.7	1.7	1.7
$P_L/P_1 =$	1.0525	1.1072	1.1643	1.2239	1.2862	1.3514	1.4197	1.4914	1.5669	1.6466
$P_U/P_1 =$	0.9497	0.9015	0.8552	0.8109	0.7684	0.7277	0.6887	0.6514	0.6157	0.5815
$C_l =$	0.0508	0.1016	0.1526	0.2037	0.2550	0.3066	0.3587	0.4112	0.4644	0.5185
$C_d =$	0.00089	0.00355	0.00800	0.01424	0.02231	0.03223	0.04404	0.05779	0.07355	0.09142
$C_{l,lin} =$	0.0508	0.1016	0.1523	0.2031	0.2539	0.3047	0.3555	0.4063	0.4570	0.5078
$C_{d,lin} =$	0.00089	0.00355	0.00798	0.01418	0.02216	0.03191	0.04343	0.05672	0.07179	0.08863
M_1	2.1	2.1	2.1	2.1	2.1	2.1	2.1	2.1	2.1	2.1
$P_L/P_1 =$	1.0597	1.1222	1.1875	1.2558	1.3272	1.4018	1.4796	1.5609	1.6457	1.7342
$P_U/P_1 =$	0.9430	0.8885	0.8366	0.7870	0.7398	0.6949	0.6521	0.6115	0.5728	0.5362
$C_l =$	0.0378	0.0756	0.1135	0.1515	0.1895	0.2277	0.2661	0.3046	0.3433	0.3822
$C_d =$	0.00066	0.00264	0.00595	0.01059	0.01658	0.02394	0.03267	0.04280	0.05437	0.06739
$C_{l,lin} =$	0.0378	0.0756	0.1134	0.1512	0.1890	0.2268	0.2646	0.3024	0.3403	0.3781
$C_{d,lin} =$	0.00066	0.00264	0.00594	0.01056	0.01650	0.02375	0.03233	0.04223	0.05345	0.06598

Example 7.2: Analysis of Thin Diamond-shaped Airfoil Comparing Linearized Flow with Shock Expansion Method

In Example 7.1 a flat plate airfoil was examined using both the linearized flow method and the shock expansion method. However, although supersonic airfoils are generally thin, they do need some thickness in order to transmit the loading due to lift. Typical supersonic airfoils have thickness to chord ratios which might be on the order of 0.05. In the present problem a diamond-shaped airfoil will be analyzed using both the linearized flow method and the shock expansion method. A thin symmetrical diamond-shaped airfoil is chosen largely due to its simplicity for analysis using the shock expansion method. In the present case, a diamond-shaped airfoil with a half wedge angle of 3° is chosen. Looking at the tangent of this angle a thickness to chord ratio of 0.0524 is found. This geometry can be visualized using the sketch in Figure 6.6. One question that arises in this analysis is how will the thickness influence the flow over the airfoil? One way to approach this would be to look at the airfoil with a zero angle of attack in a supersonic flow. In this case the initial flow encountering the symmetrical wedge with the 3° half angle would immediately cause two weak oblique shock waves to form as the flow adjusts to the concave angle to move around the airfoil. Downstream the flow must expand around the convex corners at the midpoint of the airfoil. The oblique shocks will raise the pressure on the front of the airflow while the Prandtl–Meyer expansion fans will reduce the pressure on the back side. Consequently, one new issue with the thin diamond-shaped airfoil with be wave drag due to the airfoils thickness.

The analysis will begin with an assessment using the linearized flow method. For the current analysis an angle of attack of 5° will be assumed and a free-stream Mach number of 1.7 will also be chosen. This allows a direct comparison between the present analysis and Example 7.1. Based on Section 7.3 Equation (7.10) can be used to assess the lift coefficient. This equation indicates that thickness has no significant effect on lift.

$$C_l = \frac{4\alpha}{\sqrt{M_\infty^2 - 1}} = \frac{4*5°*\dfrac{\pi}{180°}}{\sqrt{1.7^2 - 1}} = 0.2539$$

This analysis turns out to be identical to the flat plate analysis and returns the same lift coefficient. The drag coefficient due to lift will also be identical. This can be determined using Equation (7.18).

$$C_{d-l} \approx \frac{4\alpha^2}{\sqrt{M_\infty^2 - 1}} = \frac{4*\left(5°*\dfrac{\pi}{180°}\right)^2}{\sqrt{1.7^2 - 1}} = 0.02216$$

However, in addition to drag due to lift the linearized flow analysis also has a component of drag due to the airfoils thickness. This drag due to thickness can be calculated using Equation (7.20).

$$C_{d-t} \approx \frac{2}{\sqrt{M_\infty^2 - 1}} \left(\overline{\sigma_U^2} + \overline{\sigma_L^2} \right)$$

Evaluation of the drag due to thickness involves the determination of the mean of the squared slope of the airfoil surface. This mean squared surface angle can be evaluated using Equations (7.19a) and (7.19b) for the upper and lower surfaces.

$$\overline{\sigma_U^2} = \frac{1}{c} \int_0^c \left(\frac{dz_U}{dx} \right)^2 dx$$

$$\overline{\sigma_L^2} = \frac{1}{c} \int_0^c \left(\frac{dz_L}{dx} \right)^2 dx$$

The term dz/dx is simply the tangent of the half wedge angle, which is constant for this symmetrical diamond airfoil so Equation (7.20) can be evaluated as:

$$C_{d-t} \approx \frac{2}{\sqrt{M_\infty^2 - 1}} \left(\overline{\sigma_U^2} + \overline{\sigma_L^2} \right) = \frac{2}{\sqrt{1.7^2 - 1}} \left(0.0524^2 + 0.0524^2 \right) = 0.00799$$

Based on our linearized flow analysis, the drag coefficient significantly increases due to thickness resulting in an overall drag coefficient of 0.03015.

At a 5° angle of attack, similar to Figure 7.3, flow over the lower surface will cause an oblique shock due to the concave angle of the flow. Later the convex angle the flow takes at the maximum thickness causes an expansion fan and lower pressure over the last half of the bottom of the airfoil. As the wedge has a 3° half angle the initial flow over the top of the airfoil will have a weak expansion followed by another expansion on the aft section of the top of the airfoil. The flow at the trailing edge of the airfoil will not influence the airfoils lift or drag.

Sketch:

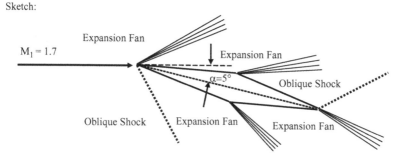

FIGURE 7.3
Sketch of a symmetrical diamond-shaped airfoil in a supersonic flow.

Sketch:

The 5° angle of attack (α) combined with the 3° wedge angle (δ) results in a concave turn of 8°. Based on the oblique shock tables in Appendix A.4, the oblique shock angle (θ) for a Mach number of 1.7 with a wedge angle of 8° is 44.4282°. This information allows the calculation of the normal and tangent components of Mach number.

$$M_{1N} = M_1 \sin\theta = 1.7 \sin 44.4282° = 1.1921$$

$$M_{1T} = M_1 \cos\theta = 1.7 \cos 44.4282° = 1.2119$$

The normal component of the Mach number can be used to find the downstream normal component and then pressure ratio. However, in this symmetrical diamond airfoil the downstream Mach number needs to be determined as well for input into the determination of the expansion fan. The normal shock information can be calculated from equations or interpolated from the normal shock table.

Normal shock: $M_{2N} = 0.8471$, $P_2/P_1 = 1.4914$, $T_2/T_1 = 1.1231$

Noting the tangential velocity does not change the downstream tangential Mach number can be determined from the upstream tangential Mach number divided by the square root of the downstream to upstream temperature ratio. The downstream Mach number can be determined from the square root of the sum of the squared components.

$$M_{2T} = M_{1T} / \sqrt{\frac{T_2}{T_1}} = \frac{1.2119}{\sqrt{1.1231}} = 1.1436$$

$$M_{2L} = \sqrt{M_{2N}^2 + M_{2T}^2} = \sqrt{.8471^2 + 1.1436^2} = 1.4232$$

Finding the Mach number downstream of the oblique shock is critical to determine the solution for the expansion fan resulting from the 6° convex expansion at the point of maximum thickness. The solution for the downstream Mach number on the bottom of the airfoil is readily solved using the Prandtl–Meyer function. The Prandtl–Meyer function at a Mach number of 1.4232 is 9.6571°. Adding 6° for the convex turn results in a Prandtl–Meyer function of 15.6571° which corresponds to a Mach number (M_{3L}) of 1.6270. The resulting pressure ratio can easily be determined using the isentropic relationships.

$$\frac{P_{3L}}{P_{2L}} = \frac{P_{3L}}{P_T}\frac{P_T}{P_{2L}} = \frac{\left(1 + \frac{k-1}{2}M_{2L}^2\right)^{\frac{k}{k-1}}}{\left(1 + \frac{k-1}{2}M_{3L}^2\right)^{\frac{k}{k-1}}} = \frac{\left(1 + \frac{0.4}{2}1.4232^2\right)^{\frac{1.4}{0.4}}}{\left(1 + \frac{0,4}{2}1.6270^2\right)^{\frac{1.4}{0.4}}} = 0.7443$$

However, this ratio references to the pressure under section 2L and the upstream ratio needs to be used to reference to the free-stream pressure.

$$\frac{P_{3L}}{P_1} = \frac{P_{3L}}{P_{2L}}\frac{P_{2L}}{P_1} = 0.7443 * 1.4914 = 1.1085$$

This last analysis provides the second pressure ratio on the bottom of the airfoil. The analysis of the airfoil top is straight forward. The angle of attack is 5° and the wedge angle is 3° so the flow takes a convex turn of 2° over the first section of the top. At a Mach number of 1.7, the Prandtl–Meyer function is 17.81° so the downstream Prandtl–Meyer function will be 19.81°. This corresponds to a downstream Mach number of 1.7684. The flow then undergoes a second expansion of 6° giving a downstream Prandtl–Meyer function of 25.81° corresponding to a downstream Mach number of 1.9794. The flow over the entire top of the airfoil is isentropic, so isentropic relationships can be used to determine pressure ratios over both upper sections.

$$\frac{P_{2u}}{P_\infty} = \frac{\left(1+\dfrac{k-1}{2}M_\infty^2\right)^{k/k-1}}{\left(1+\dfrac{k-1}{2}M_{2u}^2\right)^{k/k-1}} = \frac{\left(1+\dfrac{0.4}{2}1.7^2\right)^{1.4/0.4}}{\left(1+\dfrac{0,4}{2}1.7684^2\right)^{1.4/0.4}} = 0.9382$$

$$\frac{P_{3u}}{P_\infty} = \frac{\left(1+\dfrac{k-1}{2}M_\infty^2\right)^{k/k-1}}{\left(1+\dfrac{k-1}{2}M_{3u}^2\right)^{k/k-1}} = \frac{\left(1+\dfrac{0.4}{2}1.7^2\right)^{1.4/0.4}}{\left(1+\dfrac{0,4}{2}1.9794^2\right)^{1.4/0.4}} = 0.6779$$

All the pressure ratios over the four sections of the airfoil have now been determined. The lift and drag coefficients can now be determined. The thin diamond airfoil is a parallelogram, so the first lower surface and last upper surfaces are parallel as are the first upper surface and last lower surface. The lift coefficient is simply the lift divided by the dynamic pressure or $P_\infty * k M_\infty^2 / 2 * c$. This results in the following relationship after cancelling c, the chord length in the denominator and numerator.

$$C_l = \left[\left(\frac{P_{2L}}{P_\infty}-\frac{P_{3u}}{P_\infty}\right)*L_{2L}+\left(\frac{P_{3L}}{P_\infty}-\frac{P_{2u}}{P_\infty}\right)*L_{2u}\right]/\left(k M_\infty^2 / 2\right)$$

The horizontal lengths of these sections can easily be determined noting that each section length is equal to $c/(2 * \cos \delta)$ and its angle relative to the free-stream is $\alpha \pm \delta$. Also, in this example the angle of attack, α is 5° and the half wedge angle, δ is 3°.

$$L_{2u} = L_{3L} = \frac{c\cos(5°-3°)}{2\cos 3°} = \frac{c\,0.99939}{2*0.99863} = 0.5004 * c$$

$$L_{2L} = L_{3U} = \frac{c\cos(5° + 3°)}{2\cos 3°} = \frac{c\,0.99027}{2*0.99863} = 0.49581 * c$$

Consequently, our lift coefficient is determined to be

$$C_l = \left[(1.4914 - 0.6779) * 0.4958 + (1.1085 - 0.9382) * 0.5004\right] / \left(1.4 * 1.7^2 / 2\right)$$
$$= 0.2421$$

The drag coefficient for the airfoil can be determined similarly to the lift coefficient with the exception that the drag force is determined from the pressure force on the vertical components of the airfoil facing forward less the pressure force on the vertical components facing aft.

$$C_d = \left[\left(\frac{P_{2L}}{P_\infty} - \frac{P_{3U}}{P_\infty}\right) * H_{2L} + \left(\frac{P_{3L}}{P_\infty} - \frac{P_{2U}}{P_\infty}\right) * H_{2U}\right] / \left(k\,M_\infty^2 / 2\right)$$

The vertical components of the airfoil can be determined similarly.

$$H_{2U} = H_{3L} = \frac{c\sin(5° - 3°)}{2\cos 3°} = \frac{c\,0.01745}{2*0.99863} = 0.01747 * c$$

$$H_{2L} = H_{3U} = \frac{c\sin(5° + 3°)}{2\cos 3°} = \frac{c\,012187}{2*0.99863} = 0.06968 * c$$

Consequently, the drag coefficient is determined to be

$$C_d = \frac{\left[(1.4914 - 0.6779) * 0.0.6968 + (1.1085 - 0.9382) * 0.01747\right]}{\left(1.4 * 1.7^2 / 2\right)} = 0.02949$$

The airfoil section lift and drag coefficients were determined to be 0.2421 and 0.02949 using the shock expansion method compared with values of 0.2539 and 0.03015 using the linearized flow method. The linearized flow method produces values that are 5 percent too high in lift and 2.2 percent too high in drag. Table 7.2 provides a comparison between calculated lift and drag coefficients for the thin diamond-shaped airfoil at free-stream Mach numbers of 1.7 and 2.1 for angles of attack ranging from 1° to 10° comparing shock expansion values with linearized flow values. Generally, estimates of lift using the linearized flow method range from 1 percent too low to 6.7 percent too high and estimates of drag range from 1 percent too low to 3.8 percent too high. The linearized flow method produces reasonably accurate values in assessing supersonic airfoils.

TABLE 7.2

Comparison of Shock Expansion Lift and Drag Coefficients With Linearized Flow Lift and Drag Coefficients

α (°)	1	2	3	4	5	6	7	8	9	10
M_1	1.7	1.7	1.7	1.7	1.7	1.7	1.7	1.7	1.7	1.7
$C_l =$	0.0513	0.1026	0.1540	0.1977	0.2421	0.2873	0.3335	0.3808	0.4295	0.4799
$C_d =$	0.00891	0.01161	0.01614	0.02200	0.02949	0.03866	0.04958	0.06235	0.07711	0.09401
$C_{l,lin} =$	0.0508	0.1016	0.1523	0.2031	0.2539	0.3047	0.3555	0.4063	0.4570	0.5078
$C_{d,lin} =$	0.00888	0.01154	0.01597	0.02217	0.03015	0.03990	0.05142	0.06471	0.07978	0.09662
M_1	2.1	2.1	2.1	2.1	2.1	2.1	2.1	2.1	2.1	2.1
$C_l =$	0.0381	0.0762	0.1144	0.1453	0.1767	0.2089	0.2416	0.2751	0.3092	0.3440
$C_d =$	0.00663	0.00863	0.01198	0.01624	0.02165	0.02825	0.03608	0.04520	0.05564	0.06747
$C_{l,lin} =$	0.0378	0.0756	0.1134	0.1512	0.1890	0.2268	0.2646	0.3024	0.3403	0.3781
$C_{d,lin} =$	0.00661	0.00859	0.01189	0.01651	0.02245	0.02970	0.03828	0.04818	0.05940	0.07193

Example 7.3: Analysis of Continuously Varying Airfoil Surface Using the Linearized Flow Method

The shock expansion method was useful in evaluating relatively simple airfoil shapes such as a flat plate airfoil and a thin symmetrical diamond-shaped airfoil. However, the evaluation of lift and drag characteristics on actual airfoil shapes would be very tedious using this method. The linearized flow method is a much more straight forward approach to evaluate more complex airfoil shapes with continuously varying shapes. Based on earlier evaluation for thin airfoils at modest angles of attack, the lift coefficient is simply estimated in terms of the angle of attack. However, the drag coefficient has components of both drag due to lift and drag due to thickness. Drag due to thickness can be evaluated using Equation (7.20). One example of a continuously varying surface geometry is sketched in Figure 7.4. A simple airfoil design with a thickness to chord ratio of 0.05 can be developed using the following relationship for the top surface:

$$z\left(\frac{x}{c}\right) = 0.1\left(\frac{x}{c}\right) - 0.1\left(\frac{x}{c}\right)^2$$

The bottom surface has a similar relationship

$$z\left(\frac{x}{c}\right) = -0.1\left(\frac{x}{c}\right) + 0.1\left(\frac{x}{c}\right)^2$$

Sketch:

FIGURE 7.4
Sketch of continuous airfoil superimposed on thin diamond airfoil.

The drag due to thickness can be evaluated using Equation (7.20):

$$C_{d-t} \approx \frac{2}{\sqrt{M_\infty^2 - 1}}\left(\overline{\sigma_u^2} + \overline{\sigma_L^2}\right)$$

And the values $\overline{\sigma_u^2} + \overline{\sigma_L^2}$ can be evaluated based on the integrals:

$$\overline{\sigma_u^2} = \frac{1}{c}\int_0^c \left(\frac{dz_u}{dx}\right)^2 dx$$

and

$$\overline{\sigma_L^2} = \frac{1}{c}\int_0^c \left(\frac{dz_L}{dx}\right)^2 dx$$

Taking our derivatives:

$$\frac{dz_u}{d\left(\frac{x}{c}\right)} = 0.1 - 0.2\left(\frac{x}{c}\right)$$

And

$$\frac{dz_L}{d\left(\frac{x}{c}\right)} = -0.1 + 0.2\left(\frac{x}{c}\right)$$

then squaring, both result in

$$\left[\frac{dz}{d\left(\frac{x}{c}\right)}\right]^2 = 0.01 - 0.04\left(\frac{x}{c}\right) + 0.04\left(\frac{x}{c}\right)^2$$

Next, integrating the relationship becomes

$$\int_0^1 \left(\frac{dz}{d\left(\frac{x}{c}\right)}\right)^2 dx = \left.\left[0.01\left(\frac{x}{c}\right) - 0.02\left(\frac{x}{c}\right)^2 + 0.01333\left(\frac{x}{c}\right)^3\right]\right|_0^1 = 0.003333$$

And finally, the drag due to thickness is equal to

$$C_{d-t} \approx \frac{2}{\sqrt{M_\infty^2 - 1}}\left(\overline{\sigma_u^2} + \overline{\sigma_L^2}\right) = \frac{2}{\sqrt{M_\infty^2 - 1}}\, 0.006667$$

The drag coefficient due to thickness at a Mach number of 1.7 would be equal to 0.0097 or about 21 percent more than the diamond-shaped airfoil. This linearized flow method provides a straight forward method of evaluating the lift and drag of surfaces which are continuously varying. This method can be easily adapted to any shape using simple numerical integration. The shock expansion method is more difficult to apply to more complex airfoil shapes.

The assumptions made in this section of the chapter included that the flow was supersonic, the airfoil sections were thin and the angle of attack was modest. An airfoil section is considered to be a two-dimensional airfoil section, and issues due to tip effects, finite airfoil length, and fuselage effects are not considered. In supersonic flow, the tip of an airfoil is known to have a reduced lift component over the section affected by the Mach wave originating from the tip. Fuselage effects as well as the swept angle of the wing also have important contributions to the lift and drag characteristics of a wing. Additionally, the analysis in this section assumed that the flow was frictionless. However, over actual supersonic airfoils drag due to skin friction is typically on the order of zero lift wave drag for thin airfoils.

References

1. John, J.E.A. and Keith, T.G., 2006, *Gas Dynamics*, 3rd ed., Prentice Hall.
2. Saad, M.A., 1993, *Compressible Fluid Flow*, 2nd ed., Prentice Hall.
3. Bertin, J.J., and Cummings, R.M., 2014, *Aerodynamics for Engineers*, 6th ed., Pearson.

Chapter 7 Problems

1. Determine P/P_∞ for a supersonic flow which undergoes a 7° concave turn at a Mach number of 1.9 using linearized flow. Equation (7.6) can be used to determine the pressure coefficient and remember that the dynamic pressure, $\rho U_\infty^2 / 2g_C = P_\infty k M_\infty^2 / 2$. Check your answer using an oblique shock analysis. How close is the linearized flow approximation?

2. Determine P/P_∞ for a supersonic flow which undergoes an 8° convex turn at a Mach number of 1.8 using linearized flow. Equation (7.6) can be used to determine the pressure coefficient and remember that the dynamic pressure, $\rho U_\infty^2 / 2g_C = P_\infty k M_\infty^2 / 2$. Check your answer using a Prandtl–Meyer expansion fan analysis. How close is the linearized flow approximation?

3. Calculate the lift and drag coefficients for a flat plate airfoil at angles of attack of 3°, 5°, 7° and 9° for a free-stream Mach number of 1.8 using the linearized flow method. Check your answer at an angle of 7° using the shock expansion method. Do your results fall in the middle of results from Table 7.1 for Mach numbers of 1.7 and 2.1? How well does the linearized flow method compare with the shock expansion method?

4. Calculate the lift and drag coefficients for a flat plate airfoil at angles of attack of 4°, 6°, 8°, and 10° for a free-stream Mach number of 1.9 using the linearized flow method. Check your answer at an angle of 8° using the shock expansion method. Do your results fall in the middle of results from Table 7.1 for Mach numbers of 1.7 and 2.1? How well does the linearized flow method compare with the shock expansion method?

5. Calculate the lift and drag coefficients for a symmetrical diamond-shaped airfoil with a half wedge angle of 3° at angles of attack of 4°, 6°, 8° and 10° for a free-stream Mach number of 2.0 using the linearized flow method. Check your answer using the shock expansion method at an angle of attack of 8°. Do your results fall in the middle

of results from Table 7.2 for Mach numbers of 1.7 and 2.1? How well does the linearized flow method compare with the shock expansion method?

6. Calculate the lift and drag coefficients for a symmetrical diamond-shaped airfoil with a half wedge angle of 4° at angles of attack of 3°, 5° and 7° for a free-stream Mach number of 2.1 using the linearized flow method. Check your answer using the shock expansion method at an angle of attack of 7°. How well does the linearized flow method compare with the shock expansion method?

7. Using the linearized flow method determine the lift and drag coefficients for a parabolic airfoil with top surface relationship which can be described as $z(x/c) = 0.12\,(x/c) - 0.12\,(x/c)^2$. The bottom surface relationship which can be described as $z(x/c) = -0.12\,(x/c) + 0.12\,(x/c)^2$. The airfoil has an angle of attack of 6° and the free-stream Mach number is 2.25. How close do you believe the calculated lift and drag coefficients will be to the actual coefficients?

8

Internal Compressible Flow with Friction

Chapter 8 presents internal compressible flow with friction which is some-times called Fanno-line flow. In this chapter initially the Fanno-line equation is developed from our Tds relations, continuity, and the energy equations. The Fanno-line indicates the behavior of flows with friction which increase in entropy due to friction losses and are driven toward sonic flow for both subsonic and supersonic flows. Subsequently, the Fanno-line relationships for fL_{MAX}/D, pressure, temperature, velocity, and total pressure changes as a function of Mach number are developed. These equations are numerically integrated and provided in Appendix A.6 for specific heat ratios, k, of 1.4 and 1.31. Subsequently, example problems for flow with friction in constant area ducts are provided to show how compressible flows with friction can be solved for a variety of subsonic and supersonic flows for a range of situations. Subsequently, equations for the isothermal flow approximation are developed along with guidance restricting this simplification to low-speed flows ($M < 0.3$). Isothermal flow analyses are applied to pipeline flows and compared to Fanno-line analyses showing the very good accuracy of this approximation. Flow with friction and area change is also presented and discussed.

8.1 Introduction to Flow with Friction

The assumption of frictionless compressible flow over airfoils or in ducts is a simplification which can yield useful engineering solutions in many situations when analyzing pressure distributions and flows. Often over external flows where boundary layers are thin and attached, accurate pressure distributions and meaningful understanding can be achieved with inviscid or frictionless analyses. In short ducts with high speed and especially accelerating flows, this thin boundary layer assumption is often a reasonable and useful assumption. However, in many engineering situations when ducts are relatively long, friction can have an important impact in the analysis of flows. One area where this type of analysis is particularly important is compressible flow in pipelines. In this chapter, analyses for compressible flow with friction will be developed. Initially, adiabatic compressible flow with friction in constant area ducts will be analyzed and applied to a number of differing situations. Later, the simplification of isothermal flow will be introduced and explored. Finally, frictional flow in varying area ducts will be analyzed and discussed. A number of practical examples of frictional flow in ducts will also be solved.

8.2 Analysis of the Fanno-Line and Interpretation of Flow Behavior

An adiabatic flow has no heat transfer which means the total temperature stays constant. A steady incompressible flow with skin friction in a constant area pipe must result in a drop in pressure in the flow direction to oppose the skin friction which is opposite to the flow direction. This pressure drop will be in both the static and total pressure. A compressible flow in a constant area duct will change density with pressure changes while the flow can experience a trade-off between pressure changes and momentum changes along the duct in adjusting to the friction. Considering the mass flow rate in this steady flow is constant, a flow with friction will experience a decrease in the total pressure. In a subsonic flow, as the total pressure drops the static pressure will also drop causing a decrease in the density and a resulting increase in velocity for a constant area duct. However, earlier it was shown that a rise in static pressure can take place across a shock even though the total pressure diminishes. In this section, a Fanno-line for adiabatic flow with friction in a constant area duct will be developed to provide guidance in understanding compressible flows with friction.

An analysis of thermodynamic processes shows that in an adiabatic process the total temperature with stay constant. However, at the same time in real processes irreversibilities occur which result in an increase in the entropy and a reduction in the ability of the system to do work. Thermodynamics provides the Tds-equations which relate changes in pressure and internal energy or enthalpy to pressure or density changes. These Tds-equations will be used in the present section to provide an understanding of the path of processes along the Fanno-line for adiabatic compressible flow with friction in a constant area duct. This analysis begins by first stating the Tds equations in terms of temperature, T, specific entropy, s, pressure, P, specific volume, v, and enthalpy, h. Due to the relationship between enthalpy, h, and specific internal energy, u, where $h = u + Pv$, there are two equivalent Tds equations. One equation uses internal energy and one uses enthalpy.

$$Tds = du + Pdv = dh - vdP \tag{1.39}$$

This relation will initially be related to the change in specific entropy with velocity. Although both versions are equivalent, it is easier to work with the differential specific volume. Noting that specific volume is the reciprocal of density, the differential can be cast in terms of the density.

$$dv = -\frac{d\rho}{\rho^2} \tag{8.1}$$

Assuming an ideal gas, the ideal gas law can be used to eliminate pressure:

$$\frac{P}{\rho} = RT$$

Substituting into our first version of the *Tds* equation results in

$$Tds = du - RT\frac{d\rho}{\rho} \tag{8.2}$$

Assuming an ideal gas then $du = C_v dT$ and substituting then dividing by T:

$$ds = \frac{C_v dT}{T} - R\frac{d\rho}{\rho}$$

Dividing by the specific heat, C_v.

$$\frac{ds}{C_v} = \frac{dT}{T} - \frac{R}{C_v}\frac{d\rho}{\rho}$$

The ratio $R/C_v = k - 1$ where k is the specific heat ratio. At this point applying the continuity Equation (2.18) but assuming a constant area duct suggests that $d\rho/\rho = -dV/V$ so our entropy relationship can be recast in terms of:

$$\frac{ds}{C_v} = \frac{dT}{T} + (k-1)\frac{dV}{V} \tag{8.3}$$

It would be useful at this point to eliminate the velocity terms and put this relationship in terms of temperature. The adiabatic assumption results in the following relationship between total enthalpy and enthalpy and kinetic energy.

$$h_T = h + \frac{V^2}{2g_c} \tag{2.1}$$

Assuming a constant specific heat this relationship between velocity and the difference in total and static enthalpy can be written in terms of velocity and total and static temperatures:

$$C_p(T_T - T) = \frac{V^2}{2g_c} \ or \ V = \sqrt{2g_c C_p(T_T - T)} \tag{2.3}$$

Differentiating the relationship for V and dividing by V the relationship becomes

$$\frac{dV}{V} = \frac{-dT}{2(T_T - T)}$$

Substituting this relationship into the relationship for ds/C_V the equation becomes

$$\frac{ds}{C_v} = \frac{dT}{T} + \frac{-(k-1)dT}{2(T_T - T)} \tag{8.4}$$

At this point ds/dT can be set equal to zero and C_v can be eliminated yielding

$$\frac{ds}{dT} = 0 = \frac{1}{T} + \frac{-(k-1)}{2(T_T - T)}$$

From the earlier relationship with total temperature, $(T_T - T) = V^2/2g_cC_P$ and substitution this into the above equations yields

$$\frac{1}{T} = \frac{(k-1)g_cC_P}{V^2}$$

Noting that $C_P = kR/(k-1)$ and substituting this into the above equation while multiplying both sides by T this equation final becomes

$$1 = \frac{g_ckRT}{V^2} = \frac{a^2}{V^2} = \frac{1}{M^2}$$

This result basically shows that at a Mach number of 1 that the slope of dT/ds becomes infinite. For an adiabatic flow at a Mach number of 1 the ratio $T/T_T = 2/(k+1)$. This means when the flow is subsonic friction will cause the static pressure to drop, increasing the velocity and dropping the static temperature until the flow reaches a Mach number of 1 where it is choked. For a supersonic flow, friction will cause the velocity to decrease and the static temperature and pressure will rise in the constant area duct until a Mach number of 1 is reached. This discussion is a simple description of adiabatic flow with friction in a constant area duct. Essentially this describes the Fanno-line. Figure 8.1 below shows a simplified view of the Fanno-line in terms of temperature versus specific entropy describing the behavior of both subsonic and supersonic flows. For both subsonic and supersonic flows, friction increases the entropy and the flow moves toward a Mach number of 1.

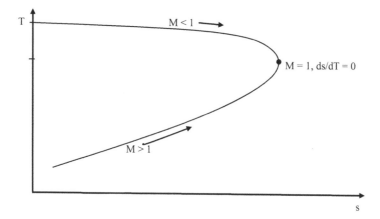

FIGURE 8.1
Schematic of Fanno-line showing increasing entropy with friction.

8.3 Adiabatic Flow with Friction in a Constant Area Duct

The analyses and problems up to this point have been inviscid or friction-less. However, real flows encounter skin friction. In short ducts, this issue of skin friction may not be a significant factor in an analysis. However, whenever the ducts are long, friction becomes an increasingly important issue in compressible flow. The influence of friction can be accounted for by using the momentum equation. Friction along a constant area duct in subsonic flow will cause a total pressure loss and a static pressure drop along with an increase in velocity and Mach number. A supersonic flow in a constant area duct will also experience a total pressure loss but the static pressure will rise as the Mach number and velocity decrease toward the sonic condition. Assuming a constant area duct, the continuity equation will relate density and velocity. Assuming the flow is adiabatic, the energy equation can relate total temperature, which is constant, to static temperature through the local velocity or Mach number. A relationship between pressure, density, and temperature can be developed from the ideal gas law. A relationship between velocity, the Mach number, and temperature can be developed from the definition of Mach number. In this current section, these relationships will be used to develop a relationship between Mach number and friction for a constant diameter adiabatic compressible flow. These relationships will also be used to develop relationships for pressure, temperature, velocity, and total pressure loss. The analysis can begin with the continuity equation or conservation of mass assuming a constant area duct. Starting with the steady-state integral equation for a control volume in the production form.

$$\dot{P}_M = 0 = \oint_{cs} \rho \vec{V} \cdot \overline{dA} \qquad (1.20)$$

The equation can be stated as the production of mass is equal to zero which equals the outflow rate less the inflow rate in a steady-state compressible flow. This relationship can be developed further by applying it to the control volume of a constant area duct as is shown below in Figure 8.2.

Since the area is constant and the flow is assumed steady-state conservation of mass can be written as the following:

$$\dot{P}_M = 0 = (\rho + d\rho)(V + dV)A - \rho V A$$

Eliminating terms, dropping second-order terms and dividing through by $\rho V A$ a relationship between changes in density and changes in velocity is developed.

$$0 = \frac{d\rho}{\rho} + \frac{dV}{V} \qquad (8.5)$$

The momentum principle provides a means to develop a relationship between frictional losses, pressure, and momentum changes. This relationship can be developed by starting with the steady-state momentum equation as applied to Figure 8.2. This development can begin by writing the steady-state integral momentum equation in the x-direction.

$$\sum F_x = \oint_{cs} V_x \rho \vec{V} \cdot \overline{dA}/g_c \qquad (1.25)$$

The forces on the control volume include the pressure forces in the positive and negative x-direction as well as the surface forces. The surface forces are related to skin friction on the fluid flow due to the no slip condition at the surface of the duct which applies a force on the fluid in the direction opposite to the flow. Using the control volume form for x-momentum and writing each of these terms out results in the following relationship:

$$PA - (P + dP)A - \tau_W \pi D dx = \rho V A (V + dV - V)/g_c$$

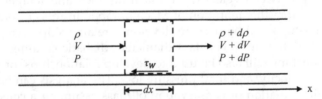

FIGURE 8.2
Schematic of constant area duct with an adiabatic compressible flow with friction.

Cancelling similar terms and dividing by the duct area, $\pi D^2/4$ this relationship simplifies to:

$$-dP - 4\tau_W \frac{dx}{D} = \rho V dV / g_C \qquad (8.6)$$

The wall shear stress can be replaced with the skin friction coefficient noting that $Cf/2 = \tau_W / (\rho V^2 / g_C)$.

$$-dP - 8Cf/2 \frac{\rho V^2}{2g_C} \frac{dx}{D} = \rho V dV / g_C \qquad (8.7)$$

However, $8\,Cf/2$ is equal to the friction factor, f, which is commonly used to account for skin friction losses in piping and ducts. Also, if the density, ρ, is replaced with the ideal gas law, $\rho = P/RT$ the denominator for the friction and momentum terms begin to look like the speed of sound squared, $a^2 = kg_C RT$. All that is needed is to multiply both terms by k/k and divide by the pressure, P.

$$-\frac{dP}{P} - f \frac{kV^2}{2kg_C RT} \frac{dx}{D} = \frac{kV^2}{kg_C RT} \frac{dV}{V}$$

This relationship now simplifies to

$$-\frac{dP}{P} - f \frac{kM^2}{2} \frac{dx}{D} = kM^2 \frac{dV}{V} \qquad (8.8)$$

The relationship, dV/V, can be evaluated based on the definition of Mach number or $V = M\sqrt{kg_C RT}$. From this relationship, the following relationship for Mach number and temperature can be developed.

$$\frac{dV}{V} = \frac{dM}{M} + \frac{dT}{2T} \qquad (8.9)$$

Based on the ideal gas in the form $P = \rho RT$, a relationship for dP/P can be developed in the form of density and temperature:

$$\frac{dP}{P} = \frac{d\rho}{\rho} + \frac{dT}{T}$$

Substituting the above equation for $-dV/V$ for $d\rho/\rho$ based on the earlier relationship for continuity results in

$$\frac{dP}{P} = -\frac{dM}{M} + \frac{dT}{2T} \qquad (8.10)$$

The momentum equations can now be written in terms of Mach number and static temperature.

$$\frac{dM}{M} - \frac{dT}{2T} - f\frac{kM^2}{2}\frac{dx}{D} = kM^2\left(\frac{dM}{M} + \frac{dT}{2T}\right) \tag{8.11}$$

The relationship for temperature can be developed based on the concept that the flow is adiabatic making the total temperature constant. An earlier relationship between static and total temperature and Mach number can be written as

$$\frac{T}{T_T} = \frac{1}{\left(1 + \dfrac{k-1}{2}M^2\right)} \tag{2.13}$$

Differentiating this relationship and dividing by the original relationship yields the following relationship for dT/T.

$$\frac{dT}{T} = \frac{-dM(k-1)M}{\left(1 + \dfrac{k-1}{2}M^2\right)} \tag{8.12}$$

Substituting the above equation into the momentum equation for dT/T yields:

$$\frac{dM}{M}\left(1 + \frac{\dfrac{(k-1)}{2}M^2}{\left(1 + \dfrac{k-1}{2}M^2\right)}\right) - f\frac{kM^2}{2}\frac{dx}{D} = kM^2\left(\frac{dM}{M}\right)\left(1 - \frac{\dfrac{(k-1)}{2}M^2}{\left(1 + \dfrac{k-1}{2}M^2\right)}\right) \tag{8.13}$$

Dividing through by $kM^2/2$ then solving for $f\, dx/D$ yields:

$$f\frac{dx}{D} = \frac{2dM}{M}\left(\frac{1 - M^2}{\left(1 + \dfrac{k-1}{2}M^2\right)kM^2}\right) \tag{8.14}$$

This equation can be integrated numerically from 1 to M, where M progresses to lower subsonic values or higher supersonic values essentially following the Fanno-line.

$$f\frac{L_{MAX}}{D} = \int_M^1 \frac{\dfrac{2}{M}(1 - M^2)dM}{\left(1 + \dfrac{k-1}{2}M^2\right)kM^2} \tag{8.15}$$

Values for $f L_{MAX}/D$ are tabulated in Appendix A.6.1 for a specific heat ratio (k) of 1.4 and in Appendix A.6.2 for a specific heat ratio of 1.31, which corresponds to the specific heat ratio of methane or natural gas. Any inaccuracy in the tabulation of results due to the numerical integration is minor compared with the uncertainty in assigning the friction factor, f, as well as the simplifications made in the development of the analysis. Finding the ratios of static pressure (P/P^*), static temperature (T/T^*), relative velocity (V/V^*), and total pressure (P_T/P_T^*) along the duct or pipe is an important part of the analysis. The "*" condition refers to the downstream choked condition. These ratios have also been tabulated in Appendices A.6.1 and A.6.2. The pressure ratio P/P^* can be developed from the ideal gas analysis, continuity, the adiabatic assumption, and velocity in terms of Mach number times the speed of sound which eventually yields:

$$\frac{dP}{P} = -\frac{dM}{M} + \frac{dT}{2T}$$

After making the substitution for dT/T as a function of Mach number:

$$\frac{dP}{P} = -\frac{dM}{M} + \frac{dT}{2T} = -\frac{dM}{M}\left(1 + \frac{\frac{(k-1)}{2}M^2}{\left(1 + \frac{k-1}{2}M^2\right)}\right)$$

Numerically integrating:

$$\int_{P^*}^{P}\frac{dP}{P} = Ln\left(\frac{P}{P^*}\right) = \int_{M}^{1}-\left(\frac{1+(k-1)M^2}{\left(1 + \frac{k-1}{2}M^2\right)M}\right)dM \tag{8.16}$$

The exponential of the natural log yields P/P^* which is tabulated in the Appendix. The adiabatic flow assumption provides a direct relationship between T/T^* noting T will be equal to the total temperature at a Mach number of 0. The resulting relationship for T/T^* can be stated simply as:

$$\frac{T}{T^*} = \frac{\left(1 + \frac{k-1}{2}\right)}{\left(1 + \frac{k-1}{2}M^2\right)} \tag{8.17}$$

The velocity ratio, V/V^* will be the reciprocal of the density ratio, ρ/ρ^*, which can be determined from the ideal gas law or

$$\frac{V}{V^*} = \frac{T}{T^*}\frac{P^*}{P} \tag{8.18}$$

The total pressure ratio can be determined from P/P^* and isentropic relationships for P/P_T and P^*/P_T^*. The isentropic relationship for P/P_T was developed from the relationship for T/T_T. In our adiabatic flow the total temperature, T_T, stays constant. Consequently, if the Mach number relationships in the isentropic relationships for P/P_T are replaced with the respective T/T_T, the total temperatures cancel and P_T/P_T^* can be determined from P/P^* and T/T^* as shown below:

$$\frac{P_T}{P_T^*} = \frac{P}{P^*} \left(\frac{T^*}{T} \right)^{\frac{k}{(k-1)}} \tag{8.19}$$

This parameter, the ratio of the local to downstream sonic total pressures, characterizes the effective total pressure loss as flow with friction develops from the local Mach number to the sonic condition at the exit of a pipe. In the next section, several examples of flow with friction in constant area ducts will be developed to provide experience in the application of the relationships tabulated in Appendix A.6.1 and A.6.2.

8.4 Application of Adiabatic Flow with Friction in a Constant Area Duct

Example 8.1: Flow with Friction in a Constant Area Duct

Consider adiabatic flow with friction in a constant area duct shown in Figure 8.1. A virtual extension is shown at the end of the duct which extends the end until the flow in this virtual section would reach a choked condition. The flow enters the pipe at a static pressure of 200 kPa and a static temperature of 300 K. The initial Mach number of the air flow is 0.3 going into a smooth duct with a diameter of 5 cm and a length of 16 m. You can assume the friction factor, f, is 0.0125.

Determine: (a) the downstream Mach number
              (b) the static pressure
              (c) the static temperature and
              (d) the upstream and downstream total pressure.

$M_1 = 0.3$

FIGURE 8.3
Schematic of constant area duct with virtual end extended to fL_{MAX}/D.

The flow can be analyzed by looking at the beginning of the tube where the Mach number is known and then looking to the point where the tube would become choked. This choked flow solution can be analyzed using our Fanno-line flow Table in Appendix A.6.1. At a Mach number of 0.3 the following information is provided in the Fanno-line table:

$$M_1 = 0.30, fL_{MAX}/D = 5.2993, P/P^* = 3.6191, T/T^* = 1.1788, V/V^* = 0.32572,$$
$$P_T/P_T^* = 2.0351$$

The flow parameter fL_{MAX}/D at the entrance to the duct is related to the maximum L/D which can be attained with an inlet Mach number of 0.30. The fL_{MAX}/D at the end of the pipe can be found by subtracting the pipes fL/D from fL_{MAX}/D at the entrance:

$$\left(f\frac{L_{MAX}}{D}\right)_2 = \left(f\frac{L_{MAX}}{D}\right)_1 - f\frac{L}{D} = 5.2993 - 0.0125\frac{16\,m}{0.05\,m} = 1.2993$$

Based on this analysis $M_2 = 0.47454$ and the following information can be interpolated from the Fanno-line table.

$$M_2 = 0.47454, P/P* = 2.2584, T/T* = 1.1483, V/V* = 0.5083, P_T/P_T* = 1.3919$$

The downstream pressure at the end of the tube can be determined from the initial static pressure and its ratio with the virtual choked flow static pressure along with the tube end static to choked flow pressure ratio. The downstream static temperature can be found similarly.

$$P_2 = P_1\frac{P^*}{P_1}\frac{P_2}{P^*} = 200\,kPa \times \frac{1}{3.6191} \times 2.2584 = 124.8\,kPa$$

$$T_2 = T_1\frac{T^*}{T_1}\frac{T_2}{T^*} = 300\,K \times \frac{1}{1.1788} \times 1.1483 = 292.2\,K$$

The flow in this example is assumed to be adiabatic. Consequently, the total temperature remains constant. However, the total pressure drops with the irreversibility of the friction. The upstream total pressure can be determined from the local static pressure and an isentropic relationship between static and total pressure at $M_1 = 0.30$.

$$P_{T1} = P_1 \times \frac{P_{T1}}{P_1} = 200\,kPa \times \frac{1}{0.93947} = 212.9\,kPa$$

The downstream total pressure can be determined using the ratio of the local total pressure to the downstream total pressure at the choked condition.

$$P_{T2} = P_{T1}\frac{P_T^*}{P_{T1}}\frac{P_{T2}}{P_T^*} = 212.9\,kPa \times \frac{1}{2.0351} \times 1.3919 = 145.6\,kPa$$

This shows that subsonic flow with friction results in an increase in Mach number and a resulting drop in static and total pressure. Static temperature decreases with Mach number in a similar manner to isentropic flow. Static temperature to total temperature ratio is entirely a function of Mach number for this adiabatic flow.

Consider a flow system with a converging nozzle followed by a constant area duct. Assume that the total pressure and temperature of the reservoir is held constant. Initially, the back pressure is equal to the reservoir pressure and no flow is entrained into the flow system (a). A schematic of the reservoir is shown below in Figure 8.4. As the back pressure downstream of the exit of the duct is lowered, flow is entrained into the converging nozzle and constant area duct system and the pressure through the nozzle and duct begin to drop (b). As the back pressure continues to drop, more flow is entrained into the nozzle and the pressure in both the nozzle and duct drop more rapidly. As the back pressure continues to drop, the pressure through the nozzle also continues to drop as does the pressure inside the pipe (c). As the back pressure continues to lower, choked flow is finally established at the exit of the nozzle (d). At that point, the maximum flow through the nozzle duct system has been established. Further, the inlet Mach number to the duct downstream from the nozzle is restricted by the flow in the duct and is limited by the fL/D of the ducting system. If the back pressure is lowered further from the choking point (d) then expansion fans can develop at the exit (e).

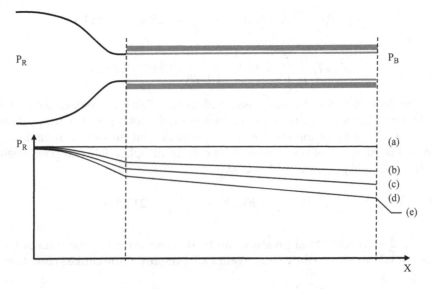

FIGURE 8.4
Flow through a nozzle and duct in series.

Example 8.2: Flow through a Converging Nozzle and a Duct in Series

Flow is entrained into a converging nozzle and a duct in series as shown in Figure 8.5. The duct is adiabatic and has a friction factor, $f = 0.02$. The pipe is 10 cm in diameter and 5 m long. The reservoir pressure is determined to be 254 kPa absolute and the exit pressure is determined to be 101.3 kPa absolute. Determine the Mach number at the entrance and exit to the duct and the exit pressure.

Analysis: The analysis can begin by assessing if the flow is choked in the ducting system. Initially, the fL/D of the duct must be found and it can initially be assumed to be the fL_{MAX}/D.

$$f\frac{L}{D} = f\frac{L_{MAX}}{D} = 0.02 \times \frac{5\,m}{0.1\,m} = 1.0$$

The flow through a converging nozzle into a duct with friction must be subsonic. The Fanno-line table brackets the fL_{MAX}/D of 1.0 between a Mach number of 0.50 and 0.51.

$$M = 0.50, fL_{MAX}/D = 1.0691, P/P* = 2.1381, T/T* = 1.1429,$$
$$V/V* = 0.53452, P_T/P_T* = 1.3398$$

$$M = 0.51, fL_{MAX}/D = 0.9904, P/P* = 2.0942, T/T* = 1.1407,$$
$$V/V* = 0.54469, P_T/P_T* = 1.3212$$

Interpolating at an fL_{MAX}/D of 1.0 produces

$$M = 0.5088, fL_{MAX}/D = 1.0000, P/P* = 2.0995, T/T* = 1.1409,$$
$$V/V* = 0.54345, P_T/P_T* = 1.3234$$

The parameter $P/P*$ provides the static pressure at the inlet to the duct divided by the static pressure at the exit of the duct when the flow is choked. By determining Pt/P at the inlet $Pt/P*$ can be determined.

$$\frac{P_T}{P} = \left(1 + \frac{k-1}{2}M^2\right)^{\frac{k}{k-1}} = \left(1 + \frac{1.4-1}{2}0.5088^2\right)^{\frac{1.4}{0.4}} = 1.1932$$

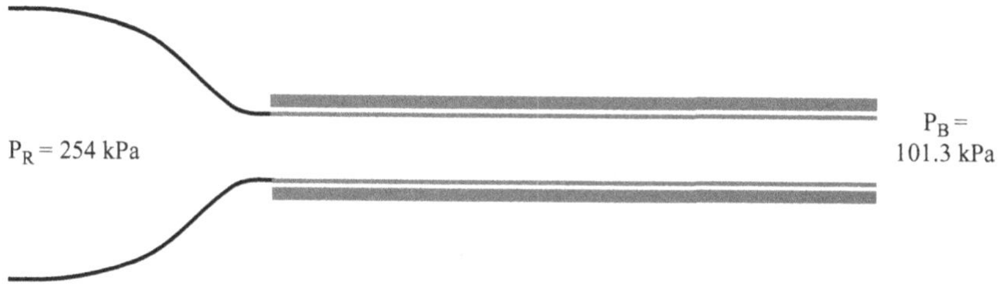

$P_R = 254$ kPa

$P_B = 101.3$ kPa

FIGURE 8.5
Compressible flow through a converging nozzle into a duct with friction.

$$\frac{P_T}{P^*} = \frac{P_T}{P}\frac{P}{P^*} = 1.1932 \times 2.0995 = 2.5052$$

The duct outlet pressure for choked flow can be determined from the ratio above.

$$P^* = \frac{P_T}{\frac{P_T}{P^*}} = \frac{254\,\text{kPa}}{2.5052} = 101.39\,\text{kPa}$$

The results of this analysis show that the flow at the end of the duct would be choked at only a tiny bit higher than the local ambient pressure. The Mach number at the entrance to the duct is 0.5088 and the flow at the exit of the duct has a Mach number of 1.0.

Example 8.3: Subsonic Flow for a Converging Nozzle and Duct in Series.

In many engineering situations, flow through a nozzle and duct in series will be subsonic. An engineer will need to determine the maximum possible flow or the pressure drop at a given flow rate for a given physical arrangement. This type of problem can be iterative in nature. As an example, let's take the previous flow situation except the pipe length will be doubled and the total pressure will be lowered. In this problem, flow enters a converging nozzle and a duct in series. The duct is adiabatic and has a friction factor, $f = 0.02$. The pipe is 10 cm in diameter and 10 m long. The reservoir pressure is determined to be 200 kPa absolute and the exit pressure is determined to be 101.3 kPa absolute. Determine the Mach number at the entrance and exit to the duct and the exit pressure.

Analysis: The analysis can begin by assessing the overall total to static pressure available and finding the fL/D of the duct. The inlet Mach number can be assessed to determine if flow at that initial Mach number through the duct is possible. First the fL/D of the duct will be found.

$$f\frac{L}{D} = 0.02 \times \frac{10\,\text{m}}{0.1\,\text{m}} = 2.0$$

Also,

$$\frac{P_T}{P} = \frac{200\,\text{kPa}}{101.3\,\text{kPa}} = 1.9739$$

At this point an analysis for a solution with a subsonic inlet and exit can be developed. The total pressure ratio will be the pressure ratio between the reservoir total and the inlet static pressure to the pipe times the ratio between the local $P/P*$ and the $P/P*$ downstream at an fL/D 2.0 lower than the upstream location. Written out this becomes:

$$\left(\frac{P_T}{P}\right)_{Avail} = \left(\frac{P_T}{P}\right)_{Inlet} \times \left(\frac{P}{P*}\right)_{Inlet} \times \left(\frac{P*}{P}\right)_{Outlet} = 1.9739$$

In this case fL_{MAX}/D at the inlet is exactly 2.0 larger than the fL_{MAX}/D at the exit. Additionally, P_T/P is based on the inlet Mach number. Solving this relationship requires finding an fL_{MAX}/D greater than 2.0 while matching the ratio in the equation above. Again, the solution requires some trial and error. Picking an upstream Mach number of 0.40.

$$M_1 = 0.40, fL_{MAX}/D = 2.3085, P/P* = 2.6958, \text{also } P_T/P = 1.11655$$

Subtracting fL/D of 2.0 from fL_{MAX}/D_1 then $fL_{MAX}/D_2 = 0.3085$. Interpolating:

$$M_2 = 0.656, fL_{MAX}/D = 0.3085, P/P* = 1.6024$$

Solving for P_T/P the result is determined to be:

$$\left(\frac{P_T}{P}\right)_{Avail} = 1.11655 \times 2.6958 / 1.6024 = 1.8784$$

Picking an upstream Mach number of 0.41.

$$M_1 = 0.41, fL_{MAX}/D = 2.1344, P/P* = 2.6280, \text{also } P_T/P = 1.1227$$

Subtracting fL/D of 2.0 from fL_{MAX}/D_1 then $fL_{MAX}/D_2 = 0.1344$. Interpolating:

$$M_2 = 0.74486, fL_{MAX}/D = 0.1344, P/P* = 1.3953$$

Solving for P_T/P the result is determined to be:

$$\left(\frac{P_T}{P}\right)_{Avail} = 1.1227 \times 2.6280 / 1.3953 = 2.1146$$

Interpolating between the two results, $M_1 = 0.404$ and $M_2 = 0.6854$.

Supersonic flow in a duct can be generated with a converging/diverging nozzle provided a low enough back pressure is achieved. Subsequently, flow downstream from the converging–diverging nozzle can discharge into a constant area duct. If the duct is shorter than the fL_{MAX}/D for that Mach number then a range of results can occur after the nozzle is choked. A schematic of this situation is shown in Figure 8.6 below. Flow through the converging–diverging portion with the downstream duct is similar to flow through a converging–diverging the nozzle without a duct through the point where a normal shock sits at the exit of the nozzle. The back pressure needed to cause a shock to sit at the exit of the nozzle is lower than the static pressure downstream of the shock, due to the pressure loss of the subsonic flow through the duct (b). As the back pressure is lowered below the value needed to cause a normal shock in the exit of the nozzle, the shock moves out into the duct. However, downstream of the shock's pressure rise a pressure loss is seen downstream in the duct due to skin friction (c). As the back pressure is further lowered in the duct the shock can move out to the exit of the duct (d). Downstream from the converging–diverging nozzle, the static pressure rises with the reduction in speed of the supersonic flow due to flow friction in the duct. If the back pressure is lowered further, an oblique shock will form at the exit of the duct (e). The back pressure can be lowered to the point where the static pressure at the exit is equal to the static pressure of the supersonic flow exiting from the duct (f). Any further lowering of the back pressure from here would cause an expansion fan to begin to form (g). If the duct is longer than the fL_{MAX}/D for that Mach number then a shock wave will form in the duct and the exit of the subsonic flow will be sonic.

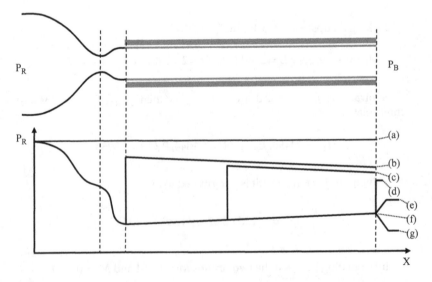

FIGURE 8.6
Compressible flow through a converging/diverging nozzle and duct with friction.

Example 8.4: Supersonic Flow for a Converging Nozzle and Duct in Series.

In this example problem, a high pressure air reservoir is connected to a converging–diverging nozzle and then an adiabatic duct of moderate length. The configuration is similar to the setup shown in Figure 8.6. The reservoir pressure is 516 kPa and the converging–diverging nozzle is designed for an exit Mach number of 2.4. The friction factor of the duct is 0.02, the diameter is 1.0 cm and the length of the duct is 0.20 m. First, assuming supersonic flow throughout the duct determine the exit pressure and the exit Mach number. Next, determine the back pressure necessary to cause a normal shock wave to sit at the exit of the duct and the corresponding exit Mach number. Finally, if the length of the duct is doubled, determine the state of the flow through the duct assuming the exit pressure is atmospheric or 101.3 kPa. The specific heat ratio, k, is assumed to be 1.4.

Analysis: The initial assumption is that the flow is supersonic at the beginning of the duct and throughout the duct. The flow through the converging–diverging nozzle can be determined from our isentropic Mach number tables. At a Mach number of 2.4 the static to total pressure ratio, P/P_T is 0.0684 and the exit to throat area ratio, A/A^* is 2.4031. The airflow enters the duct at this point and friction begins to affect the flow. Based on the Fanno-line tables at a Mach number of 2.4, $fL_{MAX}/D = 0.4099$. Also at $M = 2.4$, $P/P^* = 0.3111$. An initial check must be made to ensure that the duct can support supersonic flow for its entire length. This check can be done by subtracting fL/D of the duct from fL_{MAX}/D. The inlet to the duct can be designated as location 1 and the end of the duct can be designated as location 2.

$$\left(f\frac{L_{MAX}}{D}\right)_2 = \left(f\frac{L_{MAX}}{D}\right)_1 - f\frac{L}{D} = 0.4099 - 0.02 \times \frac{0.2\,\text{m}}{0.01\,\text{m}} = 0.0099$$

This positive result for fL_{MAX}/D indicates that the flow will still be supersonic. Further, the Fanno-line table can be used to determine the exit Mach number. The table shows a result for $fL_{MAX}/D = 0.0099$ for a Mach number of 1.1. At this same Mach number the table gives P/P^* as 0.8936. This result is consistent with Figure 8.6 which indicates that static pressure rises along the duct with supersonic flow. There is enough information now to determine the exit static pressure.

$$P_2 = P_T \frac{P_1}{P_T} \frac{P^*}{P_1} \frac{P_2}{P^*} = 516\,\text{kPa} \times 0.0684 \times \frac{1}{0.3111} \times 0.8936 = 101.38\,\text{kPa}$$

This suggests that at the chosen reservoir pressure a supersonic flow is initiated in the converging–diverging nozzle and can be maintained until the end of the duct resulting in an exit static pressure which is almost the same, only slightly above, the local atmospheric pressure. At this point the question of what a normal shock at the exit would do to the

static pressure and exit Mach number can be answered. The flow at the
end of the duct is initially supersonic with a Mach number of 1.10. Based
on the normal shock tables for $k = 1.4$, the Mach number downstream
from a normal shock with an inlet Mach number of 1.1 is 0.9118 and the
pressure ratio across the shock would be 1.245. The resulting exit static
pressure would be:

$$P_{Exit} = P_2 \frac{P_{EXIT}}{P_2} = 101.38 \text{ kPa} \times 1.245 = 126.2 \text{ kPa}$$

At this point the question becomes what would happen if the length
of the duct is doubled? In this case fL/D would now equal 0.8. Since
$fL_{MAX}/D = 0.4099$ at a Mach number of 2.4, the supersonic flow cannot
be maintained the entire distance along the duct. This increased blockage would have the effect of back pressuring the duct. This backpres-
sure would tend to cause a shockwave to form in the duct or even in
the diverging section of the nozzle. The location of the shock in the
duct relative to the duct's entrance can be determined by analyzing the
influence of a normal shock at the exit of the diffuser. Assuming that
the flow through the converging–diverging nozzle is isentropic our
first source of information will be the normal shock tables at a Mach
number of 2.4

$$N/S \text{ Table}: M_1 = 2.40, M_2 = 0.5231, P_2/P_1 = 6.5533$$

Consequently, if a normal shock existed just downstream from the
nozzle exit then Fanno-line flow would begin for this subsonic flow.
At a Mach number of 0.5231, fL_{MAX}/D and P^*/P should be interpolated
between Mach number of 0.52 and 0.53.

$$\text{Fanno} - \text{line Table}: M = 0.5231, fL_{MAX}/D = 0.8978, P*/P = 2.03928$$

Since fL_{MAX}/D is greater than fL/D for the duct (0.8) the flow at the exit
for this condition would be subsonic. The resulting exit Mach number
can now be determined.

$$\left(fL_{MAX}/D\right)_2 = \left(fL_{MAX}/D\right)_1 - fL/D = 0.8978 - 0.8 = 0.0978$$

Based on the fL_{MAX}/D of this result, the exit Mach number will be
between a Mach number of 0.77 and 0.78. Interpolating on fL_{MAX}/D the
following result is found:

$$\text{Fanno} - \text{line Table}: M = 0.77444, P/P* = 1.3367$$

With this information from the Fanno-line table the exit pressure can be determined:

$$P_{EXIT} = P_T \frac{P}{P_T} \frac{P_2}{P_1} \frac{P^*}{P_2} \frac{P_{EXIT}}{P^*}$$

$$= 516\,kPa \times 0.0684 \times 6.5533 / 2.03928 \times 1.3367 = 150.6\,kPa$$

The exit pressure determined from this analysis indicates at a back pressure of 150.6 kPa a normal shock would sit at the exit of the converging/diverging nozzle. If the back pressure was lowered, the shock wave would be expected to move down the duct. As the shock wave moved downstream, the static pressure in the duct just upstream of the shock would rise and the Mach number just upstream of the shock would decrease. The Mach number directly downstream from the shock would be higher than the previous downstream Mach number and fL_{MAX}/D would consequently decrease. The back pressure could be lowered until either choked flow appeared at the exit of the duct or subsonic flow matched the back pressure. In this case, it appears that choked flow would limit the location of the normal shock wave. The following result is arrived at by iteration.

Fanno – line Table: $M = 0.547, fL_{MAX}/D = 0.7533, P/P^* = 1.95035$

The location of the shock can be determined from the difference between the fL_{MAX}/D of the inlet Mach number and this downstream Mach number.

$$fL/D = (fL_{MAX}/D)_1 - (fL_{MAX}/D)_2 = 0.4099 - 0.3635 = 0.0464$$

The remaining fL/D of the duct will be 0.7536. The resulting normal shock will produce the following downstream conditions.

N / S Table : $M_1 = 2.21, M_2 = 0.5457, P_2/P_1 = 5.5315$

Interpolating from the Fanno-line table produces the following fL_{MAX}/D and P/P^*

Fanno–line Table: $M = 0.5457, fL_{MAX}/D = 0.7533, P/P^* = 1.95035$

The interpolation from the Fanno-line table for fL_{MAX}/D shows that a shock at the above location will produce a subsonic Mach number eventually resulting in choked flow at the exit. However, for this solution to be valid, the exit pressure from this solution much be equal to or higher than the back pressure, 101.3 kPa. Solving for the back pressure:

$$P_{EXIT} = P_T \frac{P}{P_T} \frac{P^*}{P_1} \frac{P_2}{P^*} \frac{P_3}{P_2} \frac{P^*}{P_3} = 516\,kPa \times 0.0684 \times \frac{0.3525}{0.3111} 5.5315/1.95035 = 113.4\,kPa$$

This result shows that this combination of inlet total pressure, converging/diverging nozzle, and duct length will result in a normal shock within the duct with choked flow at the exit.

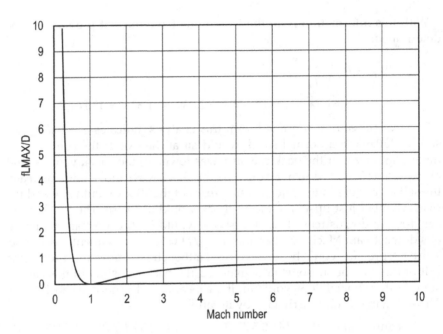

FIGURE 8.7
$f L_{MAX}/D$ versus Mach number for $k = 1.4$.

The Fanno-line Table in the Appendix tabulates values to a Mach number of 10 for a specific heat ratio of 1.4. Figure 8.7 presents the maximum $f L_{MAX}/D$ versus Mach number. At low Mach numbers $f L_{MAX}/D$ can be very high but the maximum value of above a Mach number of 1 and below a Mach number of 10 is less than 0.787. This restriction indicates that range of L/D's that can be achieved in ducts with supersonic flows is very limited. One reason for this may be that for a high-speed flow at a given total temperature there is not a large change in velocity between say Mach 10 and Mach 3. The change in velocity between Mach 10 and Mach 3 for the same T_{TOT} is only a reduction of 16 percent and the change between Mach 10 and Mach 2 is only 32 percent. Consequently, this large change in Mach number at a fixed T_{TOT} translates to a much reduced velocity decrease.

8.5 Isothermal Flow Assumption

Fanno-line flow assumes that the flow is isentropic which also means that the flow will be adiabatic. This assumption is reasonable in many situations where the ambient temperature stays constant as the total temperature and the recovery temperature, which drives heat transfer, only varies by a few degrees in subsonic flows. In supersonic flows with friction L/D's are typically small so an adiabatic flow is a reasonable assumption in these flows as well. The isothermal flow assumption suggests that the static temperature of a flow stays constant. This assumption is equally valid in long pipelines were the Mach number is low ($M < 0.3$) as in these low-speed flows the static temperature only varies a few degrees from the total temperature. This assumption is very useful as it simplifies the equations for compressible flow with friction allowing a closed form solution. Additionally, Fanno-line flow charts are relatively sparse in terms of fL_{MAX}/D at low Mach numbers and isothermal flow equations allow for simplified yet very accurate analysis at these low Mach numbers for situations where fL/D values can be large.

The analysis for isothermal flow with friction can begin with an equation developed from the momentum principle for Fanno-line flow.

$$-\frac{dP}{P} - f\frac{kM^2}{2}\frac{dx}{D} = kM^2\frac{dV}{V} \tag{8.8}$$

The goal here is to replace dP/P and dV/V in terms of Mach number relationships. The ideal gas law can be written in terms of pressure, $P = \rho\,RT$. Differentiating this equation and noting $dT = 0$.

$$\frac{dP}{P} = \frac{d\rho}{\rho} + \frac{dT}{T} = \frac{d\rho}{\rho} \tag{8.20}$$

A relationship between Mach (M) number and velocity (V) can be developed from the definition of Mach number, $M = V/a$. Differentiating the relationship between Mach number and velocity and noting that the speed of sound, a, is only a function of temperature, which is constant the following relationship between Mach number and velocity is developed.

$$\frac{dM}{M} = \frac{dV}{V} \tag{8.21}$$

Based on our analysis for continuity for our constant area duct, in Section 8.3, the change in velocity can be related to the change in density.

$$0 = \frac{d\rho}{\rho} + \frac{dV}{V}$$

These relationships for velocity, density, pressure, and Mach number can all be equated in the following way.

$$\frac{dP}{P} = \frac{d\rho}{\rho} = \frac{-dV}{V} = \frac{-dM}{M}$$

These relationships for changes in velocity and pressure in the momentum equation can be replaced with the corresponding changes in Mach number resulting in the following relationship.

$$-\frac{dM}{M} + kM^2 \frac{dM}{M} + f\frac{kM^2}{2}\frac{dx}{D} = 0$$

Rearranging,

$$f\frac{dx}{D} = \frac{\left(1 - kM^2\right)}{\dfrac{kM^2}{2}}\frac{dM}{M} \tag{8.22}$$

Integrating,

$$f\frac{L}{D} = \int_{M_1}^{M_2} \frac{\left(1 - kM^2\right)}{\dfrac{kM^2}{2}}\frac{dM}{M} \tag{8.23}$$

This integration results in the following closed form solution for isothermal flow.

$$f\frac{L}{D} = \frac{\left(1 - kM_1^2\right)}{kM_1^2} - \frac{\left(1 - kM_2^2\right)}{kM_2^2} + \ln\frac{M_1^2}{M_2^2} \tag{8.24}$$

Another simplifying feature of isothermal flow is that based on continuity $P_1 M_1 = P_2 M_2$.

The accuracy of this relationship can be evaluated at low Mach number and compared with the Fanno-line flow table. Let's compare the difference in fL_{MAX}/D from Mach numbers of 0.02 to 0.05 based on isothermal flow with the Fanno-line table, $k = 1.4$.

First, assessing using isothermal flow from Equation (8.24):

$$f\frac{L}{D} = \frac{\left(1 - 1.4 \times 0.02^2\right)}{1.4 \times 0.02^2} - \frac{\left(1 - 1.4 \times 0.05^2\right)}{1.4 \times 0.05^2} + \ln\frac{0.02^2}{0.05^2} = 1498.17$$

Next using the Fanno-line table for $k = 1.4$ with $M_1 = 0.02$ and $M_2 = 0.05$; At $M_1 = 0.02$, $fL_{MAX}/D = 1778.45$ and at $M_2 = 0.05$, $fL_{MAX}/D = 280.02$. Consequently, the difference in fL_{MAX}/D from the Fanno-line table amounts to 1498.43. This difference in the result is less than 0.02 percent, which is only a minor fraction of the uncertainty in the friction factor. In this region of the Fanno-line table where Mach numbers are less than 0.1, the change in fL_{MAX}/D with Mach number is highly non-linear and any interpolation which is not based on some type of higher order method can be expected to be much more inaccurate than the isothermal flow assumption. Consequently, using the isothermal flow assumption for flow in ducts or pipes where the peak Mach number is less than 0.1 should be considered.

Example 8.5: Use of Isothermal Flow for Natural Gas Pipeline Analysis

The maximum distance between compressor stations is to be estimated using the isothermal flow analysis. The pipeline diameter is set at 0.75 m. The maximum pressure (P_1) is determined to be 100 bar and the minimum pressure (P_2) is set at 20 bar. The nominal summer temperature in the pipeline is estimated at 300 K and the maximum allowable pipeline Mach number is set at 0.1. A typical pipeline friction factor, f, is estimated to be 0.012. Natural gas has a specific heat ratio of about 1.31. The desired result from Equation (8.24) is the pipeline length to diameter ratio. However, the inlet Mach number is unknown. However, based on continuity, $P_1 M_1 = P_2 M_2$. This means that the inlet Mach number, M_1 should be equal to the maximum Mach number divided by the pressure ratio or $M_2 = 0.02$. Equation (8.24) can now be solved for fL/D.

$$f \frac{L}{D} = \frac{\left(1 - 1.31 \times 0.02^2\right)}{1.31 \times 0.02^2} - \frac{\left(1 - 1.31 \times 0.1^2\right)}{1.31 \times 0.1^2} + \ln \frac{0.02^2}{0.1^2} = 1828.8$$

Noting that $f = 0.012$, L/D becomes

$$\frac{L}{D} = \left(f \frac{L}{D}\right) \Big/ f = \frac{1828.8}{0.012} = 152{,}351$$

Noting that $D = 0.75$ m then $L_{MAX} = 114.3$ km. This distance is equal to about 71 miles. Typical distances between compressor stations for natural gas transmission lines are between about 40 and 100 miles.

8.6 Adiabatic Flow with Friction and Area Change

In the preceding sections, equations were developed for adiabatic flow with friction for constant area ducts. At low speeds the effects of friction caused a gradual pressure loss accompanied with a rise in velocity. However, at high-speeds subsonic flows quickly became choked and supersonic flows quickly decreased in velocity and moved toward sonic flow. When low-speed flows include area change, contracting area quickly accelerates flow and the influence of skin friction is reduced. However, when flow area expands in a low-speed flow, flow is susceptible to separation which can have an adverse influence of flow uniformity and which can negate any significant pressure recovery. In high-speed flows, acceleration will tend to significantly reduce the influence of skin friction for both high subsonic velocities and accelerating supersonic flows. However, when a high-speed flow is decelerated, it becomes very susceptible to separation. Moreover, at transonic velocities, small changes in flow area can have a very significant influence on velocity. Consequently, area changes in high-speed decelerating flows are best dealt with using higher order methods such as highly resolved computational fluid dynamics. In the present section the equations of motion for adiabatic flow with friction and area change are developed. However, the analysis will focus on the situation where the Mach number remains constant.

Assuming a varying area duct, the continuity equation relates changes in density and velocity to changes in area. Assuming the flow is adiabatic, the energy equation can relate total temperature, which is constant, to static temperature through the local velocity or Mach number. A relationship between pressure, density, and temperature can be developed from the ideal gas law. A relationship between velocity, the Mach number, and temperature can be developed from the definition of Mach number. These relationships will be used to develop a relationship between Mach number, friction, and area change for a varying area adiabatic compressible flow. The analysis can begin with the continuity equation or conservation of mass. Starting with the steady-state integral equation for a control volume in the production form.

$$\dot{P}_M = 0 = \oint_{CS} \rho \bar{V} \cdot \overline{dA} \tag{1.20}$$

Here, the production of mass is equal to zero which equals the outflow rate less the inflow rate in a steady-state compressible flow. This relationship is developed by applying it to the control volume of a varying area duct as is shown below in Figure 8.8.

Applying the control volume formulation of conservation of mass to our varying area duct shown above.

$$\dot{P}_M = 0 = \oint_{CS} \rho V \cdot dA = (\rho + d\rho)(V + dV)(A + dA) - \rho V A$$

Multiplying each term keeping only first-order terms or higher then dividing through by $\rho V A$ the relationship previously obtained.

$$\frac{d\rho}{\rho} + \frac{dV}{V} + \frac{dA}{A} = 0 \tag{2.18}$$

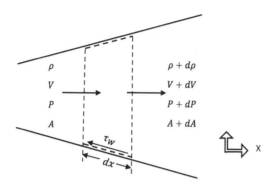

FIGURE 8.8
Schematic of control volume for compressible flow with friction and area change.

This development can begin by writing the steady-state integral momentum equation in the x-direction.

$$\sum F_x = \oint_{CS} V_x \rho \vec{V} \cdot \overrightarrow{dA}/g_C \qquad (1.25)$$

The forces include pressure forces and shear forces. The net pressure force on this volume with increasing area is $(-dPA)$ while the net shear force is shear stress, τ_W, times the surface area, A_S. The change in X-momentum is just the mass flow rate times the change in velocity across the control volume. Assuming a circular duct and estimating the shear stress from the skin friction coefficient times the dynamic pressure, plus noting $Cf = f/4$ the resulting equation is actually the same as Equation (8.8).

$$-\frac{dP}{P} - f\frac{kM^2}{2}\frac{dx}{D} = kM^2\frac{dV}{V} \qquad (8.8)$$

The pressure term can be eliminated once again using the ideal gas law. However, now when the density term is eliminated using continuity, the area term is added and as a result this analysis varies from the analysis for a constant area duct. The velocity terms can again be eliminated using Equation (8.9) and temperature can be eliminated using Equation (8.12). The resulting equation becomes:

$$f\frac{dx}{D}\frac{kM^2}{2} = \frac{dA}{A} + \frac{dM}{M}\left(\frac{1-M^2}{\left(1+\frac{k-1}{2}M^2\right)}\right) \qquad (8.25)$$

This equation is actually Equation (8.14) but multiplied by $kM^2/2$ and with the addition of dA/A on the right-hand side. Solving this equation is not possible except for the case where $dA/A = 0$ or $dM/M = 0$. Fanno-line flow was flow with a constant area duct or $dA/A = 0$ and this has been addressed fairly comprehensively in this chapter. The case of constant Mach number ($dM/M = 0$) results in the following equation.

$$\frac{f}{D}\frac{kM^2}{2} = \frac{1}{A}\frac{dA}{dx} \tag{8.26}$$

Equation (8.26) can be rewritten for a circular duct as:

$$f\frac{kM^2}{4} = \frac{dD}{dx} \tag{8.27}$$

This equation simply indicates that the diameter or area of a duct would need to gradually expand to accommodate a constant Mach number in a compressible flow.

High-speed flows with area change were generally discussed at the beginning of this section. There are many examples of high-speed flows with area change in engineering. For example in gas turbines including modern turbofan engines as well as large industrial gas turbines, high-speed flow with area change occurs in virtually every component in the engines. The tips of fan blades and first stage compressor blades often run at supersonic Mach numbers relative to the blades while the flow near the hub is typically subsonic but in the transonic regime. The gas turbine industry has been a leading driver of computation fluid dynamics methods dating back at least into the 70s due to the complex nature of flows with friction and area change. Additionally, these turbomachinery flows included surface curvature and often a significant influence of secondary flows and flow field turbulence. The use of Fanno-line flow is very useful in the design and analysis of compressible flow pipelines such as natural gas transmission lines. However, in most high-speed flows, the analysis of the development of boundary layers is a key aspect of a useful flow field solution. In high-speed converging–diverging nozzles, boundary layers are expected to stay thin so boundary layer analysis is most critical in helping to determine the flow blockage due to the boundary layer. Boundary layer analysis for the assessment of heat load is often another important issue. However, in high-speed diffusers where flow is purposely slowed to help the recovery of pressure, the rate of deceleration is critical to the assessment of boundary layer development. In these flows, boundary layers are very susceptible to separation so accurate CFD results require a well-resolved grid and the choice of a turbulence model which accurately predicts boundary layer development in adverse pressure gradients.

References

1. John, J.E.A. and Keith, T.G., 2006, *Gas Dynamics*, 3rd Ed., Prentice Hall.
2. Saad, M.A., 1993, *Compressible Fluid Flow*, 2nd Ed., Prentice Hall.

Chapter 8 Problems

1. Flow moves through a 5 cm diameter pipe at an entrance Mach number of 0.5. The initial static pressure is 500 kPa and the initial static temperature is 295 K. The length of the pipe is 1.75 m. You can assume that the friction factor is 0.02. Find the exit Mach number and exit static pressure and temperature.

2. Compressed air leaves a power house at a flow rate of 4 kg/s and at a static pressure of 860 kPa and a static temperature of 306 K moving through a 0.125 m diameter pipe. The distance from the power house to the experiment is 100 m. If the piping system friction factor is 0.02, estimate the pressure drop between the power house and the experiment. Also determine the local Mach number and static temperature at the entrance of the experiment.

3. A 3 cm diameter pipe is connected to the exit of a converging nozzle which is connected to a reservoir with a total pressure of 900 kPa and a total temperature of 400 K. The length of the pipe is 45 cm long and the friction factor, f, is assumed to be 0.03. Determine the maximum flow rate for the system. Also determine the exit static temperature and pressure. Also determine the maximum flow rate if the pipe is removed from the nozzle so the converging nozzle now has a 3 cm exit diameter.

4. A reservoir has a total pressure of 600 kPa and it is attached to a converging–diverging nozzle with a throat to exit area ratio of 1.7. The nozzle attached to a duct has an L/D of 12 and a friction factor $f = 0.015$. Determine the exit Mach number and exit static pressure for flow exiting the duct at a back pressure of 0 kPa (absolute). Also determine the range of back pressure that would result in a normal shock within the duct.

5. A 2.5 cm diameter pipe is connected through a converging nozzle to a reservoir with a total pressure (P_R) of 250 kPa as shown in Figure 8.9 below. The pipe is 4 m long and the friction factor $f = 0.0198$. The pipe contracts to 2 cm in diameter at the end of the pipe and prior to the discharge. Assume the back pressure is 101 kPa. Find the mass flow rate through the system.

FIGURE 8.9
Flow from a reservoir into a constant area duct with friction and then through a converging nozzle.

6. A natural gas pipeline is designed to flow 158 kg/s of methane (MW = 16.04 kg/kmol). The initial static pressure is 75 bar (7500 kPa) and the pipeline static temperature is believed to be a constant value of 285 K (isothermal). The pipeline is designed to be 50 km in length between compressor stations and the friction factor is estimated to be $f = 0.012$. Determine the downstream Mach number and pressure drop at the end of the station. Compare your answer assuming instead that your analysis is for Fanno-line flow.

7. A carbon dioxide pipeline (CO_2, MW = 44.01 kg/kmol, $k = 1.28$) is sent from central North Dakota to the oil fields in western North Dakota and Saskatchewan. The pipeline is designed to span 100 km between stations and is proposed to be 0.9 m in diameter. If the maximum inlet pressure is limited to 8000 kPa and the minimum pressure is proposed to be 2000 kPa, determine the maximum flow rate in the pipe if the static temperature is 282 K and the friction factor, $f = 0.012$. You can assume that this flow is isothermal.

9

Internal Compressible Flow with Heat Addition

Chapter 9 introduces internal compressible flow with heat transfer. The introduction provided in Section 9.1 discusses some examples of when heat addition and its effects are important in compressible flows. Section 9.2 develops the governing equations for constant area frictionless compressible flow with heat addition. Section 9.3 introduces the Rayleigh line. The Rayleigh line provides a physical understanding of how heating and cooling changes subsonic and supersonic flows. Section 9.4 introduces frictionless compressible flow with heat transfer and area change. Section 9.5 presents constant area compressible flow with both heat transfer and skin friction. This chapter also provides a number of examples and some end of the chapter problems to help support student learning through application.

9.1 Introduction

Compressible flows are influenced by heat addition to a system as well as heat rejection from a system. Heat transfer can occur not only due to temperature differences between the wall of a channel and the flow but also due to thermal energy release caused by combustion. The process of condensation as well as evaporation can also influence compressible flows. However, these phenomena are more complex as condensation is associated with an increase in density without temperature change while evaporation results in a decrease in density. The most significant effects associated with heat addition in compressible flows are due to combustion. Compressible flow systems with chemical energy release include combustion systems used for energy conversion as well as afterburners. The influence of thermal energy release on pressure drop is much more critical at higher Mach numbers so determining the effects of heat release on compressible flow in systems such as the afterburners or even scramjets (supersonic combustion ramjets) is very important. Combustion systems can also include area change; this issue will be addressed in this chapter. Also the combination of heat transfer and skin friction effects can be important in heat exchangers where a gas is the working fluid. In this chapter, the equations which quantify the influence of heat release and rejection on compressible flows will be developed and applied to a number of engineering situations.

9.2 Constant Area Frictionless Flow with Heat Addition

The ideal gas law implies that the addition of thermal energy to a gas will cause an expansion. In a constant mass flow rate gas through a constant area duct, this addition of thermal energy will cause an increase in velocity. In this section, the principles of conservation of mass, the momentum principle, and conservation of energy will be applied to a compressible flow along with the ideal gas law and other relationships to develop equations relating pressure and temperature to Mach number and heat addition or rejection. The general approach is similar to the development in other chapters and similar to Chapters 2, 3, 5, and 8, tables relating these changes as a function of Mach number have been developed and are presented in Appendix A.7.

The most direct approach to the development of relationships governing compressible flow with heat transfer is with a control volume analysis showing heat addition to a steady, constant area frictionless flow as shown below in Figure 9.1. The principles of conservation of mass, momentum and conservation of energy are applied to the control volume below.

Applying the steady-state conservation of mass principle for our control volume, Equation (1.20) results in the outflow minus the inflow of mass is zero written simply in terms of one-dimensional uniform flow.

$$\dot{P}_M = 0 = \oint_{CS} \rho V \cdot dA = \rho_2 V_2 A_2 - \rho_1 V_1 A_1 \qquad (1.20)$$

Noting that the area is constant, or $A_1 = A_2$, this equation reduces to the result that the mass flux, ρV, is constant across the control volume.

$$\rho_2 V_2 = \rho_1 V_1 \qquad (9.1)$$

This relationship, Equation (9.1), will be used together with the results of the momentum principle to develop a relationship for static temperature changes. The momentum principle for steady-state flow can now be applied to our control volume. The analysis is begun with the steady-state version of the x-momentum equation, Equation (1.25).

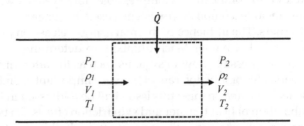

FIGURE 9.1
Analysis of a steady compressible flow with heat addition.

$$\sum F_X = F_{Sx} + F_{Bx} + R_X = \oint_{cs} V_X \rho V \cdot dA/g_C \qquad (1.25)$$

Assuming the horizontal flow of a gas, body forces can be eliminated. Additionally, noting this is a frictionless flow, the control surface includes only the fluid, a one-dimensional uniform flow, which is moving through the control volume eliminating the possibility of a reaction force. The resulting surface forces are from pressure times the cross-sectional area. The momentum forces constitute the outflow less inflow of x-momentum. The initial result is shown below:

$$P_1 A_1 - P_2 A_2 = \rho_2 V_2^2 A_2/g_C - \rho_1 V_1^2 A_1/g_C \qquad (9.2)$$

Noting that the duct area is constant and then separating the inflow from the outflow.

$$P_1 + \rho_1 V_1^2/g_C = P_2 + \rho_2 V_2^2/g_C \qquad (9.3)$$

Equations (9.2) and (9.3) are shown to be the same as Equations (3.1) and (3.2) for a normal shock wave. Consequently, as the ideal gas law is substituted for density in developing an equation for pressure changes from the momentum equation, it is not surprising that the result is the same as Equation (3.3).

$$P_1 + \frac{P_1 V_1^2}{R T_1 g_C} = P_2 + \frac{P_2 V_2^2}{R T_2 g_C} \qquad (9.4)$$

Similarly, when pressure is factored out from each side the result is the same as Equation (3.4).

$$P_1 \left(1 + \frac{V_1^2}{R T_1 g_C} \right) = P_2 \left(1 + \frac{V_2^2}{R T_2 g_C} \right) \qquad (9.5)$$

If the velocity squared term is multiplied by (k/k) then the denominator under that term becomes the speed of sound squared and that term can be written as the specific heat ratio, k, times the Mach number squared. This substitution results in an equation similar to Equation (3.5).

$$P_1 \left(1 + k M_1^2 \right) = P_2 \left(1 + k M_2^2 \right) \qquad (9.6)$$

Consequently, pressure changes can be related to Mach number changes when heat addition occurs in this steady, frictionless, constant area flow. Heat addition will be shown to cause an increase in Mach number and a drop in static pressure while flow across a normal shockwave is known to cause a decrease in Mach number from supersonic flow to subsonic flow with a corresponding rise in pressure. However, Equation (9.6) can be applied to either subsonic or supersonic flows with heat addition while Equation (3.5) can only be applied to a flow which is initially supersonic.

An equation for the relationship between temperature and Mach number for this steady, constant area, frictionless flow with heat addition can be developed using Equation (9.1). However, this equation is transformed using ideal gas for density and Mach number times the speed of sound for velocity as shown below in Equation (9.7), which is also identical to Equation (3.8).

$$\frac{P_2}{RT_2} M_2 \sqrt{kg_C RT_2} = \frac{P_1}{RT_1} M_1 \sqrt{kg_C RT_1} \tag{9.7}$$

Noting the gas constant, R, the specific heat ratio, and g_C are constant, this equation can be reduced to a relationship between pressure, Mach number, and temperature.

$$\frac{P_2}{P_1} \frac{M_2}{M_1} = \sqrt{\frac{T_2}{T_1}} \tag{9.8}$$

This Equation (9.8) is identical to Equation (3.9). However, at this point our development varies from the development for a normal shock in that, at this point, Equation (9.6) is substituted for the pressure ratio, P_2/P_1. The substitution of Equation (9.6) eliminates the pressure ratio.

$$\sqrt{\frac{T_2}{T_1}} = \frac{\left(1 + k M_1^2\right)}{\left(1 + k M_2^2\right)} \frac{M_2}{M_1} \tag{9.9}$$

Squaring both sides and rearranging the following relationship provides a relationship for Mach number and static temperature ratio for heat addition for our steady, constant area, frictionless flow.

$$\frac{T_2}{T_1} = \left[\frac{M_2}{M_1}\right]^2 \left[\frac{\left(1 + k M_1^2\right)}{\left(1 + k M_2^2\right)}\right]^2 \tag{9.10}$$

The final principle used to analyze this flow is conservation of energy. Conservation of energy for a steady-state control volume flow for a single stream can most simply be written as:

$$\dot{P}_E = 0 = \dot{M}_{OUT}\left(h_2 + \frac{V_2^2}{2g_C} + \frac{g z_2}{g_C}\right) + \dot{W}_{CV} - \dot{M}_{IN}\left(h_1 + \frac{V_1^2}{2g_C} + \frac{g z_1}{g_C}\right) - \dot{Q}_{CV} \tag{1.28}$$

This equation can be simplified for this flow by eliminating the work term, dividing the equation by $\dot{M}$, and then eliminating the potential energy terms, $g z/g_C$. Solving for $\dot{Q}/\dot{M}$ results in the following equation.

$$\frac{\dot{Q}}{\dot{M}} = \left(h_2 + \frac{V_2^2}{2g_C}\right) - \left(h_1 + \frac{V_1^2}{2g_C}\right) \tag{9.11}$$

Previously, the static enthalpy, h, plus the kinetic energy was defined as the total enthalpy, h_T, in Equation (2.1). Also in Equation (2.2), the total temperature was defined as the static temperature plus the kinetic energy divided by the specific heat at constant pressure, C_P.

$$T_T = T + \frac{V^2}{2g_c C_p}$$

Consequently, using Equation (9.11) with the definition of stagnation temperature, a relationship between thermal energy addition and total temperature change can be developed.

$$T_{T2} = T_{T1} + \frac{\dot{Q}}{\dot{M} C_P} \tag{9.12}$$

In previous chapters isentropic, flow was shown to achieve a limit of a Mach number of one in a converging nozzle. Heat addition is somewhat similar to a converging duct in which both subsonic and supersonic flows will move toward sonic flow, $M = 1$, when thermal energy is added. Consequently, choosing the reference state for $M = 1$ is practical in the development of Mach number relationships for steady, constant area, frictionless flow with heat addition. Starting with Equation (9.6) and referencing to the situation of $M = 1$ where $P = P^*$ this pressure relationship can be rewritten in terms of P/P^* as a function of Mach number.

$$\frac{P}{P^*} = \frac{1+k}{1+k M^2} \tag{9.13}$$

A similar transformation can be applied to the temperature relationship in Equation (9.10).

$$\frac{T}{T^*} = \frac{M^2 (1+k)^2}{\left(1+k M^2\right)^2} \tag{9.14}$$

Based on Equation (9.1), for conservation of mass in a constant area duct, a relationship for V/V^* can be developed when combined with the two equations above and the ideal gas law.

$$\frac{V}{V^*} = \frac{\rho^*}{\rho} = \frac{P^*}{P} \frac{T}{T^*} = \frac{M^2 (1+k)}{\left(1+k M^2\right)} \tag{9.15}$$

Relationships for total temperature and total pressure can also be developed based on the relationships developed for static temperature and pressure and the local relationships between total and static temperature and total and static pressure. The total temperature ratios along with the energy equation can be used to determine the change in the Mach number. Subsequently, the static pressure, temperature, and total pressure loss can be determined. T_T/T_T^* can be determined by first starting from T/T^*.

$$\frac{T_T}{T_T^*} = \frac{T}{T^*}\frac{T_T}{T}\frac{T^*}{T_T^*} = \frac{M^2(1+k)^2}{(1+kM^2)^2}\frac{\left(1+\dfrac{k-1}{2}M^2\right)}{\left(1+\dfrac{k-1}{2}\right)} \tag{9.16}$$

Similarly, P_T/P_T^* can be determined by first starting from P/P^*.

$$\frac{P_T}{P_T^*} = \frac{P}{P^*}\frac{P_T}{P}\frac{P^*}{P_T^*} = \frac{1+k}{1+kM^2}\left[\frac{\left(1+\dfrac{k-1}{2}M^2\right)}{\left(1+\dfrac{k-1}{2}\right)}\right]^{\!k/_{k-1}} \tag{9.17}$$

Equations (9.13) through (9.17) comprise the relationships which make up the Rayleigh-line flow tables which are tabulated in Appendix A.7 for specific heat ratios, k, of 1.4 and 1.3. These relationships are plotted as a function of Mach number in Figure 9.2 shown below. One significant issue in heating a compressible flow is the resulting total pressure loss. The P_T/P_T^* line indicates that heating a subsonic flow causes a notable total pressure loss when $M > 0.2$ and it suggests heating a supersonic flow is always problematic in terms of the total pressure loss.

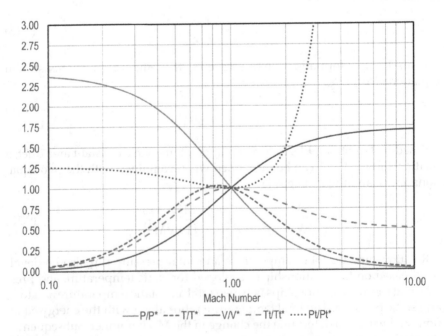

FIGURE 9.2
Plot of Rayleigh line flow variables as a function of Mach number, $k = 1.4$.

Example 9.1:   Steady Frictionless Flow with Heat Transfer in a Constant Area Duct

Airflow enters a combustion system at 300 K and a velocity of 35 m/s as shown in Figure 9.3. The air is mixed with methane and burned, increasing the enthalpy by 980 kJ/kg. The C_p of these products can be assumed to average 1.09 kJ/kg/K for this temperature rise in the gas. You can assume that the specific heat ratio is 1.4. Determine the downstream conditions including the total temperature and pressure changes.

Analysis: The downstream condition can be determined using Equation (9.12) along with the Rayleigh line table. However, initially the inlet total temperature needs to be determined. Additionally, the Rayleigh line table is based on Mach number so the example can be solved most easily by determining the Mach number.

$$M_1 = V_1 / a_1$$

$$a_1 = \sqrt{k\, g_C\, R\, T_1} = \sqrt{1.4\,(1)\, 287\; \frac{J}{kg\,K}\, 300\, K} = 347.2\; \frac{m}{s}$$

$$M_1 = \frac{35\; \frac{m}{s}}{347.2\; \frac{m}{s}} = 0.1008$$

The Mach number of this flow has been found to be essentially, $M_1 = 0.10$. Based on Table A.7:

$$M_1 = 0.10, P\,/\,P^* = 2.36686, T\,/\,T^* = 0.05602, V\,/\,V^* = 0.02367,$$
$$T_T\,/\,T_T{}^* = 0.04678, P_T\,/\,P_T{}^* = 1.25915$$

The total temperature can be determined using Equation (2.12).

$$T_{T1} = T_1 \left(1 + \frac{k-1}{2} M_1{}^2\right) = 300\, K \left(1 + \frac{1.4-1}{2}\, 0.1^2\right) = 300.6\, K$$

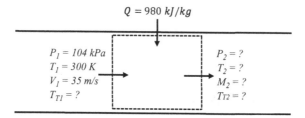

$$Q = 980\ kJ/kg$$

$P_1 = 104\ kPa$
$T_1 = 300\ K$
$V_1 = 35\ m/s$
$T_{T1} = ?$

$P_2 = ?$
$T_2 = ?$
$M_2 = ?$
$T_{T2} = ?$

FIGURE 9.3
Analysis of the effect of heat addition on steady compressible frictionless flow.

Now the downstream total temperature can be found using the energy equation of Equation (9.12).

$$T_{T2} = T_{T1} + \frac{Q}{C_P} = 300.6 \text{ K} + \frac{980 \,^{kJ}\!\big/\!_{kg}}{1.09 \,^{kJ}\!\big/\!_{kg \, K}} = 1199.7 \text{ K}$$

Noting that T_T/T_T^* denotes the total temperature ratio between the current point and the choking point, the initial ratio at condition 1 can be updated for the ratio at condition 2 using the initial T_T/T_T^* ratio and the total temperature ratio.

$$\frac{T_{T2}}{T_T^*} = \frac{T_{T1}}{T_T^*}\frac{T_{T2}}{T_{T1}} = 0.04678 \times \frac{1198.7 \text{ K}}{300.6 \text{ K}} = 0.1867$$

The total temperature ratio, T_{T2}/T_T^*, allows the use of Appendix A.7 for determining the Mach number. Interpolating on T_T/T_T^*, the Mach number is determined to be 0.2083.

$$M_2 = 0.2083, \, P/P^* = 2.26256, \, T/T^* = 0.22212, \, V/V^* = 0.09817,$$
$$T_T/T_T^* = 0.1867, \, P_T/P_T^* = 1.23197$$

The table allows the determination of the static temperature and pressure and total pressure drop.

$$T_2 = T_1 \frac{T^*}{T_1}\frac{T_2}{T^*} = 300 \text{ K} \times \frac{0.22212}{0.05602} = 1189.5 \text{ K}$$

$$P_2 = P_1 \frac{P^*}{P_1}\frac{P_2}{P^*} = 104 \text{ kPa} \times \frac{2.26256}{2.36686} = 99.42 \text{ kPa}$$

$$P_{T2} = P_{T1} \frac{P_{T2}}{P_T^*}\frac{P_T^*}{P_{T1}} = 104.73 \text{ kPa} \times \frac{1.23197}{1.25915} = 102.47 \text{ kPa}$$

This problem shows that even at this relatively low inlet Mach number that the significant rate of heat addition produces a total pressure drop of 2.26 kPa. This amounts to a significant total pressure drop of around 2.2 percent. If reducing the total pressure loss in this combustion system was critical, then reducing the inlet Mach number to the combustion system would be an approach.

9.3 Rayleigh Line Analysis

The Rayleigh line relates changes in $s - s^*$ to changes in T/T^*. This relationship can be developed from an intermediate equation in the development of a relationship for the changes of entropy for an ideal gas with a constant specific heat.

$$ds = C_P \frac{dT}{T} - R \frac{dP}{P}$$

If this equation is integrated from the local value of T to the value at $M = 1$, T^*, the following results is found.

$$s - s^* = C_P \ln \frac{T}{T^*} - R \ln \frac{P}{P^*}$$

Dividing this equation by R, the gas constant results in

$$\frac{s - s^*}{R} = \frac{C_P}{R} \ln \frac{T}{T^*} - \ln \frac{P}{P^*} \tag{9.18}$$

If Equations (9.13) and (9.14) are substituted into Equation (9.18), then $(s - s^*)/R$ can be determined as a function of Mach number. Next, T/T^* can be plotted as a function of $(s - s^*)/R$ which is one method to present the Rayleigh line.

The shape of the Rayleigh line in terms of T/T^* as a function of $(s - s^*)/R + C$ is presented in Figure 9.4. The constant C is added to $(s - s^*)/R$ as s^* is the maximum entropy for a given Rayleigh line and as a result $(s - s^*)$ will be negative. Note that for a given Rayleigh line both the mass flow rate and T_T^* will be the same for both the subsonic and sonic flow lines. The plot shows

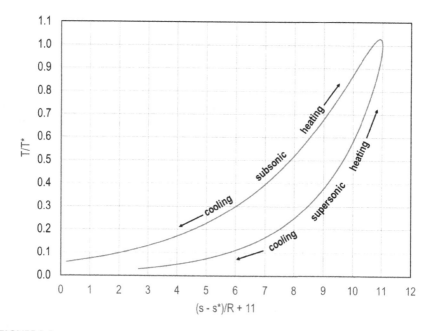

FIGURE 9.4
Plot of Raleigh line in terms of *T/T* versus $(s - s^*)/R$ showing the influence of heating and cooling as well as subsonic and supersonic flow.

two points of tangency. One point is at the maximum for the T/T^* line where T/T^* is greater than one and $d(T/T^*)/d(s - s^*) = 0$. This just indicates that in this region the static temperature is more influenced by increasing Mach number than increasing T_T^*. The other point of tangency occurs where $d(s - s^*)/d(T/T^*) = 0$. This occurs at a Mach number of 1 as will be shown presently. Once a Rayleigh line reaches this second tangency any further increase in heating will result is a reduced flow rate. A subsonic flow cannot increase past this line to a supersonic flow and a supersonic flow cannot move past this location to a subsonic flow. However, it is possible for a normal shock to occur under certain conditions resulting in the supersonic flow becoming a subsonic flow but at a higher entropy level.

The maximum for T/T^* on the upper curve can be shown by differentiation. Noting by differentiation using the change rule.

$$\frac{d\left(T/T^*\right)}{ds} = \frac{d\left(T/T^*\right)}{dM}\frac{dM}{ds}$$

Differentiating $d(T/T^*)/dM$ produces the following result.

$$\frac{d\left(T/T^*\right)}{dM} = \frac{2M\left(1+k\right)^2\left(1+kM^2\right)^2 - 2\left(2kM\right)\left(1+kM^2\right)M^2\left(1+k\right)^2}{\left(1+kM^2\right)^4} \tag{9.19}$$

Simplifying and setting this relationship to equal zero indicates that the maximum value of T/T^* occurs at a Mach number of $M = 1/\sqrt{k}$. The Mach number of the peak in T/T^* is 0.84515 and the peak in T/T^* is determined to be 1.02857.

The other interesting point of tangency occurs when $d(s - s^*)/d(T/T^*) = 0$. This result can be found in a similar manner to the other point of tangency. Using differentiation by the chain rule.

$$\frac{d\dfrac{\left(s-s^*\right)}{R}}{d\left(T/T^*\right)} = \frac{d\dfrac{\left(s-s^*\right)}{R}}{dM}\frac{dM}{d\left(T/T^*\right)}$$

The term $(s - s^*)/R$ can be written in terms of M and then differentiated starting with Equation (9.18) substituting in Equations (9.14) and (9.13).

$$\frac{s-s^*}{R} = \frac{C_P}{R}\ln\frac{M^2\left(1+k\right)^2}{\left(1+kM^2\right)^2} - \ln\frac{1+k}{1+kM^2} \tag{9.20}$$

Differentiating the right-hand side noting the $C_P/R = k/(k-1)$, the result is shown below:

$$\frac{d\frac{\left(s-s^*\right)}{R}}{dM} = \frac{k}{(k-1)}\frac{2}{M} - \frac{k}{(k-1)}\frac{4kM}{\left(1+kM^2\right)} + \frac{2kM}{1+kM^2} \qquad (9.21)$$

Simplifying the relationship after finding a common denominator and setting this equal to zero, the result becomes

$$\frac{d\frac{\left(s-s^*\right)}{R}}{dM} = 0 = \frac{2k-2kM^2}{(k-1)M\left(1+kM^2\right)} \qquad (9.22)$$

The term $(2k-2kM^2)$ can only be zero when the Mach number, $M = 1$. This demonstrates that the maximum entropy occurs at a Mach number equal to one. A subsonic flow accelerating to a supersonic flow due to heat addition would violate the second law as would a supersonic flow decelerating to a subsonic flow due to heat addition. However, this does not preclude a supersonic flow from becoming a subsonic flow through a normal shock.

What happens when additional thermal energy is added after a flow becomes choked due to heat addition? Assuming that the inlet total pressure and temperature have not changed, additional heat addition will cause the flow rate to decrease essentially putting the new flow rate on a different Rayleigh line with a lower mass flow rate.

Example 9.2: Heat Addition with Choked Exit Flow

A reservoir of air has a total pressure of 400 kPa and a total temperature of 296 K. The reservoir is attached to a converging nozzle contracting into a 0.025 m diameter tube. The back pressure of the system is 100 kPa. Heat addition rates ranging from 100 kJ/kg and 250 kJ/kg are to be added to the flow. The specific heat ratio for the air can is assumed to be 1.005 kJ/kg/K. Determine the mass flow rates, inlet Mach number, and exit static pressure for each case. Note that the exit static pressure must equal or exceed the back pressure for choked flow ($M = 1$) at the exit. A schematic of the flow setup is shown in Figure 9.5 below.

Analysis for Q_1:

The initial assumption made in this analysis is that flow through the pipe is subsonic, and it is choked at the exit. This situation makes $T_{T,EXIT} = T_T^*$. The inlet total temperature in both cases is 296 K. Initially, the exit total temperatures can be determined from the heat addition.

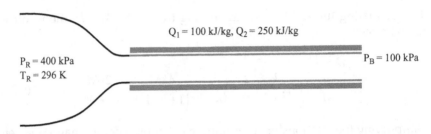

FIGURE 9.5
Converging nozzle and frictionless flow in pipe with heat addition in series.

$$T_{T2} = T_{T1} + \frac{Q}{C_p} = 296\text{ K} + \frac{100\,\dfrac{\text{kJ}}{\text{kg}}}{1.005\,\dfrac{\text{kJ}}{\text{kg K}}} = 395.5\text{ K} = T_T^*$$

Based on the value of T_{T1}/T_T^*, the inlet Mach number to the duct can be determined.

$$\frac{T_{T1}}{T_T^*} = \frac{296\text{ K}}{395.5\text{ K}} = 0.74842$$

The Rayleigh line table for a specific heat ratio, $k = 1.4$ shows that at a Mach number of 0.54.

$$M_1 = 0.54, P/P^* = 1.70425, T/T^* = 0.84695, V/V^* = 0.49696,$$
$$T_T/T_T^* = 0.74695, P_T/P_T^* = 1.09789$$

The initial check must show that the back pressure is low enough to allow for choked flow. This can be determined by checking P^* against the back pressure. Starting with the total pressure, the local static pressure can be determined from the Mach number and the isentropic tables $(P/P_T = 0.814165)$. Next, the exit static pressure can be determined from the ratio, P/P^* in the Rayleigh line table as tabulated above.

$$P^* = P_T \frac{P}{P_T} \frac{P^*}{P} = 400\text{ kPa}\,\frac{0.82005}{1.70425} = 192.47\text{ kPs}$$

This analysis shows that choked flow is possible for a heat addition of 100 kJ/kg. The mass flow rate can be easily calculated based on the inlet Mach number, M_1, and the static conditions there using the one-dimensional uniform flow assumption.

$$\dot{m} = \rho V A = \frac{P}{RT} M\sqrt{kg_cRT}\,A$$

The static temperature and pressure can be determined at the exit of the converging nozzle using either the isentropic tables or the relationships that they are based on. In the analysis below, the static temperature is determined from total temperature and Mach number while the static pressure is determined from the static to total pressure ratio from the isentropic tables.

$$T = \frac{T_T}{\left[1 + \frac{(k-1)}{2}M^2\right]} = \frac{296\ \text{K}}{\left[1 + \frac{0.4}{2}0.54^2\right]} = \frac{296\ \text{K}}{1.0583} = 279.7\ \text{K}$$

$$P = P_T\ \frac{P}{P_T} = 400\ \text{kPa}\ 0.82005 = 328.02\ \text{kPa}$$

The mass flow rate can now be determined based on one-dimensional uniform flow with the density determined using the ideal gas law, the velocity determined from the Mach number times the speed of sound along with the cross-sectional area.

$$\dot{m} = \frac{328{,}020\ \text{Pa}}{287\ \frac{\text{J}}{\text{kg K}}\ 279.7\ \text{K}}\ 0.54\sqrt{1.4(1)287\ \frac{\text{J}}{\text{kg K}}\ 279.7\text{K}}\ \frac{\pi}{4}(0.025\ \text{m})^2$$

$$= 0.363\ \text{kg/s}$$

Analysis for Q_2:

The assumption made in this analysis is that flow through the pipe is subsonic and it is choked at the exit. This situation will again make $T_{T,EXIT} = T_T^*$. The inlet total temperature is 296 K. Initially, the exit total temperatures can be determined.

$$T_{T2} = T_{T1} + \frac{Q}{C_P} = 296\ \text{K} + \frac{250\ \frac{\text{kJ}}{\text{kg}}}{1.005\ \frac{\text{kJ}}{\text{kg K}}} = 544.8\ \text{K} = T_T^*$$

Based on the value of T_{T1}/T_T^*, the inlet Mach number to the duct can be determined.

$$\frac{T_{T1}}{T_T^*} = \frac{296\ \text{K}}{395.5\ \text{K}} = 0.54336$$

The Rayleigh line table for a specific heat ratio, $k = 1.4$, can be interpolated to show that at a Mach number of 0.4082 the following values are determined.

$$M_1 = 0.4082, P\ /\ P^* = 1.946025, T\ /\ T^* = 0.631, V\ /\ V^* = 0.32437,$$
$$T_T\ /\ T_T^* = 0.54336, P_T\ /\ P_T^* = 1.15304$$

Again, the initial check must show that the back pressure is low enough to allow for choked flow. The validity of the result can be determined by checking P^* against the back pressure. Starting with the total pressure and temperature, the local static pressure and temperature can be determined from the Mach number using the isentropic tables ($P/P_T = 0.89194$, $T/T_T = 0.967747$). The exit static pressure can be determined from the ratio, P/P^* in the Rayleigh line table as tabulated above and the static to total pressure ratio.

$$P^* = P_T\ \frac{P}{P_T}\ \frac{P^*}{P} = 400\ \text{kPa}\ \frac{0.89194}{1.946025} = 183.26\ \text{kPa}$$

This demonstrates that the flow with the higher heat addition to the air (Q_2 = 250 kJ/kg) can still achieve sonic flow at the pipe exit. The mass flow rate can be determined in a manner similar to the value for Q_1 as shown above and has been determined to be 0.285 kg/s.

The previous two examples involved heat addition. One relevant question would be what happens when a flow had heat rejection instead of heat addition? Also when might a compressible flow encounter this heat rejection?

Many engineering systems reject heat from high temperature exhaust flows in order to improve the thermodynamic efficiency of a cycle. Two examples include a waste heat recovery steam generator and a gas turbine regenerator. In both cases very hot but low pressure exhaust gas leaves a power turbine at a relatively high Mach number. A Mach number roughly in the range of 0.3 to 0.4 would be reasonable to expect at the exit of a power turbine. Assuming the flow is then directed to a diffuser, a device used to slow the flow down, the diffuser exit Mach number would likely be in the range of 0.2 to 0.27. The exit temperature of a power turbine can easily be as high as 900 K or higher while the exit temperature of a waste heat recovery steam generator is likely to be on the order of 430 K.

Example 9.3: Analysis of Frictionless Pressure Changes During Heat Rejection

In this analysis assume that air enters a heat exchanger at a total temperature of 900 K and a total pressure of 110 kPa as shown schematically in Figure 9.6. Assume that the inlet Mach number is 0.22. Find the exit Mach number and exit total and static pressure if the exit total temperature is 445 K. Assume that the specific heat ratio, k = 1.4, and that the specific heat at constant pressure is 1.005 kJ/kg/K. Find Q.

Analysis: In this analysis, thermal energy is being removed from the system shown in the schematic above. The Rayleigh line figure indicates that for cooling below a Mach number of 0.845 that the static temperature decreases. If Table A.7 is then viewed for guidance it is clear that along with the lower static and total temperatures the Mach number can also be expected to decrease. Based on the table, at the same time, the static and total pressure show an increase. Can this be possible without violating the second law of thermodynamics? The answer of course is that it is possible for the static and total pressure to rise as thermal energy is being rejected to the environment which produces a positive contribution to the production of entropy.

This analysis can start by looking at the tabulated values of the Rayleigh line table at a Mach number of 0.22.

$$M_1 = 0.22, P/P^* = 2.2477, T/T^* = 0.24452, V/V^* = 0.10879,$$
$$T_T/T_T^* = 0.20574, P_T/P_T^* = 1.22814$$

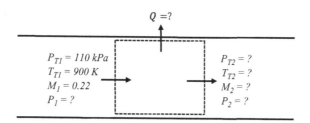

FIGURE 9.6
Analysis of the effect of heat rejection on pressure in steady compressible flow.

The T_T/T_T^* at the exit can be determined using the total temperature ratio.

$$\frac{T_{T2}}{T_T^*} = \frac{T_{T1}}{T_T^*}\frac{T_{T2}}{T_{T1}} = 0.20574 \times \frac{445\,K}{900\,K} = 0.10173$$

Based on the Rayleigh line table, the total temperature ratio suggests a downstream Mach number of approximately 0.15.

$$M_2 = 0.15, P/P^* = 2.2477, T/T^* = 0.12181, V/V^* = 0.05235,$$
$$T_T/T_T^* = 0.10196, P_T/P_T^* = 1.24863$$

The downstream total pressure is very straight forward to find using the total pressure ratios.

$$P_{T2} = P_{T1}\frac{P_T^*}{P_{T1}}\frac{P_{T2}}{P_T^*} = 110\,kPa\,\frac{1.24863}{1.22814} = 111.84\,kPa$$

The downstream static pressure can be determined from the local static to total pressure ratio at a Mach number of 0.15.

$$M_2 = 0.15, P/P_T = 0.984408, T/T_T = 0.99552$$

$$P_2 = P_{T2}\frac{P_2}{P_{T2}} = 111.84\,kPa * 0.984408 = 110.09\,kPa$$

The heat transfer rate per unit mass of flow can be easily determined from the first Law of Thermodynamics. Equation (9.12) can be adapted for our purposes.

$$\frac{\dot{Q}}{\dot{M}} = Q = C_P(T_{T2} - T_{T1}) = 1.005\frac{kJ}{kg\,K}(445\,K - 900\,K) = -457.3\frac{kJ}{kg}$$

Conclusion
The results of this analysis using the concepts of Rayleigh line flow suggest that if a frictionless heat exchanger could be developed that the transfer of thermal energy out of a compressible flow results in the slight

increase of the total pressure. The heat addition process due to combustion can act similarly to a frictionless flow in which very high levels of heat release can occur in a relatively short distance. Chemical heat absorption occurs much less commonly. Largely, thermal energy loss from a system occurs due to heat transfer processes such as in a heat exchanger. In our example, our heat exchanger passed the thermal energy from the hot, low density exhaust gas to the cooler, higher pressure working fluid. Typically, the heat exchanger area on this hot side is very large, which means that the fluid must travel over a very large surface area. Consequently, any heat transfer process rejecting heat from a hot, low pressure exhaust gas is also accompanied by substantial pressure loss due to skin friction. In Section 9.5 the combined influence of heat gain or loss to a system with skin friction is examined.

9.4 Frictionless Flow with Heat Addition and Area Change

Constant area frictionless flow with heat addition was analyzed in Section 9.2. In this section, frictionless flow with heat transfer and area change is examined as shown schematically in Figure 9.7. In the previous development, area was constant which allowed a simplified development. In the current analysis, the relationships of conservation of mass, momentum, energy, the ideal gas law, and the determination of velocity from Mach number and the speed of sound can be employed in differential form. The goal of the analysis is to develop a relationship which relates Mach number area change and heat addition. However, based on the first law, heat addition can easily be described in terms of the change in total temperature. The result of this analysis will be posed in terms of the total temperature.

This analysis can begin with differential conservation of mass which was previously developed in Chapter 2 for a varying area duct as Equation (2.18).

$$\frac{d\rho}{\rho} + \frac{dV}{V} + \frac{dA}{A} = 0 \tag{2.18}$$

Differential momentum was also developed in Chapter 2 as Equation (2.19).

$$dP + \frac{\rho V dV}{g_c} = 0 \tag{2.19}$$

Conservation of energy is written in this chapter in two forms and it can be developed from Equation (9.12) into the following differential form.

$$dq = C_p dT_T \tag{9.23}$$

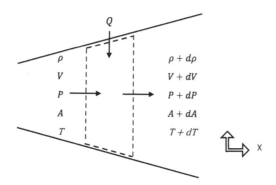

FIGURE 9.7
Control volume for compressible flow with heat addition and area change.

The differentiation of velocity in terms of the Mach number times the speed of sound yields Equation (8.9) from Chapter 8.

$$\frac{dV}{V} = \frac{dM}{M} + \frac{dT}{2T} \qquad (8.9)$$

The ideal gas law in terms of $P = \rho RT$ can also be differentiated yielding the following equation from Chapter 8.

$$\frac{dP}{P} = \frac{d\rho}{\rho} + \frac{dT}{T}$$

The density term $d\rho/\rho$ can be solved for in the above equation and is substituted into our continuity equation.

$$\frac{dP}{P} - \frac{dT}{T} + \frac{dV}{V} + \frac{dA}{A} = 0$$

If the momentum Equation (2.19) is divided by P and solved for dP/P, the pressure term can be eliminated resulting in the following relationship.

$$-\frac{\rho V^2 dV / V}{Pg_c} - \frac{dT}{T} + \frac{dV}{V} + \frac{dA}{A} = 0 \qquad (9.24)$$

From the ideal gas law $\rho/P = 1/RT$ so substituting this into the first term in the equation above and multiplying it by k/k yields

$$-\frac{kM^2 dV}{V} - \frac{dT}{T} + \frac{dV}{V} + \frac{dA}{A} = 0$$

The terms involving velocity can be combined and dV/V can be replaced by Equation (8.9).

$$\frac{(1-kM^2)dM}{M} - \frac{(kM^2+1)dT}{2T} + \frac{dA}{A} = 0 \qquad (9.25)$$

The static temperature term can most ideally be replaced with a similar term with total temperature. Noting that

$$T_T = T\left(1 + \frac{k-1}{2}M^2\right)$$

Differentiating results in

$$\frac{dT_T}{T_T} = \frac{dT}{T} + \frac{(k-1)M^2 \dfrac{dM}{M}}{\left(1 + \dfrac{k-1}{2}M^2\right)}$$

Solving for dT/T and substituting in the Equation (9.25), the relation for the influence on area change and heat transfer is determined.

$$\frac{(1-M^2)\dfrac{dM}{M}}{\left(1 + \dfrac{k-1}{2}M^2\right)} - \frac{(kM^2+1)dT_T}{2T_T} + \frac{dA}{A} = 0 \qquad (9.26)$$

This equation can easily be posed in terms of heat addition, dq, by substituting Equation (9.23) for T_T. However, perhaps the most critical issue related to combustion at high velocities is the loss of total pressure. A useful understanding can be developed from studying Figure 9.3. Looking at the line for P_T/P_T^* at an increasing Mach number the curve disappears above the value of 3.0 before a Mach number of 3 is reached. This means if thermal energy or heat release is added in a constant area duct and driven toward choking more than 2/3rds of the total pressure will be lost in the process. The question in this section becomes how does area change influence total pressure loss with heat addition in high-speed flows? The simplest analysis would be for a flow with a constant Mach number resulting in the following equation.

$$\frac{dA}{A} = \frac{(1+kM^2)dT_T}{2T_T} \qquad (9.27)$$

The momentum equation can provide some insight into pressure change. Dividing the momentum Equation (2.19) by P the equation can be rewritten.

$$\frac{dP}{P} + kM^2\frac{dV}{V} = 0 \qquad (9.28)$$

Using the relationship that velocity is equal to the Mach number times the speed of sound for a constant Mach number:

$$\frac{dV}{V} = \frac{dT}{2T}$$

When the Mach number is constant then:

$$\frac{dT_T}{T_T} = \frac{dT}{T} \text{ and also } \frac{dP_T}{P_T} = \frac{dP}{P}$$

Substituting these relationships into Equation (9.28) the following result is arrived at:

$$\frac{dP_T}{P_T} = -\frac{kM^2}{2}\frac{dT_T}{T_T} \tag{9.29}$$

This relationship suggests that total pressure loss is increasingly parasitic with increasing Mach number at a constant Mach number which is problematic for high-speed flows with heat addition. One flow situation with high-speed heat addition would be combustion in a scramjet or supersonic combustion ram jet. This propulsion method has often been proposed for hypersonic vehicles. This issue of significant pressure drop is a critical one for scramjet development. The key concept behind scramjets is to reduce the total pressure loss associated with slowing down the flow for inlets of more conventional ramjets or low bypass turbofan engines. However, noting that heat addition at higher Mach numbers is also associated with very significant total pressure loss, the challenge of designing an efficient scramjet is substantial. Additionally, although efforts to develop pressure gain combustion in turbomachinery has been a topic of research for decades, the development of a practical system has remained elusive.

Example 9.4: Constant Mach Number Pressure Loss in a Scramjet

A scramjet has been designed to fly at Mach number of 5.5 with combustion designed to occur at a constant Mach number of 2.0. If the limiting combustion temperature is 2000 K, determine the total pressure loss associated with combustion. Assume that that standard day temperature at altitude is 220 K.

Analysis: Integrating Equation (9.29) assuming a constant Mach number then taking the exponential of both sides results in the following equation.

$$\frac{P_{T2}}{P_{T1}} = \left(\frac{T_{T2}}{T_{T1}}\right)^{-kM^2/2} \tag{9.30}$$

This equation can be applied to the present situation by first determining the inlet total temperature.

$$T_{T1} = T_1\left(1 + \frac{k-1}{2}M^2\right) = 220\,\text{K}\left(1 + \frac{0.4}{2}5.5^2\right) = 1551\,\text{K}$$

Noting $T_{T2} = 2000\,\text{K}$ the final to initial total pressure in the combustion system can be determined.

$$\frac{P_{T2}}{P_{T1}} = \left(\frac{T_{T2}}{T_{T1}}\right)^{-kM^2\!/2} = \left(\frac{2000\,\text{K}}{1551\,\text{K}}\right)^{-1.4 \cdot 2^2\!/2} = 0.4907$$

Consequently, the combustion at this constant Mach number of 2 results in a total pressure loss of over 50 percent of the combustor inlet total pressure. This pressure loss does not include the pressure loss of the inlet diffuser. This problem highlights one of the significant challenges related to high-speed flight and the required propulsion.

9.5 Constant Area Flow with Heat Transfer and Friction

Previous sections examine frictionless flow with heat addition and rejection. In many cases where heat addition is relatively rapid due to chemical heat release, the frictionless flow approximation has useful validity. However, in many cases where convective heat transfer occurs through the wall of a tube, flow with friction, in addition to heat transfer, is present. In Chapter 8, flow with friction in constant area ducts was analyzed and equations derived from control volumes were developed. These equations were integrated to develop the Fanno-line flow. The equations for constant area flow with heat transfer and skin friction have similarities to the equations developed in Chapter 8 as well as equations developed in this chapter for heat addition. The analysis can begin with the schematic shown in Figure 9.8, showing a constant area flow with both heat transfer and skin friction. In the current analysis, the heat transfer occurs due to convection. However, this analysis

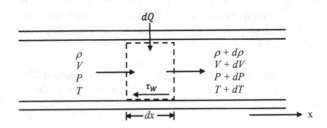

FIGURE 9.8
Compressible flow in a constant area duct with heat transfer and friction.

can begin in a manner similar to the development of flow with skin friction in a constant area duct.

This analysis can begin with conservation of mass in a constant area duct and the momentum equation with skin friction by borrowing Equations (8.5) and (8.6) from Chapter 8.

$$0 = \frac{d\rho}{\rho} + \frac{dV}{V} \tag{8.5}$$

$$dP + 4\tau_W \frac{dx}{D} + \frac{\rho V dV}{g_C} = 0 \tag{8.6}$$

The momentum equation in Chapter 8 is updated to replace the shear stress, τ_W, with the friction factor, f, replacing the velocity squared terms with Mach number resulting in Equation (8.8).

$$\frac{dP}{P} + f\frac{kM^2}{2}\frac{dx}{D} + kM^2\frac{dV}{V} = 0 \tag{8.8}$$

The pressure term can be eliminated by differentiating ideal gas and substituting, and the velocity term can be eliminated using the relationship for velocity as Mach number times the speed of sound as shown in Equation (8.9). These substitutions result in Equation (8.11).

$$\left(kM^2 - 1\right)\frac{dM}{M} + \frac{\left(kM^2 + 1\right)}{2}\frac{dT}{T} + f\frac{kM^2}{2}\frac{dx}{D} = 0 \tag{8.11}$$

The influence of the heat addition term can be developed by taking the definition of total temperature and differentiating to develop the relationship in terms of total temperature and Mach number to substitute for the static temperature term above. This approach was shown in the previous section. With this substitution Equation (8.11) becomes

$$\frac{\left(M^2 - 1\right)}{\left(1 + \frac{k-1}{2}M^2\right)}\frac{dM}{M} + \frac{\left(kM^2 + 1\right)}{2}\frac{dT_T}{T_T} + f\frac{kM^2}{2}\frac{dx}{D} = 0 \tag{9.31}$$

The use of the total temperature allows a means to account for heat addition. Heat addition in the current analysis is due to convection. The appropriate substitution can be developed starting with Equation (9.23).

$$dT_T = \frac{dq}{C_P} \tag{9.23}$$

The heat addition rate needs to be developed in terms of convection heat transfer. Based on the control volume in Figure 9.7, the heat transfer rate is based on heat addition to the fluid from the wall.

$$\dot{q} = hA_S\left(T_W - T_T\right) \tag{9.32}$$

Here, A_S is the internal surface area of the tube which is the same area that is associated with skin friction. The heat transfer coefficient, h, (not to be confused with enthalpy) times the temperature difference is used to estimate the heat flux. However, this heat addition is shown as a heat addition rate and needs to be developed in terms of a differential addition.

$$dq = \frac{d\dot{q}}{\dot{m}} = \frac{h\pi D dx\left(T_W - T_T\right)}{\rho A V} \tag{9.33}$$

Here, the surface area is written as the perimeter of the tube, πD, times the differential distance along the tube, dx, providing a differential heat addition. In terms of dT_T, this relationship can be written as

$$dT_T = \frac{d\dot{q}}{\dot{m}C_P} = \frac{h\pi D dx\left(T_W - T_T\right)}{\rho V C_P \pi D^2\big/4} = \frac{h}{\rho C_P V}\frac{\pi D dx}{\pi D^2\big/4}\left(T_W - T_T\right) \tag{9.34}$$

The term $h/\rho V C_P$ is the Stanton number, St. In the case of fully developed flow in pipes for gases where the Prandtl number is close to one, the Stanton number is typically very close to the skin friction coefficient, $Cf/2$. Also, the skin friction coefficient, $Cf/2$, is equal to the friction factor divided by 8, $Cf/2 = f/8$. Consequently, if the Stanton number ($h/\rho V C_P$) is replaced by the friction factor divided by 8 ($f/8$), then Equation (9.34) can be rewritten in the following terms.

$$dT_T = \frac{f}{2}\frac{dx}{D}\left(T_W - T_T\right) \tag{9.35}$$

This relationship can be substituted into Equation (9.31) with the following results.

$$\frac{\left(M^2 - 1\right)}{\left(1 + \frac{k-1}{2}M^2\right)}\frac{dM}{M} + \frac{\left(kM^2 + 1\right)}{4}f\frac{dx}{D}\frac{\left(T_W - T_T\right)}{T_T} + f\frac{kM^2}{2}\frac{dx}{D} = 0 \tag{9.36}$$

It should be noted that if the heat transfer term is eliminated, Equation (9.36) can be shown to be equivalent to Equation (8.14). The benefit of Equation (9.36) is that it shows the relative importance of the heat transfer and skin friction terms. Also it should be noted that both heat addition and skin friction cause a total pressure drop and drive both a subsonic and a supersonic flow toward a Mach number of one, while heat rejection should drive the flow in the opposite direction. As an example, take a flow with a Mach number of 0.05, a $(T_W\text{-}T_T) = 20$ K, and a $T_T = 300$ K. Looking at the value of the term multiplying $f\,dx/D$ for the skin friction term (last), $kM^2/2 = 0.00175$. The heat transfer term multiplier $[kM^2/4^*(T_W\text{-}T_T)/T_T]$ is equal to 0.0000583. The $[1/4^*(T_W\text{-}T_T)/T_T]$ term associated with the heat transfer term does not typically have a significant effect on pressure drop. This comparison suggests that for this situation, the skin friction term would have a much larger influence on pressure drop and the increase in velocity of a flow. Also one issue to keep in mind is that as the heat transfer process is occurring, if the wall temperature is constant, then the term $(T_W\text{-}T_T)/T_T$ would decrease as flow moved through the tube further diminishing the overall impact of heat transfer on the flow relative to skin friction. Consequently, this simple analysis suggests that when a convection process takes place, the skin friction terms dominates unless the term $(T_W\text{-}T_T)/T_T$ is order one or larger. However, elevated temperature differences in heat exchangers equate to a significant level of exergy destruction. As a result in heat exchanger design, velocities are typically kept relatively low to reduce pressure losses.

Example 9.5: Comparison between Pressure Drop Due to Skin Friction and Pressure Drop Due to Heat Transfer

A small heat exchanger is constructed from a tube within a tube. The air flows within the inner tube while the wall of the tube is maintained using low pressure steam at atmospheric pressure. This maintains the wall at 373.15 K. The air flows into the tube at a Mach number of 0.07 at a temperature of 300 K. Determine the pressure drop independently due to skin friction and due to heat transfer. The pipe is 1.0 m long with an internal diameter of 0.025 m.

Analysis: The Mach number is quite low and so is fL/D. An accurate and straight forward method in this situation to calculate the pressure loss is using the Darcy–Weisbach equation. It can be written in compressible form as

$$\Delta P = f\frac{L}{D}\frac{kM^2}{2}P_\infty \tag{9.37}$$

Note the similarity to the last term in Equation (9.31) which represents the friction loss. In this problem the friction factor, f, can be assumed to be 0.022. The friction factor generally depends on the flow Reynolds number and roughness and most fluid mechanic textbooks provide straight forward methods to determine f.

$$\Delta P = f\frac{L}{D}\frac{kM^2}{2}P_\infty = 0.022\frac{1.0\,\text{m}}{0.025\,\text{m}}\frac{1.4\,0.07^2}{2}101325\,\text{Pa} = 305.8\,\text{Pa}$$

The pressure drop due to heat addition can be determined using the Rayleigh line flow table. In this problem, the tube wall can be taken to be the condensing temperature of the low pressure steam, 373.15 K. The heat transfer coefficient (h) can be estimated using the Dittus–Boelter correlation [3]. For heating this equation can be described as:

$$Nu_D = \frac{hD}{k} = 0.023\, Re_D^{0.8} Pr^{0.4}$$

The Prandtl number (Pr) for air is about 0.71. The thermal conductivity (k) of air is around 0.02614 W/m/K at 300 K. The viscosity (μ) of air is 0.00001846 Pa/s at 300 K. At 300 K the speed of sound (a) is 347.2 m/s so at a Mach number of 0.07 the velocity would be 24.3 m/s. At 1 atmosphere and 300 K the density (ρ) of air is 1.177 kg/m³. The resulting Reynolds number can be determined:

$$Re_D = \frac{\rho VD}{\mu} = \frac{1.177\,\frac{kg}{m^3}\,24.3\,\frac{m}{s}\,0.025\,m}{0.00001846\,Pa\,s} = 38{,}733$$

Calculating the Nusselt number:

$$Nu_D = 0.023\, Re_D^{0.8} Pr^{0.8} = 0.023 \quad 38{,}733^{0.8}\; 0.71^4 = 93.9$$

The resulting heat transfer coefficient becomes

$$h = \frac{k}{D}\, Nu_D = \frac{0.02614\,\frac{W}{m\,K}}{0.025\,m}\,93.9 = 98.15\,\frac{W}{m^2K}$$

Note that as the air moves through the tube it increases in temperature and the wall to fluid temperature diminishes reducing the rate of heat transfer. This rising temperature must be accounted for in determining the air temperature rise. The temperature rise is related to the heat transfer rate times the area as well as the fluid flow thermal capacitance. The ratio of these two values is called NTU or the number of heat transfer units.

$$NTU = \frac{hA_s}{\dot{m}C_P} = \frac{h\pi DL}{\rho V \frac{\pi D^2}{4}} = \frac{98.15\,\frac{W}{m^2K}\,\pi\,0.025\,m\,1.0\,m}{1.177\,\frac{kg}{m^3}\,24.3\,\frac{m}{s}\,\pi\,\frac{(0.025\,m)^2}{4}} = 0.5464$$

In the current case, the tube wall temperature would be expected to stay constant as the heat transfer coefficient on the steam side is expected to be much higher than the air side. This heat exchanger would be called an infinite capacitance heat exchanger and the effectiveness can be determined as shown below:

$$\varepsilon = \frac{T_{T,OUT} - T_{T,IN}}{T_W - T_{T,IN}} = 1 - exp(-NTU) = 1 - exp(-0.5464) = 0.4209$$

The temperature rise then is the effectiveness, ε, times the wall to fluid inlet temperature difference.

$$T_{T,OUT} - T_{T,IN} = \varepsilon (T_W - T_{T,IN}) = 0.4209(373.15\,\text{K} - 300\,\text{K}) = 30.79\,\text{K}$$

This temperature difference then increases the total temperature. At this point, the inlet static temperature has been given. At this low Mach number, there is little difference between the static and total temperatures but for completeness, $T_{T,IN}$, is calculated below.

$$T_{T,IN} = T_{IN}\left(1 + \frac{k-1}{2}M^2\right) = 300\,\text{K} \times \left(1 + \frac{k-1}{2}0.07^2\right) = 300.29\,\text{K}$$

The outlet total temperature is simply $T_{T,IN}$ plus the temperature rise in the heat exchanger.

$$T_{T,OUT} = T_{T,IN} + \varepsilon (T_W - T_{T,IN}) = 300.29\,\text{K} + 30.79\,\text{K} = 331.08\,\text{K}$$

Noting the inlet Mach number is 0.07 into the heat exchanger, the Rayleigh line table provides a $T_T/T_T^* = 0.02322$ and a P_T/P_T^* of 1.26536. Noting the inlet and outlet total temperature of the heat exchanger:

$$\frac{T_{T2}}{T_T^*} = \frac{T_{T1}}{T_T^*}\frac{T_{T2}}{T_{T1}} = 0.02322 \times \frac{331.08\,\text{K}}{300.29\,\text{K}} = 0.02560$$

Based on interpolation, $M_2 = 0.07356$ resulting in P_{T2}/P_T^* of 1.26312. The total pressure loss can now be determined. However, first the inlet total pressure needs to be calculated.

$$P_{T,IN} = P_{IN}\left(1 + \frac{k-1}{2}M^2\right)^{k-\frac{1}{k}} = 101325\,\text{Pa} \times \left(1 + \frac{k-1}{2}0.07^2\right)^{\frac{1.4}{0.4}} = 101673\,\text{Pa}$$

At this point the downstream total pressure can be determined based on the upstream and downstream total pressure ratios.

$$P_{T2} = P_{T1}\frac{P_T^*}{P_{T1}}\frac{P_{T2}}{P_T^*} = 101637\,\text{Pa}\frac{1.26312}{1.26536} = 101637\,\text{Pa}$$

The total pressure loss due to the heat addition is only 36 Pa, while the total pressure loss due to skin friction is about 306 Pa. Equation (9.36) suggested that the pressure loss due to skin friction is on the order of $f\,kM^2/2$ while the loss due to heat transfer should be on the order of $[fkM^2/4*(T_W-T_T)/T_T]$. Earlier, the compressible flow version of Darcy–Weisbach equation was expressed as:

$$\Delta P = f \frac{L}{D} \frac{kM^2}{2} P_\infty$$

However, note that this formula is only strictly applicable to Mach numbers below 0.2 to 0.3. Above these Mach numbers, Fanno-line flow or CFD should be applied. Earlier this formula was used to estimate the pressure losses due to skin friction. If differs from the term above by L/DP_∞. If this addition is applied to the heat addition term the pressure loss estimate is

$$\frac{\left(kM^2\right)}{4} f \frac{L}{D} \frac{\left(T_W - T_T\right)}{T_T} P_\infty = \frac{\left(1.4*0.07^2\right)}{4} 0.022 \frac{1.0\,\text{m}}{0.025\,\text{m}} \frac{73.15\,\text{K}}{300.29\,\text{K}} 101325\,\text{Pa} = 37\,\text{Pa}$$

This estimate is close to the calculated value of 36 Pa. Note that Equation (9.36) needs to be solved as an integral so the estimate above is only an estimate and should not be thought of as an accurate approach.

References

1. John, J.E.A. and Keith, T.G., 2006, *Gas Dynamics*, 3rd ed., Prentice Hall.
2. Saad, M.A., 1993, *Compressible Fluid Flow*, 2nd ed., Prentice Hall.
3. Kreith, F., and Bohn, M.S., 2001, *Principles of Heat Transfer*, 6th ed., Brooks/Cole.

Chapter 9 Problems

1. Airflow enters a combustion system at 504 K and a velocity of 65 m/s. The air is mixed with propane and burned increasing the enthalpy of by 790 kJ/kg. The C_P of these products can be assumed to average 1.08 kJ/kg/K for this temperature rise in the gas. You can assume that the specific heat ratio is 1.4. Determine the downstream conditions $(T_2, P_2, M_2, P_{T2}, T_{T2})$.

2. A compressor system supplies air at a total pressure of 350 kPa and a total temperature of 340 K. The reservoir is attached to a converging nozzle contracting into a 0.05 m diameter tube. The back pressure of the system is 101 kPa. Heat addition of 300 kJ/kg is added to the flow. The specific heat ratio for the air is assumed to be 1.02 kJ/kg/K. Calculate the mass flow rate, inlet Mach number, and exit

static pressure for this case. The exit static pressure must equal or exceed the back pressure for choked flow ($M = 1$) at the exit.

3. A compressor system supplies air at a total pressure of 200 kPa and a total temperature of 450 K. The reservoir is attached to a converging nozzle contracting into a 0.035 m diameter tube. The back pressure of the system is 100 kPa. Heat rejection of 150 kJ/kg is removed from the flow. The specific heat ratio for the air can is assumed to be 1.02 kJ/kg/K. Determine the mass flow rate, exit Mach number, and exit static pressure for this case. Refer to the Rayleigh line in thinking about this problem.

4. Air enters a heat exchanger at a total temperature of 1000 K and a total pressure of 200 kPa. Assume that the inlet Mach number is 0.27. Find the exit Mach number and exit total and static pressure if the exit total temperature is 525 K assuming the flow is frictionless. Assume that the specific heat ratio, $k = 1.4$, and that the specific heat at constant pressure is 1.05 kJ/kg/K. Find Q.

5. A scramjet has been designed to fly at Mach 5.6 with combustion design to occur at a constant Mach number of 1.8 due to the shape of the system in the heat addition region. If the limiting combustion temperature is 2100 K, determine the total pressure loss associated with combustion. Assume that that standard day temperature at altitude is 220 K.

6. A small heat exchanger is constructed from a tube within a tube. Gas flows through a small heat exchanger consisting of a tube held at constant temperature due to the condensation of low pressure steam. This maintains the wall at 400 K. The air flows into the tube at a Mach number of 0.08 at a temperature of 320 K. Determine the pressure drop independently due to skin friction and due to heat transfer. The pipe is 1.0 m long with an internal diameter of 0.02 m.

Appendix

A.1 Isentropic Mach Number Tables

TABLE A.1.1
Isentropic Mach number, k = 1.4

M	P/Pt	T/Tt	A/A*	M	P/Pt	T/Tt	A/A*	M	P/Pt	T/Tt	A/A*
0.00	1.00000	1.00000	infinity	0.50	0.84302	0.95238	1.33984	1.00	0.52828	0.83333	1.00000
0.01	0.99993	0.99998	57.8738	0.51	0.83737	0.95055	1.32117	1.01	0.52213	0.83055	1.00008
0.02	0.99972	0.99992	28.9421	0.52	0.83165	0.94869	1.30339	1.02	0.51602	0.82776	1.00033
0.03	0.99937	0.99982	19.3005	0.53	0.82588	0.94681	1.28645	1.03	0.50994	0.82496	1.00074
0.04	0.99888	0.99968	14.4815	0.54	0.82005	0.94489	1.27032	1.04	0.50389	0.82215	1.00131
0.05	0.99825	0.99950	11.5914	0.55	0.81417	0.94295	1.25495	1.05	0.49787	0.81934	1.00203
0.06	0.99748	0.99928	9.66591	0.56	0.80823	0.94098	1.24029	1.06	0.49189	0.81651	1.00291
0.07	0.99658	0.99902	8.29153	0.57	0.80224	0.93898	1.22633	1.07	0.48595	0.81368	1.00394
0.08	0.99553	0.99872	7.26161	0.58	0.79621	0.93696	1.21301	1.08	0.48005	0.81085	1.00512
0.09	0.99435	0.99838	6.46134	0.59	0.79013	0.93491	1.20031	1.09	0.47418	0.80800	1.00645
0.10	0.99303	0.99800	5.82183	0.60	0.78400	0.93284	1.18820	1.10	0.46835	0.80515	1.00793
0.11	0.99158	0.99759	5.29923	0.61	0.77784	0.93073	1.17665	1.11	0.46257	0.80230	1.00955
0.12	0.98998	0.99713	4.86432	0.62	0.77164	0.92861	1.16565	1.12	0.45682	0.79944	1.01131
0.13	0.98826	0.99663	4.49686	0.63	0.76540	0.92646	1.15515	1.13	0.45111	0.79657	1.01322
0.14	0.98640	0.99610	4.18240	0.64	0.75913	0.92428	1.14515	1.14	0.44545	0.79370	1.01527
0.15	0.98441	0.99552	3.91034	0.65	0.75283	0.92208	1.13562	1.15	0.43983	0.79083	1.01745
0.16	0.98228	0.99491	3.67274	0.66	0.74650	0.91986	1.12654	1.16	0.43425	0.78795	1.01978
0.17	0.98003	0.99425	3.46351	0.67	0.74014	0.91762	1.11789	1.17	0.42872	0.78506	1.02224
0.18	0.97765	0.99356	3.27793	0.68	0.73376	0.91535	1.10965	1.18	0.42322	0.78218	1.02484
0.19	0.97514	0.99283	3.11226	0.69	0.72735	0.91306	1.10182	1.19	0.41778	0.77929	1.02757
0.20	0.97250	0.99206	2.96352	0.70	0.72093	0.91075	1.09437	1.20	0.41238	0.77640	1.03044
0.21	0.96973	0.99126	2.82929	0.71	0.71448	0.90841	1.08729	1.21	0.40702	0.77350	1.03344
0.22	0.96685	0.99041	2.70760	0.72	0.70803	0.90606	1.08057	1.22	0.40171	0.77061	1.03657
0.23	0.96383	0.98953	2.59681	0.73	0.70155	0.90369	1.07419	1.23	0.39645	0.76771	1.03983
0.24	0.96070	0.98861	2.49556	0.74	0.69507	0.90129	1.06814	1.24	0.39123	0.76481	1.04323
0.25	0.95745	0.98765	2.40271	0.75	0.68857	0.89888	1.06242	1.25	0.38606	0.76190	1.04675
0.26	0.95408	0.98666	2.31729	0.76	0.68207	0.89644	1.05700	1.26	0.38093	0.75900	1.05041
0.27	0.95060	0.98563	2.23847	0.77	0.67556	0.89399	1.05188	1.27	0.37586	0.75610	1.05419
0.28	0.94700	0.98456	2.16555	0.78	0.66905	0.89152	1.04705	1.28	0.37083	0.75319	1.05810
0.29	0.94329	0.98346	2.09793	0.79	0.66254	0.88903	1.04251	1.29	0.36585	0.75029	1.06214
0.30	0.93947	0.98232	2.03507	0.80	0.65602	0.88652	1.03823	1.30	0.36091	0.74738	1.06630
0.31	0.93554	0.98114	1.97651	0.81	0.64951	0.88400	1.03422	1.31	0.35603	0.74448	1.07060
0.32	0.93150	0.97993	1.92185	0.82	0.64300	0.88146	1.03046	1.32	0.35119	0.74158	1.07502
0.33	0.92736	0.97868	1.87074	0.83	0.63650	0.87890	1.02696	1.33	0.34640	0.73867	1.07957
0.34	0.92312	0.97740	1.82288	0.84	0.63000	0.87633	1.02370	1.34	0.34166	0.73577	1.08424
0.35	0.91877	0.97609	1.77797	0.85	0.62351	0.87374	1.02067	1.35	0.33697	0.73287	1.08904
0.36	0.91433	0.97473	1.73578	0.86	0.61703	0.87114	1.01787	1.36	0.33233	0.72997	1.09396
0.37	0.90979	0.97335	1.69609	0.87	0.61057	0.86852	1.01530	1.37	0.32773	0.72707	1.09902
0.38	0.90516	0.97193	1.65870	0.88	0.60412	0.86589	1.01294	1.38	0.32319	0.72418	1.10419
0.39	0.90043	0.97048	1.62343	0.89	0.59768	0.86324	1.01080	1.39	0.31869	0.72128	1.10950
0.40	0.89561	0.96899	1.59014	0.90	0.59126	0.86059	1.00886	1.40	0.31424	0.71839	1.11493
0.41	0.89071	0.96747	1.55867	0.91	0.58486	0.85791	1.00713	1.41	0.30984	0.71550	1.12048
0.42	0.88572	0.96592	1.52890	0.92	0.57848	0.85523	1.00560	1.42	0.30549	0.71262	1.12616
0.43	0.88065	0.96434	1.50072	0.93	0.57211	0.85253	1.00426	1.43	0.30118	0.70973	1.13197
0.44	0.87550	0.96272	1.47401	0.94	0.56578	0.84982	1.00311	1.44	0.29693	0.70685	1.13790
0.45	0.87027	0.96108	1.44867	0.95	0.55946	0.84710	1.00215	1.45	0.29272	0.70398	1.14396
0.46	0.86496	0.95940	1.42463	0.96	0.55317	0.84437	1.00136	1.46	0.28856	0.70110	1.15015
0.47	0.85958	0.95769	1.40180	0.97	0.54691	0.84162	1.00076	1.47	0.28445	0.69824	1.15646
0.48	0.85413	0.95595	1.38010	0.98	0.54067	0.83887	1.00034	1.48	0.28039	0.69537	1.16290
0.49	0.84861	0.95418	1.35947	0.99	0.53446	0.83611	1.00008	1.49	0.27637	0.69251	1.16947

(Continued)

TABLE A.1.1 (*Continued*)

Isentropic Mach number, k = 1.4

M	P/Pt	T/Tt	A/A*	M	P/Pt	T/Tt	A/A*	M	P/Pt	T/Tt	A/A*
1.50	0.27240	0.68966	1.17617	2.00	0.12780	0.55556	1.68750	2.50	0.05853	0.44444	2.63672
1.51	0.26848	0.68680	1.18299	2.01	0.12583	0.55309	1.70165	2.51	0.05762	0.44247	2.66146
1.52	0.26461	0.68396	1.18994	2.02	0.12389	0.55064	1.71597	2.52	0.05674	0.44051	2.68645
1.53	0.26078	0.68112	1.19702	2.03	0.12197	0.54819	1.73047	2.53	0.05586	0.43856	2.71171
1.54	0.25700	0.67828	1.20423	2.04	0.12009	0.54576	1.74514	2.54	0.05500	0.43662	2.73723
1.55	0.25326	0.67545	1.21157	2.05	0.11823	0.54333	1.75999	2.55	0.05415	0.43469	2.76301
1.56	0.24957	0.67262	1.21904	2.06	0.11640	0.54091	1.77502	2.56	0.05332	0.43277	2.78906
1.57	0.24593	0.66980	1.22664	2.07	0.11460	0.53851	1.79022	2.57	0.05250	0.43085	2.81538
1.58	0.24233	0.66699	1.23438	2.08	0.11282	0.53611	1.80561	2.58	0.05169	0.42895	2.84197
1.59	0.23878	0.66418	1.24224	2.09	0.11107	0.53373	1.82119	2.59	0.05090	0.42705	2.86884
1.60	0.23527	0.66138	1.25024	2.10	0.10935	0.53135	1.83694	2.60	0.05012	0.42517	2.89598
1.61	0.23181	0.65858	1.25836	2.11	0.10766	0.52898	1.85289	2.61	0.04935	0.42329	2.92339
1.62	0.22839	0.65579	1.26663	2.12	0.10599	0.52663	1.86902	2.62	0.04859	0.42143	2.95109
1.63	0.22501	0.65301	1.27502	2.13	0.10434	0.52428	1.88533	2.63	0.04784	0.41957	2.97907
1.64	0.22168	0.65023	1.28355	2.14	0.10273	0.52194	1.90184	2.64	0.04711	0.41772	3.00733
1.65	0.21839	0.64746	1.29222	2.15	0.10113	0.51962	1.91854	2.65	0.04639	0.41589	3.03588
1.66	0.21515	0.64470	1.30102	2.16	0.09956	0.51730	1.93544	2.66	0.04568	0.41406	3.06472
1.67	0.21195	0.64194	1.30996	2.17	0.09802	0.51499	1.95252	2.67	0.04498	0.41224	3.09385
1.68	0.20879	0.63919	1.31904	2.18	0.09649	0.51269	1.96981	2.68	0.04429	0.41043	3.12327
1.69	0.20567	0.63645	1.32825	2.19	0.09500	0.51041	1.98729	2.69	0.04362	0.40863	3.15299
1.70	0.20259	0.63371	1.33761	2.20	0.09352	0.50813	2.00497	2.70	0.04295	0.40683	3.18301
1.71	0.19956	0.63099	1.34710	2.21	0.09207	0.50586	2.02286	2.71	0.04229	0.40505	3.21333
1.72	0.19656	0.62827	1.35674	2.22	0.09064	0.50361	2.04094	2.72	0.04165	0.40328	3.24395
1.73	0.19361	0.62556	1.36651	2.23	0.08923	0.50136	2.05923	2.73	0.04102	0.40151	3.27488
1.74	0.19070	0.62285	1.37643	2.24	0.08785	0.49912	2.07773	2.74	0.04039	0.39976	3.30611
1.75	0.18782	0.62016	1.38649	2.25	0.08648	0.49689	2.09644	2.75	0.03978	0.39801	3.33766
1.76	0.18499	0.61747	1.39670	2.26	0.08514	0.49468	2.11535	2.76	0.03917	0.39627	3.36952
1.77	0.18219	0.61479	1.40705	2.27	0.08382	0.49247	2.13447	2.77	0.03858	0.39454	3.40169
1.78	0.17944	0.61211	1.41755	2.28	0.08251	0.49027	2.15381	2.78	0.03799	0.39282	3.43418
1.79	0.17672	0.60945	1.42819	2.29	0.08123	0.48809	2.17336	2.79	0.03742	0.39111	3.46699
1.80	0.17404	0.60680	1.43898	2.30	0.07997	0.48591	2.19313	2.80	0.03685	0.38941	3.50012
1.81	0.17140	0.60415	1.44992	2.31	0.07873	0.48374	2.21312	2.81	0.03629	0.38771	3.53358
1.82	0.16879	0.60151	1.46101	2.32	0.07751	0.48158	2.23332	2.82	0.03574	0.38603	3.56737
1.83	0.16622	0.59888	1.47225	2.33	0.07631	0.47944	2.25375	2.83	0.03520	0.38435	3.60148
1.84	0.16369	0.59626	1.48365	2.34	0.07512	0.47730	2.27440	2.84	0.03467	0.38268	3.63593
1.85	0.16119	0.59365	1.49519	2.35	0.07396	0.47517	2.29528	2.85	0.03415	0.38102	3.67072
1.86	0.15873	0.59104	1.50689	2.36	0.07281	0.47305	2.31638	2.86	0.03363	0.37937	3.70584
1.87	0.15631	0.58845	1.51875	2.37	0.07168	0.47095	2.33771	2.87	0.03312	0.37773	3.74131
1.88	0.15392	0.58586	1.53076	2.38	0.07057	0.46885	2.35928	2.88	0.03263	0.37610	3.77711
1.89	0.15156	0.58329	1.54293	2.39	0.06948	0.46676	2.38107	2.89	0.03213	0.37447	3.81327
1.90	0.14924	0.58072	1.55526	2.40	0.06840	0.46468	2.40310	2.90	0.03165	0.37286	3.84977
1.91	0.14695	0.57816	1.56774	2.41	0.06734	0.46262	2.42537	2.91	0.03118	0.37125	3.88662
1.92	0.14470	0.57561	1.58039	2.42	0.06630	0.46056	2.44787	2.92	0.03071	0.36965	3.92383
1.93	0.14247	0.57307	1.59320	2.43	0.06527	0.45851	2.47061	2.93	0.03025	0.36806	3.96139
1.94	0.14028	0.57054	1.60617	2.44	0.06426	0.45647	2.49360	2.94	0.02980	0.36647	3.99932
1.95	0.13813	0.56802	1.61931	2.45	0.06327	0.45444	2.51683	2.95	0.02935	0.36490	4.03760
1.96	0.13600	0.56551	1.63261	2.46	0.06229	0.45242	2.54031	2.96	0.02891	0.36333	4.07625
1.97	0.13390	0.56301	1.64608	2.47	0.06133	0.45041	2.56403	2.97	0.02848	0.36177	4.11527
1.98	0.13184	0.56051	1.65972	2.48	0.06038	0.44841	2.58801	2.98	0.02805	0.36022	4.15466
1.99	0.12981	0.55803	1.67352	2.49	0.05945	0.44642	2.61224	2.99	0.02764	0.35868	4.19443

TABLE A.1.2

Isentropic Mach number, k = 1.3

M	P/Pt	T/Tt	A/A*	M	P/Pt	T/Tt	A/A*	M	P/Pt	T/Tt	A/A*
0.00	1.00000	1.00000	infinity	0.50	0.85255	0.96386	1.34785	1.00	0.54573	0.86957	1.00000
0.01	0.99994	0.99999	58.5261	0.51	0.84717	0.96245	1.32884	1.01	0.53957	0.86729	1.00009
0.02	0.99974	0.99994	29.2681	0.52	0.84174	0.96102	1.31073	1.02	0.53344	0.86501	1.00034
0.03	0.99942	0.99987	19.5177	0.53	0.83624	0.95957	1.29347	1.03	0.52733	0.86271	1.00077
0.04	0.99896	0.99976	14.6442	0.54	0.83068	0.95809	1.27703	1.04	0.52126	0.86041	1.00136
0.05	0.99838	0.99963	11.7214	0.55	0.82506	0.95659	1.26136	1.05	0.51521	0.85809	1.00212
0.06	0.99766	0.99946	9.77400	0.56	0.81939	0.95507	1.24642	1.06	0.50919	0.85577	1.00304
0.07	0.99682	0.99927	8.38398	0.57	0.81367	0.95353	1.23216	1.07	0.50320	0.85344	1.00412
0.08	0.99585	0.99904	7.34230	0.58	0.80790	0.95196	1.21857	1.08	0.49724	0.85109	1.00536
0.09	0.99475	0.99879	6.53287	0.59	0.80207	0.95038	1.20561	1.09	0.49132	0.84874	1.00676
0.10	0.99353	0.99850	5.88600	0.60	0.79620	0.94877	1.19324	1.10	0.48542	0.84638	1.00831
0.11	0.99217	0.99819	5.35736	0.61	0.79029	0.94714	1.18145	1.11	0.47957	0.84401	1.01002
0.12	0.99069	0.99784	4.91740	0.62	0.78433	0.94548	1.17020	1.12	0.47374	0.84164	1.01188
0.13	0.98909	0.99747	4.54566	0.63	0.77834	0.94381	1.15947	1.13	0.46795	0.83925	1.01389
0.14	0.98736	0.99707	4.22751	0.64	0.77230	0.94212	1.14924	1.14	0.46220	0.83686	1.01605
0.15	0.98551	0.99664	3.95224	0.65	0.76623	0.94040	1.13949	1.15	0.45649	0.83446	1.01836
0.16	0.98353	0.99617	3.71181	0.66	0.76012	0.93867	1.13019	1.16	0.45081	0.83206	1.02081
0.17	0.98143	0.99568	3.50007	0.67	0.75399	0.93691	1.12134	1.17	0.44518	0.82964	1.02342
0.18	0.97921	0.99516	3.31225	0.68	0.74782	0.93514	1.11290	1.18	0.43958	0.82723	1.02617
0.19	0.97687	0.99461	3.14457	0.69	0.74162	0.93335	1.10488	1.19	0.43402	0.82480	1.02906
0.20	0.97441	0.99404	2.99401	0.70	0.73540	0.93153	1.09724	1.20	0.42850	0.82237	1.03210
0.21	0.97183	0.99343	2.85813	0.71	0.72916	0.92970	1.08998	1.21	0.42303	0.81993	1.03529
0.22	0.96914	0.99279	2.73492	0.72	0.72289	0.92785	1.08308	1.22	0.41759	0.81749	1.03861
0.23	0.96633	0.99213	2.62274	0.73	0.71660	0.92598	1.07653	1.23	0.41220	0.81504	1.04208
0.24	0.96341	0.99143	2.52020	0.74	0.71029	0.92409	1.07032	1.24	0.40685	0.81259	1.04570
0.25	0.96037	0.99071	2.42616	0.75	0.70397	0.92219	1.06443	1.25	0.40154	0.81013	1.04945
0.26	0.95722	0.98996	2.33963	0.76	0.69764	0.92027	1.05886	1.26	0.39628	0.80766	1.05335
0.27	0.95397	0.98918	2.25978	0.77	0.69129	0.91833	1.05360	1.27	0.39106	0.80520	1.05739
0.28	0.95060	0.98838	2.18590	0.78	0.68493	0.91637	1.04863	1.28	0.38588	0.80272	1.06157
0.29	0.94713	0.98754	2.11737	0.79	0.67856	0.91440	1.04395	1.29	0.38075	0.80025	1.06589
0.30	0.94355	0.98668	2.05366	0.80	0.67218	0.91241	1.03954	1.30	0.37566	0.79777	1.07035
0.31	0.93986	0.98579	1.99430	0.81	0.66580	0.91040	1.03541	1.31	0.37062	0.79528	1.07495
0.32	0.93608	0.98487	1.93888	0.82	0.65942	0.90838	1.03154	1.32	0.36562	0.79280	1.07969
0.33	0.93220	0.98393	1.88706	0.83	0.65303	0.90634	1.02792	1.33	0.36067	0.79030	1.08458
0.34	0.92821	0.98296	1.83851	0.84	0.64665	0.90429	1.02455	1.34	0.35576	0.78781	1.08960
0.35	0.92413	0.98196	1.79296	0.85	0.64026	0.90222	1.02142	1.35	0.35090	0.78531	1.09477
0.36	0.91995	0.98093	1.75015	0.86	0.63388	0.90014	1.01853	1.36	0.34609	0.78282	1.10008
0.37	0.91568	0.97988	1.70987	0.87	0.62750	0.89804	1.01587	1.37	0.34132	0.78031	1.10553
0.38	0.91132	0.97880	1.67192	0.88	0.62113	0.89593	1.01343	1.38	0.33660	0.77781	1.11112
0.39	0.90687	0.97769	1.63612	0.89	0.61477	0.89380	1.01121	1.39	0.33193	0.77530	1.11686
0.40	0.90233	0.97656	1.60232	0.90	0.60842	0.89166	1.00921	1.40	0.32730	0.77280	1.12274
0.41	0.89771	0.97541	1.57036	0.91	0.60208	0.88951	1.00741	1.41	0.32272	0.77029	1.12876
0.42	0.89300	0.97422	1.54012	0.92	0.59575	0.88734	1.00582	1.42	0.31819	0.76778	1.13492
0.43	0.88821	0.97301	1.51147	0.93	0.58943	0.88516	1.00443	1.43	0.31370	0.76527	1.14123
0.44	0.88334	0.97178	1.48433	0.94	0.58313	0.88297	1.00323	1.44	0.30927	0.76275	1.14769
0.45	0.87839	0.97052	1.45857	0.95	0.57685	0.88077	1.00223	1.45	0.30487	0.76024	1.15429
0.46	0.87336	0.96924	1.43412	0.96	0.57058	0.87855	1.00142	1.46	0.30053	0.75773	1.16103
0.47	0.86827	0.96793	1.41090	0.97	0.56434	0.87632	1.00080	1.47	0.29623	0.75521	1.16792
0.48	0.86310	0.96659	1.38882	0.98	0.55811	0.87408	1.00035	1.48	0.29198	0.75269	1.17496
0.49	0.85785	0.96524	1.36783	0.99	0.55191	0.87183	1.00009	1.49	0.28777	0.75018	1.18215

(Continued)

TABLE A.1.2 (*Continued*)

Isentropic Mach number, k = 1.3

M	P/Pt	T/Tt	A/A*	M	P/Pt	T/Tt	A/A*	M	P/Pt	T/Tt	A/A*
1.50	0.28361	0.74766	1.18949	2.00	0.13046	0.62500	1.77319	2.50	0.05692	0.51613	2.95446
1.51	0.27950	0.74515	1.19697	2.01	0.12836	0.62266	1.78993	2.51	0.05598	0.51413	2.98668
1.52	0.27544	0.74263	1.20461	2.02	0.12628	0.62032	1.80690	2.52	0.05504	0.51215	3.01931
1.53	0.27142	0.74012	1.21240	2.03	0.12424	0.61800	1.82411	2.53	0.05413	0.51017	3.05235
1.54	0.26745	0.73760	1.22034	2.04	0.12223	0.61567	1.84157	2.54	0.05323	0.50820	3.08580
1.55	0.26352	0.73509	1.22843	2.05	0.12025	0.61336	1.85926	2.55	0.05234	0.50623	3.11967
1.56	0.25964	0.73258	1.23668	2.06	0.11830	0.61105	1.87720	2.56	0.05147	0.50428	3.15396
1.57	0.25580	0.73007	1.24508	2.07	0.11638	0.60874	1.89538	2.57	0.05061	0.50233	3.18867
1.58	0.25202	0.72756	1.25364	2.08	0.11449	0.60644	1.91382	2.58	0.04977	0.50039	3.22382
1.59	0.24827	0.72505	1.26236	2.09	0.11262	0.60415	1.93250	2.59	0.04894	0.49845	3.25941
1.60	0.24457	0.72254	1.27124	2.10	0.11079	0.60187	1.95145	2.60	0.04813	0.49652	3.29544
1.61	0.24092	0.72004	1.28027	2.11	0.10898	0.59959	1.97065	2.61	0.04733	0.49461	3.33191
1.62	0.23731	0.71754	1.28947	2.12	0.10720	0.59731	1.99011	2.62	0.04654	0.49269	3.36883
1.63	0.23375	0.71503	1.29883	2.13	0.10545	0.59505	2.00984	2.63	0.04577	0.49079	3.40621
1.64	0.23023	0.71253	1.30835	2.14	0.10373	0.59279	2.02983	2.64	0.04500	0.48889	3.44406
1.65	0.22675	0.71004	1.31804	2.15	0.10203	0.59054	2.05009	2.65	0.04426	0.48700	3.48236
1.66	0.22332	0.70754	1.32789	2.16	0.10036	0.58829	2.07063	2.66	0.04352	0.48512	3.52114
1.67	0.21993	0.70505	1.33791	2.17	0.09872	0.58605	2.09144	2.67	0.04280	0.48325	3.56040
1.68	0.21659	0.70256	1.34810	2.18	0.09710	0.58382	2.11253	2.68	0.04208	0.48138	3.60013
1.69	0.21329	0.70008	1.35846	2.19	0.09550	0.58159	2.13390	2.69	0.04138	0.47952	3.64036
1.70	0.21003	0.69759	1.36899	2.20	0.09393	0.57937	2.15555	2.70	0.04070	0.47767	3.68107
1.71	0.20681	0.69511	1.37969	2.21	0.09239	0.57716	2.17750	2.71	0.04002	0.47582	3.72228
1.72	0.20364	0.69264	1.39057	2.22	0.09087	0.57496	2.19973	2.72	0.03935	0.47399	3.76400
1.73	0.20050	0.69016	1.40163	2.23	0.08937	0.57276	2.22226	2.73	0.03870	0.47216	3.80622
1.74	0.19741	0.68769	1.41286	2.24	0.08790	0.57057	2.24508	2.74	0.03806	0.47034	3.84896
1.75	0.19436	0.68522	1.42427	2.25	0.08645	0.56838	2.26821	2.75	0.03742	0.46852	3.89221
1.76	0.19135	0.68276	1.43587	2.26	0.08503	0.56621	2.29164	2.76	0.03680	0.46671	3.93599
1.77	0.18838	0.68030	1.44764	2.27	0.08362	0.56404	2.31538	2.77	0.03619	0.46491	3.98031
1.78	0.18545	0.67785	1.45960	2.28	0.08224	0.56187	2.33942	2.78	0.03559	0.46312	4.02515
1.79	0.18256	0.67540	1.47175	2.29	0.08088	0.55972	2.36379	2.79	0.03500	0.46134	4.07054
1.80	0.17972	0.67295	1.48408	2.30	0.07955	0.55757	2.38847	2.80	0.03442	0.45956	4.11648
1.81	0.17690	0.67050	1.49661	2.31	0.07823	0.55543	2.41347	2.81	0.03385	0.45779	4.16297
1.82	0.17413	0.66807	1.50932	2.32	0.07694	0.55329	2.43879	2.82	0.03329	0.45603	4.21002
1.83	0.17140	0.66563	1.52223	2.33	0.07566	0.55117	2.46445	2.83	0.03274	0.45427	4.25764
1.84	0.16870	0.66320	1.53533	2.34	0.07441	0.54905	2.49043	2.84	0.03219	0.45252	4.30583
1.85	0.16605	0.66077	1.54863	2.35	0.07318	0.54693	2.51675	2.85	0.03166	0.45078	4.35459
1.86	0.16343	0.65835	1.56213	2.36	0.07197	0.54483	2.54341	2.86	0.03114	0.44905	4.40394
1.87	0.16084	0.65594	1.57583	2.37	0.07077	0.54273	2.57042	2.87	0.03062	0.44732	4.45388
1.88	0.15830	0.65353	1.58974	2.38	0.06960	0.54064	2.59777	2.88	0.03011	0.44560	4.50442
1.89	0.15579	0.65112	1.60384	2.39	0.06844	0.53856	2.62547	2.89	0.02962	0.44389	4.55556
1.90	0.15331	0.64872	1.61816	2.40	0.06731	0.53648	2.65352	2.90	0.02913	0.44218	4.60731
1.91	0.15087	0.64632	1.63269	2.41	0.06619	0.53441	2.68194	2.91	0.02864	0.44049	4.65967
1.92	0.14847	0.64393	1.64742	2.42	0.06509	0.53235	2.71071	2.92	0.02817	0.43880	4.71266
1.93	0.14610	0.64155	1.66237	2.43	0.06401	0.53030	2.73986	2.93	0.02771	0.43711	4.76628
1.94	0.14377	0.63917	1.67754	2.44	0.06295	0.52825	2.76937	2.94	0.02725	0.43544	4.82053
1.95	0.14147	0.63679	1.69292	2.45	0.06190	0.52621	2.79926	2.95	0.02680	0.43377	4.87542
1.96	0.13920	0.63442	1.70852	2.46	0.06087	0.52418	2.82952	2.96	0.02636	0.43211	4.93097
1.97	0.13697	0.63206	1.72435	2.47	0.05986	0.52216	2.86017	2.97	0.02592	0.43045	4.98717
1.98	0.13476	0.62970	1.74040	2.48	0.05886	0.52014	2.89121	2.98	0.02550	0.42881	5.04403
1.99	0.13260	0.62735	1.75668	2.49	0.05789	0.51813	2.92264	2.99	0.02508	0.42717	5.10156

A.2 Normal Shock Tables

TABLE A.2.1

Normal Shock Table, k = 1.4

M_1	M_2	P_2/P_1	T_2/T_1	P_{T2}/P_{T1}	ρ_2/ρ_1	M_1	M_2	P_2/P_1	T_2/T_1	P_{T2}/P_{T1}	ρ_2/ρ_1
1.00	1.00000	1.00000	1.00000	1.00000	1.00000	1.50	0.70109	2.45833	1.32022	0.92979	1.86207
1.01	0.99013	1.02345	1.00664	1.00000	1.01669	1.51	0.69758	2.49345	1.32688	0.92659	1.87918
1.02	0.98052	1.04713	1.01325	0.99999	1.03344	1.52	0.69413	2.52880	1.33357	0.92332	1.89626
1.03	0.97115	1.07105	1.01981	0.99997	1.05024	1.53	0.69073	2.56438	1.34029	0.92000	1.91331
1.04	0.96203	1.09520	1.02634	0.99992	1.06709	1.54	0.68739	2.60020	1.34703	0.91662	1.93033
1.05	0.95313	1.11958	1.03284	0.99985	1.08398	1.55	0.68410	2.63625	1.35379	0.91319	1.94732
1.06	0.94445	1.14420	1.03931	0.99975	1.10092	1.56	0.68087	2.67253	1.36057	0.90970	1.96427
1.07	0.93598	1.16905	1.04575	0.99961	1.11790	1.57	0.67768	2.70905	1.36738	0.90615	1.98119
1.08	0.92771	1.19413	1.05217	0.99943	1.13492	1.58	0.67455	2.74580	1.37422	0.90255	1.99808
1.09	0.91965	1.21945	1.05856	0.99920	1.15199	1.59	0.67147	2.78278	1.38108	0.89890	2.01493
1.10	0.91177	1.24500	1.06494	0.99893	1.16908	1.60	0.66844	2.82000	1.38797	0.89520	2.03175
1.11	0.90408	1.27078	1.07129	0.99860	1.18621	1.61	0.66545	2.85745	1.39488	0.89145	2.04852
1.12	0.89656	1.29680	1.07763	0.99821	1.20338	1.62	0.66251	2.89513	1.40182	0.88765	2.06526
1.13	0.88922	1.32305	1.08396	0.99777	1.22057	1.63	0.65962	2.93305	1.40879	0.88381	2.08197
1.14	0.88204	1.34953	1.09027	0.99726	1.23779	1.64	0.65677	2.97120	1.41578	0.87992	2.09863
1.15	0.87502	1.37625	1.09658	0.99669	1.25504	1.65	0.65396	3.00958	1.42280	0.87599	2.11525
1.16	0.86816	1.40320	1.10287	0.99605	1.27231	1.66	0.65119	3.04820	1.42985	0.87201	2.13183
1.17	0.86145	1.43038	1.10916	0.99535	1.28961	1.67	0.64847	3.08705	1.43693	0.86800	2.14836
1.18	0.85488	1.45780	1.11544	0.99457	1.30693	1.68	0.64579	3.12613	1.44403	0.86394	2.16486
1.19	0.84846	1.48545	1.12172	0.99372	1.32426	1.69	0.64315	3.16545	1.45117	0.85985	2.18131
1.20	0.84217	1.51333	1.12799	0.99280	1.34161	1.70	0.64054	3.20500	1.45833	0.85572	2.19772
1.21	0.83601	1.54145	1.13427	0.99180	1.35898	1.71	0.63798	3.24478	1.46552	0.85156	2.21408
1.22	0.82999	1.56980	1.14054	0.99073	1.37636	1.72	0.63545	3.28480	1.47274	0.84736	2.23040
1.23	0.82408	1.59838	1.14682	0.98958	1.39376	1.73	0.63296	3.32505	1.47999	0.84312	2.24667
1.24	0.81830	1.62720	1.15309	0.98836	1.41116	1.74	0.63051	3.36553	1.48727	0.83886	2.26289
1.25	0.81264	1.65625	1.15938	0.98706	1.42857	1.75	0.62809	3.40625	1.49458	0.83457	2.27907
1.26	0.80709	1.68553	1.16566	0.98568	1.44599	1.76	0.62570	3.44720	1.50192	0.83024	2.29520
1.27	0.80164	1.71505	1.17195	0.98422	1.46341	1.77	0.62335	3.48838	1.50929	0.82589	2.31128
1.28	0.79631	1.74480	1.17825	0.98268	1.48084	1.78	0.62104	3.52980	1.51669	0.82151	2.32731
1.29	0.79108	1.77478	1.18456	0.98107	1.49827	1.79	0.61875	3.57145	1.52412	0.81711	2.34329
1.30	0.78596	1.80500	1.19087	0.97937	1.51570	1.80	0.61650	3.61333	1.53158	0.81268	2.35922
1.31	0.78093	1.83545	1.19720	0.97760	1.53312	1.81	0.61428	3.65545	1.53907	0.80823	2.37510
1.32	0.77600	1.86613	1.20353	0.97575	1.55055	1.82	0.61209	3.69780	1.54659	0.80376	2.39093
1.33	0.77116	1.89705	1.20988	0.97382	1.56797	1.83	0.60993	3.74038	1.55415	0.79927	2.40671
1.34	0.76641	1.92820	1.21624	0.97182	1.58538	1.84	0.60780	3.78320	1.56173	0.79476	2.42244
1.35	0.76175	1.95958	1.22261	0.96974	1.60278	1.85	0.60570	3.82625	1.56935	0.79023	2.43811
1.36	0.75718	1.99120	1.22900	0.96758	1.62018	1.86	0.60363	3.86953	1.57700	0.78569	2.45373
1.37	0.75269	2.02305	1.23540	0.96534	1.63757	1.87	0.60158	3.91305	1.58468	0.78112	2.46930
1.38	0.74829	2.05513	1.24181	0.96304	1.65494	1.88	0.59957	3.95680	1.59239	0.77655	2.48481
1.39	0.74396	2.08745	1.24825	0.96065	1.67231	1.89	0.59758	4.00078	1.60014	0.77196	2.50027
1.40	0.73971	2.12000	1.25469	0.95819	1.68966	1.90	0.59562	4.04500	1.60792	0.76736	2.51568
1.41	0.73554	2.15278	1.26116	0.95566	1.70699	1.91	0.59368	4.08945	1.61573	0.76274	2.53103
1.42	0.73144	2.18580	1.26764	0.95306	1.72430	1.92	0.59177	4.13413	1.62357	0.75812	2.54633
1.43	0.72741	2.21905	1.27414	0.95039	1.74160	1.93	0.58988	4.17905	1.63144	0.75349	2.56157
1.44	0.72345	2.25253	1.28066	0.94765	1.75888	1.94	0.58802	4.22420	1.63935	0.74884	2.57675
1.45	0.71956	2.28625	1.28720	0.94484	1.77614	1.95	0.58618	4.26958	1.64729	0.74420	2.59188
1.46	0.71574	2.32020	1.29377	0.94196	1.79337	1.96	0.58437	4.31520	1.65527	0.73954	2.60695
1.47	0.71198	2.35438	1.30035	0.93901	1.81058	1.97	0.58258	4.36105	1.66328	0.73488	2.62196
1.48	0.70829	2.38880	1.30695	0.93600	1.82777	1.98	0.58082	4.40713	1.67132	0.73021	2.63692
1.49	0.70466	2.42345	1.31357	0.93293	1.84493	1.99	0.57907	4.45345	1.67939	0.72555	2.65182

(Continued)

TABLE A.2.1 (*Continued*)

Normal Shock Table, k = 1.4

M₁	M₂	P₂/P₁	T₂/T₁	P$_{T2}$/P$_{T1}$	ρ₂/ρ₁	M₁	M₂	P₂/P₁	T₂/T₁	P$_{T2}$/P$_{T1}$	ρ₂/ρ₁
2.00	0.57735	4.50000	1.68750	0.72087	2.66667	2.50	0.51299	7.12500	2.13750	0.49901	3.33333
2.01	0.57565	4.54678	1.69564	0.71620	2.68145	2.51	0.51203	7.18345	2.14742	0.49502	3.34516
2.02	0.57397	4.59380	1.70382	0.71153	2.69618	2.52	0.51109	7.24213	2.15737	0.49105	3.35692
2.03	0.57231	4.64105	1.71203	0.70685	2.71085	2.53	0.51015	7.30105	2.16737	0.48711	3.36863
2.04	0.57068	4.68853	1.72027	0.70218	2.72546	2.54	0.50923	7.36020	2.17739	0.48318	3.38028
2.05	0.56906	4.73625	1.72855	0.69751	2.74002	2.55	0.50831	7.41958	2.18746	0.47928	3.39187
2.06	0.56747	4.78420	1.73686	0.69284	2.75451	2.56	0.50741	7.47920	2.19756	0.47540	3.40341
2.07	0.56589	4.83238	1.74521	0.68817	2.76895	2.57	0.50651	7.53905	2.20770	0.47155	3.41489
2.08	0.56433	4.88080	1.75359	0.68351	2.78332	2.58	0.50562	7.59913	2.21788	0.46772	3.42631
2.09	0.56280	4.92945	1.76200	0.67885	2.79764	2.59	0.50474	7.65945	2.22809	0.46391	3.43767
2.10	0.56128	4.97833	1.77045	0.67420	2.81190	2.60	0.50387	7.72000	2.23834	0.46012	3.44898
2.11	0.55978	5.02745	1.77893	0.66956	2.82610	2.61	0.50301	7.78078	2.24863	0.45636	3.46023
2.12	0.55829	5.07680	1.78745	0.66492	2.84024	2.62	0.50216	7.84180	2.25896	0.45263	3.47143
2.13	0.55683	5.12638	1.79601	0.66029	2.85432	2.63	0.50131	7.90305	2.26932	0.44891	3.48257
2.14	0.55538	5.17620	1.80459	0.65567	2.86835	2.64	0.50048	7.96453	2.27972	0.44522	3.49365
2.15	0.55395	5.22625	1.81322	0.65105	2.88231	2.65	0.49965	8.02625	2.29015	0.44156	3.50468
2.16	0.55254	5.27653	1.82188	0.64645	2.89621	2.66	0.49883	8.08820	2.30063	0.43792	3.51565
2.17	0.55115	5.32705	1.83057	0.64185	2.91005	2.67	0.49802	8.15038	2.31114	0.43430	3.52657
2.18	0.54977	5.37780	1.83930	0.63727	2.92383	2.68	0.49722	8.21280	2.32168	0.43070	3.53743
2.19	0.54840	5.42878	1.84806	0.63270	2.93756	2.69	0.49642	8.27545	2.33227	0.42714	3.54824
2.20	0.54706	5.48000	1.85686	0.62814	2.95122	2.70	0.49563	8.33833	2.34289	0.42359	3.55899
2.21	0.54572	5.53145	1.86569	0.62359	2.96482	2.71	0.49485	8.40145	2.35355	0.42007	3.56969
2.22	0.54441	5.58313	1.87456	0.61905	2.97837	2.72	0.49408	8.46480	2.36425	0.41657	3.58033
2.23	0.54311	5.63505	1.88347	0.61453	2.99185	2.73	0.49332	8.52838	2.37498	0.41310	3.59092
2.24	0.54182	5.68720	1.89241	0.61002	3.00527	2.74	0.49256	8.59220	2.38576	0.40965	3.60146
2.25	0.54055	5.73958	1.90138	0.60553	3.01863	2.75	0.49181	8.65625	2.39657	0.40623	3.61194
2.26	0.53930	5.79220	1.91040	0.60105	3.03194	2.76	0.49107	8.72053	2.40741	0.40283	3.62237
2.27	0.53805	5.84505	1.91944	0.59659	3.04518	2.77	0.49033	8.78505	2.41830	0.39945	3.63274
2.28	0.53683	5.89813	1.92853	0.59214	3.05836	2.78	0.48960	8.84980	2.42922	0.39610	3.64307
2.29	0.53561	5.95145	1.93765	0.58771	3.07149	2.79	0.48888	8.91478	2.44018	0.39277	3.65334
2.30	0.53441	6.00500	1.94680	0.58329	3.08455	2.80	0.48817	8.98000	2.45117	0.38946	3.66355
2.31	0.53322	6.05878	1.95599	0.57890	3.09755	2.81	0.48746	9.04545	2.46221	0.38618	3.67372
2.32	0.53205	6.11280	1.96522	0.57452	3.11049	2.82	0.48676	9.11113	2.47328	0.38293	3.68383
2.33	0.53089	6.16705	1.97448	0.57015	3.12338	2.83	0.48606	9.17705	2.48439	0.37969	3.69389
2.34	0.52974	6.22153	1.98378	0.56581	3.13620	2.84	0.48538	9.24320	2.49554	0.37649	3.70389
2.35	0.52861	6.27625	1.99311	0.56148	3.14897	2.85	0.48469	9.30958	2.50672	0.37330	3.71385
2.36	0.52749	6.33120	2.00249	0.55718	3.16167	2.86	0.48402	9.37620	2.51794	0.37014	3.72375
2.37	0.52638	6.38638	2.01189	0.55289	3.17432	2.87	0.48335	9.44305	2.52920	0.36700	3.73361
2.38	0.52528	6.44180	2.02134	0.54862	3.18690	2.88	0.48269	9.51013	2.54050	0.36389	3.74341
2.39	0.52419	6.49745	2.03082	0.54437	3.19943	2.89	0.48203	9.57745	2.55183	0.36080	3.75316
2.40	0.52312	6.55333	2.04033	0.54014	3.21190	2.90	0.48138	9.64500	2.56321	0.35773	3.76286
2.41	0.52206	6.60945	2.04988	0.53594	3.22430	2.91	0.48073	9.71278	2.57462	0.35469	3.77251
2.42	0.52100	6.66580	2.05947	0.53175	3.23665	2.92	0.48010	9.78080	2.58607	0.35167	3.78211
2.43	0.51996	6.72238	2.06910	0.52758	3.24894	2.93	0.47946	9.84905	2.59755	0.34867	3.79167
2.44	0.51894	6.77920	2.07876	0.52344	3.26117	2.94	0.47884	9.91753	2.60908	0.34570	3.80117
2.45	0.51792	6.83625	2.08846	0.51931	3.27335	2.95	0.47821	9.98625	2.62064	0.34275	3.81062
2.46	0.51691	6.89353	2.09819	0.51521	3.28546	2.96	0.47760	10.05520	2.63224	0.33982	3.82002
2.47	0.51592	6.95105	2.10797	0.51113	3.29752	2.97	0.47699	10.12438	2.64387	0.33692	3.82937
2.48	0.51493	7.00880	2.11777	0.50707	3.30951	2.98	0.47638	10.19380	2.65555	0.33404	3.83868
2.49	0.51395	7.06678	2.12762	0.50303	3.32145	2.99	0.47578	10.26345	2.66726	0.33118	3.84794

TABLE A.2.2
Normal Shock Table, k = 1.3

M₁	M₂	P₂/P₁	T₂/T₁	P_T2/P_T1	ρ₂/ρ₁	M₁	M₂	P₂/P₁	T₂/T₁	P_T2/P_T1	ρ₂/ρ₁
1.00	1.00000	1.00000	1.00000	1.00000	1.00000	1.50	0.69425	2.41304	1.24732	0.92610	1.93458
1.01	0.99012	1.02272	1.00520	1.00000	1.01743	1.51	0.69057	2.44707	1.25243	0.92269	1.95386
1.02	0.98049	1.04567	1.01036	0.99999	1.03495	1.52	0.68695	2.48132	1.25755	0.91921	1.97315
1.03	0.97109	1.06884	1.01549	0.99997	1.05254	1.53	0.68338	2.51580	1.26268	0.91566	1.99242
1.04	0.96192	1.09224	1.02059	0.99992	1.07021	1.54	0.67987	2.55050	1.26784	0.91205	2.01170
1.05	0.95297	1.11587	1.02566	0.99985	1.08795	1.55	0.67642	2.58543	1.27301	0.90838	2.03097
1.06	0.94422	1.13972	1.03070	0.99975	1.10577	1.56	0.67301	2.62059	1.27820	0.90464	2.05023
1.07	0.93568	1.16380	1.03572	0.99960	1.12366	1.57	0.66966	2.65597	1.28340	0.90084	2.06948
1.08	0.92733	1.18810	1.04072	0.99942	1.14162	1.58	0.66637	2.69158	1.28863	0.89699	2.08872
1.09	0.91918	1.21263	1.04569	0.99919	1.15965	1.59	0.66312	2.72742	1.29387	0.89308	2.10795
1.10	0.91120	1.23739	1.05065	0.99891	1.17774	1.60	0.65992	2.76348	1.29914	0.88911	2.12717
1.11	0.90340	1.26237	1.05559	0.99857	1.19590	1.61	0.65677	2.79977	1.30442	0.88508	2.14637
1.12	0.89578	1.28758	1.06051	0.99817	1.21411	1.62	0.65366	2.83628	1.30972	0.88100	2.16556
1.13	0.88832	1.31302	1.06542	0.99771	1.23239	1.63	0.65061	2.87302	1.31504	0.87688	2.18474
1.14	0.88102	1.33868	1.07032	0.99719	1.25072	1.64	0.64759	2.90998	1.32038	0.87270	2.20390
1.15	0.87388	1.36457	1.07521	0.99660	1.26911	1.65	0.64463	2.94717	1.32574	0.86847	2.22304
1.16	0.86688	1.39068	1.08009	0.99595	1.28756	1.66	0.64170	2.98459	1.33112	0.86419	2.24216
1.17	0.86004	1.41702	1.08496	0.99522	1.30606	1.67	0.63882	3.02223	1.33652	0.85987	2.26127
1.18	0.85333	1.44358	1.08982	0.99441	1.32460	1.68	0.63598	3.06010	1.34194	0.85550	2.28035
1.19	0.84676	1.47037	1.09468	0.99353	1.34320	1.69	0.63317	3.09820	1.34739	0.85109	2.29941
1.20	0.84033	1.49739	1.09953	0.99258	1.36184	1.70	0.63041	3.13652	1.35285	0.84664	2.31845
1.21	0.83403	1.52463	1.10438	0.99155	1.38053	1.71	0.62769	3.17507	1.35834	0.84214	2.33747
1.22	0.82785	1.55210	1.10923	0.99043	1.39926	1.72	0.62501	3.21384	1.36384	0.83761	2.35646
1.23	0.82179	1.57980	1.11408	0.98924	1.41803	1.73	0.62236	3.25284	1.36937	0.83304	2.37542
1.24	0.81585	1.60772	1.11892	0.98796	1.43685	1.74	0.61975	3.29207	1.37492	0.82844	2.39436
1.25	0.81003	1.63587	1.12377	0.98661	1.45570	1.75	0.61718	3.33152	1.38050	0.82380	2.41328
1.26	0.80432	1.66424	1.12862	0.98517	1.47458	1.76	0.61464	3.37120	1.38609	0.81913	2.43216
1.27	0.79872	1.69284	1.13347	0.98365	1.49350	1.77	0.61214	3.41110	1.39171	0.81442	2.45102
1.28	0.79322	1.72167	1.13833	0.98204	1.51246	1.78	0.60967	3.45123	1.39735	0.80969	2.46984
1.29	0.78783	1.75072	1.14318	0.98035	1.53144	1.79	0.60724	3.49159	1.40301	0.80493	2.48864
1.30	0.78253	1.78000	1.14805	0.97858	1.55046	1.80	0.60484	3.53217	1.40870	0.80014	2.50740
1.31	0.77734	1.80950	1.15292	0.97672	1.56950	1.81	0.60247	3.57298	1.41441	0.79532	2.52613
1.32	0.77224	1.83923	1.15779	0.97478	1.58857	1.82	0.60013	3.61402	1.42014	0.79048	2.54483
1.33	0.76723	1.86919	1.16267	0.97275	1.60767	1.83	0.59783	3.65528	1.42589	0.78561	2.56350
1.34	0.76232	1.89937	1.16756	0.97064	1.62678	1.84	0.59555	3.69677	1.43167	0.78073	2.58213
1.35	0.75749	1.92978	1.17246	0.96845	1.64592	1.85	0.59330	3.73848	1.43747	0.77582	2.60073
1.36	0.75274	1.96042	1.17737	0.96618	1.66508	1.86	0.59109	3.78042	1.44330	0.77089	2.61929
1.37	0.74808	1.99128	1.18229	0.96382	1.68426	1.87	0.58890	3.82258	1.44915	0.76595	2.63781
1.38	0.74351	2.02237	1.18722	0.96139	1.70345	1.88	0.58674	3.86497	1.45502	0.76099	2.65630
1.39	0.73901	2.05368	1.19215	0.95887	1.72266	1.89	0.58461	3.90759	1.46092	0.75601	2.67475
1.40	0.73459	2.08522	1.19710	0.95627	1.74189	1.90	0.58251	3.95043	1.46684	0.75102	2.69316
1.41	0.73024	2.11698	1.20206	0.95359	1.76112	1.91	0.58043	3.99350	1.47279	0.74601	2.71153
1.42	0.72597	2.14897	1.20704	0.95084	1.78037	1.92	0.57838	4.03680	1.47876	0.74100	2.72986
1.43	0.72177	2.18119	1.21202	0.94801	1.79963	1.93	0.57635	4.08032	1.48475	0.73597	2.74815
1.44	0.71764	2.21363	1.21702	0.94510	1.81889	1.94	0.57436	4.12407	1.49077	0.73093	2.76640
1.45	0.71358	2.24630	1.22204	0.94211	1.83816	1.95	0.57238	4.16804	1.49682	0.72589	2.78461
1.46	0.70958	2.27920	1.22706	0.93905	1.85744	1.96	0.57043	4.21224	1.50289	0.72083	2.80277
1.47	0.70566	2.31232	1.23211	0.93592	1.87672	1.97	0.56851	4.25667	1.50898	0.71577	2.82089
1.48	0.70179	2.34567	1.23716	0.93272	1.89601	1.98	0.56661	4.30132	1.51510	0.71071	2.83897
1.49	0.69799	2.37924	1.24223	0.92944	1.91529	1.99	0.56473	4.34620	1.52124	0.70564	2.85701

(Continued)

TABLE A.2.2 (*Continued*)

Normal Shock Table, k = 1.3

M₁	M₂	P₂/P₁	T₂/T₁	P_{T2}/P_{T1}	ρ₂/ρ₁	M₁	M₂	P₂/P₁	T₂/T₁	P_{T2}/P_{T1}	ρ₂/ρ₁
2.00	0.56288	4.39130	1.52741	0.70057	2.87500	2.50	0.49290	6.93478	1.86938	0.46098	3.70968
2.01	0.56105	4.43663	1.53360	0.69550	2.89295	2.51	0.49185	6.99142	1.87691	0.45672	3.72497
2.02	0.55924	4.48219	1.53982	0.69042	2.91085	2.52	0.49081	7.04828	1.88447	0.45248	3.74020
2.03	0.55745	4.52797	1.54607	0.68535	2.92870	2.53	0.48978	7.10537	1.89205	0.44827	3.75537
2.04	0.55569	4.57398	1.55234	0.68028	2.94651	2.54	0.48876	7.16268	1.89967	0.44409	3.77049
2.05	0.55394	4.62022	1.55863	0.67521	2.96427	2.55	0.48775	7.22022	1.90731	0.43994	3.78555
2.06	0.55222	4.66668	1.56496	0.67014	2.98199	2.56	0.48676	7.27798	1.91498	0.43581	3.80055
2.07	0.55052	4.71337	1.57130	0.66508	2.99965	2.57	0.48577	7.33597	1.92268	0.43171	3.81549
2.08	0.54883	4.76028	1.57768	0.66002	3.01727	2.58	0.48479	7.39419	1.93041	0.42764	3.83038
2.09	0.54717	4.80742	1.58408	0.65496	3.03484	2.59	0.48382	7.45263	1.93816	0.42360	3.84521
2.10	0.54553	4.85478	1.59050	0.64992	3.05236	2.60	0.48286	7.51130	1.94594	0.41958	3.85998
2.11	0.54390	4.90237	1.59695	0.64488	3.06983	2.61	0.48191	7.57020	1.95375	0.41560	3.87469
2.12	0.54230	4.95019	1.60343	0.63984	3.08726	2.62	0.48097	7.62932	1.96159	0.41164	3.88935
2.13	0.54071	4.99823	1.60993	0.63482	3.10463	2.63	0.48004	7.68867	1.96946	0.40771	3.90395
2.14	0.53914	5.04650	1.61646	0.62981	3.12195	2.64	0.47912	7.74824	1.97735	0.40380	3.91849
2.15	0.53760	5.09500	1.62302	0.62481	3.13922	2.65	0.47820	7.80804	1.98528	0.39993	3.93298
2.16	0.53606	5.14372	1.62960	0.61982	3.15644	2.66	0.47730	7.86807	1.99323	0.39609	3.94740
2.17	0.53455	5.19267	1.63620	0.61484	3.17361	2.67	0.47640	7.92832	2.00121	0.39227	3.96177
2.18	0.53305	5.24184	1.64284	0.60987	3.19072	2.68	0.47552	7.98880	2.00921	0.38848	3.97609
2.19	0.53157	5.29124	1.64950	0.60492	3.20779	2.69	0.47464	8.04950	2.01725	0.38472	3.99034
2.20	0.53011	5.34087	1.65619	0.59998	3.22480	2.70	0.47377	8.11043	2.02531	0.38099	4.00454
2.21	0.52866	5.39072	1.66290	0.59506	3.24176	2.71	0.47291	8.17159	2.03340	0.37729	4.01868
2.22	0.52723	5.44080	1.66964	0.59015	3.25866	2.72	0.47205	8.23297	2.04152	0.37361	4.03276
2.23	0.52582	5.49110	1.67641	0.58526	3.27551	2.73	0.47121	8.29458	2.04967	0.36997	4.04679
2.24	0.52442	5.54163	1.68320	0.58038	3.29231	2.74	0.47037	8.35642	2.05785	0.36635	4.06076
2.25	0.52304	5.59239	1.69002	0.57552	3.30906	2.75	0.46954	8.41848	2.06605	0.36276	4.07467
2.26	0.52167	5.64337	1.69687	0.57068	3.32575	2.76	0.46871	8.48077	2.07428	0.35920	4.08853
2.27	0.52032	5.69458	1.70375	0.56586	3.34239	2.77	0.46790	8.54328	2.08255	0.35567	4.10233
2.28	0.51898	5.74602	1.71065	0.56105	3.35897	2.78	0.46709	8.60602	2.09083	0.35217	4.11607
2.29	0.51765	5.79768	1.71758	0.55627	3.37550	2.79	0.46629	8.66898	2.09915	0.34869	4.12975
2.30	0.51635	5.84957	1.72453	0.55150	3.39197	2.80	0.46550	8.73217	2.10750	0.34525	4.14338
2.31	0.51505	5.90168	1.73152	0.54676	3.40839	2.81	0.46471	8.79559	2.11587	0.34183	4.15696
2.32	0.51377	5.95402	1.73853	0.54203	3.42475	2.82	0.46394	8.85923	2.12428	0.33844	4.17047
2.33	0.51250	6.00658	1.74556	0.53733	3.44106	2.83	0.46317	8.92310	2.13271	0.33508	4.18393
2.34	0.51125	6.05937	1.75263	0.53265	3.45731	2.84	0.46240	8.98720	2.14117	0.33175	4.19734
2.35	0.51001	6.11239	1.75972	0.52799	3.47351	2.85	0.46164	9.05152	2.14966	0.32844	4.21068
2.36	0.50878	6.16563	1.76684	0.52335	3.48965	2.86	0.46089	9.11607	2.15817	0.32516	4.22398
2.37	0.50757	6.21910	1.77398	0.51874	3.50573	2.87	0.46015	9.18084	2.16672	0.32191	4.23721
2.38	0.50637	6.27280	1.78115	0.51415	3.52176	2.88	0.45941	9.24584	2.17529	0.31869	4.25039
2.39	0.50518	6.32672	1.78835	0.50958	3.53773	2.89	0.45868	9.31107	2.18389	0.31550	4.26352
2.40	0.50400	6.38087	1.79558	0.50504	3.55365	2.90	0.45796	9.37652	2.19252	0.31233	4.27659
2.41	0.50284	6.43524	1.80284	0.50052	3.56951	2.91	0.45724	9.44220	2.20118	0.30919	4.28960
2.42	0.50169	6.48984	1.81012	0.49602	3.58531	2.92	0.45653	9.50810	2.20987	0.30608	4.30256
2.43	0.50055	6.54467	1.81743	0.49155	3.60105	2.93	0.45583	9.57423	2.21859	0.30300	4.31546
2.44	0.49942	6.59972	1.82477	0.48711	3.61674	2.94	0.45513	9.64059	2.22733	0.29994	4.32831
2.45	0.49831	6.65500	1.83213	0.48269	3.63238	2.95	0.45444	9.70717	2.23611	0.29691	4.34111
2.46	0.49720	6.71050	1.83953	0.47829	3.64795	2.96	0.45375	9.77398	2.24491	0.29391	4.35384
2.47	0.49611	6.76623	1.84695	0.47393	3.66347	2.97	0.45307	9.84102	2.25374	0.29093	4.36653
2.48	0.49503	6.82219	1.85440	0.46958	3.67893	2.98	0.45240	9.90828	2.26260	0.28798	4.37916
2.49	0.49396	6.87837	1.86187	0.46527	3.69433	2.99	0.45173	9.97577	2.27149	0.28506	4.39173

A.3 Shock Tube Tables

TABLE A.3.1

Shock Tube Table, $k = 1.4$, $a_1 = a_4$

P_2/P_1	P_4/P_1	P_2/P_1	P_4/P_1	P_2/P_1	P_4/P_1	P_2/P_1	P_4/P_1	P_2/P_1	P_4/P_1	P_2/P_1	P_4/P_1
1.05	1.1029	3.65	18.8899	6.25	92.774	8.85	328.923	11.45	1015.54	14.05	2946.74
1.10	1.2116	3.70	19.5893	6.30	95.263	8.90	336.458	11.50	1036.99	14.10	3006.83
1.15	1.3263	3.75	20.3081	6.35	97.807	8.95	344.149	11.55	1058.87	14.15	3068.12
1.20	1.4471	3.80	21.0466	6.40	100.409	9.00	351.999	11.60	1081.19	14.20	3130.65
1.25	1.5742	3.85	21.8052	6.45	103.069	9.05	360.011	11.65	1103.95	14.25	3194.43
1.30	1.7078	3.90	22.5845	6.50	105.789	9.10	368.189	11.70	1127.17	14.30	3259.49
1.35	1.8478	3.95	23.3848	6.55	108.570	9.15	376.536	11.75	1150.86	14.35	3325.85
1.40	1.9946	4.00	24.2066	6.60	111.413	9.20	385.055	11.80	1175.02	14.40	3393.55
1.45	2.1483	4.05	25.0503	6.65	114.320	9.25	393.749	11.85	1199.67	14.45	3462.61
1.50	2.3090	4.10	25.9165	6.70	117.291	9.30	402.622	11.90	1224.81	14.50	3533.06
1.55	2.4769	4.15	26.8055	6.75	120.327	9.35	411.678	11.95	1250.45	14.55	3604.92
1.60	2.6522	4.20	27.7179	6.80	123.431	9.40	420.920	12.00	1276.61	14.60	3678.23
1.65	2.8351	4.25	28.6542	6.85	126.604	9.45	430.351	12.05	1303.29	14.65	3753.01
1.70	3.0257	4.30	29.6149	6.90	129.847	9.50	439.976	12.10	1330.50	14.70	3829.30
1.75	3.2241	4.35	30.6005	6.95	133.161	9.55	449.798	12.15	1358.26	14.75	3907.12
1.80	3.4307	4.40	31.6115	7.00	136.547	9.60	459.821	12.20	1386.57	14.80	3986.51
1.85	3.6456	4.45	32.6485	7.05	140.008	9.65	470.049	12.25	1415.45	14.85	4067.50
1.90	3.8690	4.50	33.7119	7.10	143.545	9.70	480.486	12.30	1444.90	14.90	4150.12
1.95	4.1010	4.55	34.8025	7.15	147.159	9.75	491.137	12.35	1474.95	14.95	4234.41
2.00	4.3420	4.60	35.9206	7.20	150.851	9.80	502.004	12.40	1505.59	15.00	4320.40
2.05	4.5920	4.65	37.0670	7.25	154.624	9.85	513.094	12.45	1536.85	15.05	4408.12
2.10	4.8513	4.70	38.2422	7.30	158.479	9.90	524.409	12.50	1568.73	15.10	4497.61
2.15	5.1202	4.75	39.4468	7.35	162.417	9.95	535.955	12.55	1601.25	15.15	4588.91
2.20	5.3988	4.80	40.6813	7.40	166.441	10.00	547.735	12.60	1634.42	15.20	4682.06
2.25	5.6873	4.85	41.9466	7.45	170.551	10.05	559.755	12.65	1668.25	15.25	4777.09
2.30	5.9861	4.90	43.2431	7.50	174.750	10.10	572.020	12.70	1702.76	15.30	4874.03
2.35	6.2952	4.95	44.5715	7.55	179.039	10.15	584.533	12.75	1737.96	15.35	4972.94
2.40	6.6151	5.00	45.9325	7.60	183.420	10.20	597.301	12.80	1773.86	15.40	5073.86
2.45	6.9459	5.05	47.3268	7.65	187.896	10.25	610.327	12.85	1810.48	15.45	5176.81
2.50	7.2878	5.10	48.7551	7.70	192.467	10.30	623.618	12.90	1847.83	15.50	5281.85
2.55	7.6411	5.15	50.2179	7.75	197.136	10.35	637.178	12.95	1885.93	15.55	5389.02
2.60	8.0062	5.20	51.7162	7.80	201.905	10.40	651.012	13.00	1924.79	15.60	5498.37
2.65	8.3831	5.25	53.2505	7.85	206.775	10.45	665.126	13.05	1964.42	15.65	5609.93
2.70	8.7723	5.30	54.8217	7.90	211.749	10.50	679.526	13.10	2004.85	15.70	5723.76
2.75	9.1740	5.35	56.4305	7.95	216.828	10.55	694.216	13.15	2046.08	15.75	5839.90
2.80	9.5885	5.40	58.0777	8.00	222.016	10.60	709.204	13.20	2088.14	15.80	5958.39
2.85	10.0160	5.45	59.7640	8.05	227.313	10.65	724.494	13.25	2131.04	15.85	6079.30
2.90	10.4568	5.50	61.4903	8.10	232.722	10.70	740.092	13.30	2174.80	15.90	6202.67
2.95	10.9113	5.55	63.2573	8.15	238.246	10.75	756.006	13.35	2219.44	15.95	6328.55
3.00	11.3798	5.60	65.0659	8.20	243.886	10.80	772.240	13.40	2264.96	16.00	6456.99
3.05	11.8625	5.65	66.9170	8.25	249.645	10.85	788.801	13.45	2311.40	16.05	6588.05
3.10	12.3598	5.70	68.8115	8.30	255.525	10.90	805.696	13.50	2358.77	16.10	6721.78
3.15	12.8720	5.75	70.7501	8.35	261.529	10.95	822.931	13.55	2407.08	16.15	6858.24
3.20	13.3995	5.80	72.7338	8.40	267.659	11.00	840.513	13.60	2456.36	16.20	6997.49
3.25	13.9425	5.85	74.7636	8.45	273.917	11.05	858.449	13.65	2506.63	16.25	7139.58
3.30	14.5015	5.90	76.8403	8.50	280.306	11.10	876.745	13.70	2557.90	16.30	7284.57
3.35	15.0767	5.95	78.9649	8.55	286.829	11.15	895.409	13.75	2610.20	16.35	7432.53
3.40	15.6685	6.00	81.1384	8.60	293.488	11.20	914.449	13.80	2663.54	16.40	7583.52
3.45	16.2773	6.05	83.3618	8.65	300.287	11.25	933.870	13.85	2717.96	16.45	7737.60
3.50	16.9034	6.10	85.6361	8.70	307.226	11.30	953.682	13.90	2773.46	16.50	7894.84
3.55	17.5473	6.15	87.9622	8.75	314.310	11.35	973.892	13.95	2830.08	16.55	8055.30
3.60	18.2093	6.20	90.3413	8.80	321.542	11.40	994.507	14.00	2887.83	16.60	8219.05

TABLE A.3.2

Shock Tube Table, $k = 1.3$, $a_1 = a_4$

P_2/P_1	P_4/P_1	P_2/P_1	P_4/P_1	P_2/P_1	P_4/P_1	P_2/P_1	P_4/P_1	P_2/P_1	P_4/P_1	P_2/P_1	P_4/P_1
1.05	1.1029	3.65	17.8220	6.25	80.122	8.85	254.722	11.45	689.05	14.05	1704.12
1.10	1.2116	3.70	18.4541	6.30	82.115	8.90	259.956	11.50	701.64	14.10	1733.00
1.15	1.3263	3.75	19.1023	6.35	84.148	8.95	265.281	11.55	714.43	14.15	1762.34
1.20	1.4471	3.80	19.7669	6.40	86.221	9.00	270.701	11.60	727.44	14.20	1792.14
1.25	1.5742	3.85	20.4481	6.45	88.336	9.05	276.216	11.65	740.66	14.25	1822.41
1.30	1.7078	3.90	21.1463	6.50	90.493	9.10	281.829	11.70	754.10	14.30	1853.16
1.35	1.8478	3.95	21.8619	6.55	92.692	9.15	287.540	11.75	767.76	14.35	1884.40
1.40	1.9946	4.00	22.5950	6.60	94.934	9.20	293.351	11.80	781.64	14.40	1916.12
1.45	2.1483	4.05	23.3461	6.65	97.221	9.25	299.264	11.85	795.75	14.45	1948.35
1.50	2.3090	4.10	24.1154	6.70	99.552	9.30	305.280	11.90	810.09	14.50	1981.08
1.55	2.4769	4.15	24.9034	6.75	101.929	9.35	311.401	11.95	824.66	14.55	2014.32
1.60	2.6522	4.20	25.7103	6.80	104.352	9.40	317.628	12.00	839.47	14.60	2048.09
1.65	2.8351	4.25	26.5365	6.85	106.823	9.45	323.964	12.05	854.53	14.65	2082.39
1.70	3.0257	4.30	27.3823	6.90	109.341	9.50	330.410	12.10	869.82	14.70	2117.23
1.75	3.2241	4.35	28.2482	6.95	111.908	9.55	336.968	12.15	885.37	14.75	2152.61
1.80	3.4307	4.40	29.1344	7.00	114.524	9.60	343.639	12.20	901.17	14.80	2188.55
1.85	3.6456	4.45	30.0413	7.05	117.191	9.65	350.426	12.25	917.23	14.85	2225.05
1.90	3.8690	4.50	30.9693	7.10	119.909	9.70	357.329	12.30	933.54	14.90	2262.12
1.95	4.1010	4.55	31.9188	7.15	122.678	9.75	364.352	12.35	950.12	14.95	2299.77
2.00	4.3420	4.60	32.8901	7.20	125.501	9.80	371.495	12.40	966.97	15.00	2338.01
2.05	4.5920	4.65	33.8837	7.25	128.377	9.85	378.761	12.45	984.10	15.05	2376.85
2.10	4.8513	4.70	34.8999	7.30	131.308	9.90	386.152	12.50	1001.50	15.10	2416.30
2.15	5.1202	4.75	35.9392	7.35	134.294	9.95	393.669	12.55	1019.18	15.15	2456.36
2.20	5.3988	4.80	37.0020	7.40	137.337	10.00	401.315	12.60	1037.14	15.20	2497.05
2.25	5.6873	4.85	38.0886	7.45	140.437	10.05	409.092	12.65	1055.40	15.25	2538.38
2.30	5.9861	4.90	39.1996	7.50	143.595	10.10	417.001	12.70	1073.95	15.30	2580.34
2.35	6.2952	4.95	40.3352	7.55	146.812	10.15	425.044	12.75	1092.79	15.35	2622.97
2.40	6.6151	5.00	41.4961	7.60	150.090	10.20	433.225	12.80	1111.94	15.40	2666.26
2.45	6.9459	5.05	42.6825	7.65	153.428	10.25	441.544	12.85	1131.40	15.45	2710.22
2.50	7.2878	5.10	43.8950	7.70	156.829	10.30	450.005	12.90	1151.18	15.50	2754.87
2.55	7.6411	5.15	45.1340	7.75	160.293	10.35	458.609	12.95	1171.26	15.55	2800.21
2.60	8.0062	5.20	46.4001	7.80	163.821	10.40	467.358	13.00	1191.68	15.60	2846.26
2.65	8.3831	5.25	47.6935	7.85	167.414	10.45	476.255	13.05	1212.41	15.65	2893.03
2.70	8.7723	5.30	49.0149	7.90	171.073	10.50	485.303	13.10	1233.48	15.70	2940.52
2.75	9.1740	5.35	50.3647	7.95	174.800	10.55	494.502	13.15	1254.89	15.75	2988.75
2.80	9.5885	5.40	51.7434	8.00	178.595	10.60	503.857	13.20	1276.64	15.80	3037.73
2.85	10.0160	5.45	53.1515	8.05	182.460	10.65	513.368	13.25	1298.74	15.85	3087.47
2.90	10.4568	5.50	54.5896	8.10	186.396	10.70	523.040	13.30	1321.19	15.90	3137.99
2.95	10.9113	5.55	56.0581	8.15	190.403	10.75	532.873	13.35	1343.99	15.95	3189.28
3.00	11.3798	5.60	57.5575	8.20	194.483	10.80	542.872	13.40	1367.17	16.00	3241.38
3.05	11.8625	5.65	59.0885	8.25	198.637	10.85	553.037	13.45	1390.71	16.05	3294.28
3.10	12.3598	5.70	60.6515	8.30	202.867	10.90	563.372	13.50	1414.62	16.10	3348.01
3.15	12.8720	5.75	62.2470	8.35	207.173	10.95	573.880	13.55	1438.92	16.15	3402.56
3.20	13.3995	5.80	63.8757	8.40	211.557	11.00	584.562	13.60	1463.60	16.20	3457.96
3.25	13.9425	5.85	65.5382	8.45	216.020	11.05	595.423	13.65	1488.67	16.25	3514.23
3.30	14.5015	5.90	67.2349	8.50	220.563	11.10	606.464	13.70	1514.15	16.30	3571.36
3.35	15.0767	5.95	68.9665	8.55	225.187	11.15	617.688	13.75	1540.02	16.35	3629.37
3.40	15.6685	6.00	70.7336	8.60	229.895	11.20	629.099	13.80	1566.31	16.40	3688.29
3.45	16.2773	6.05	72.5367	8.65	234.687	11.25	640.698	13.85	1593.01	16.45	3748.11
3.50	16.9034	6.10	74.3766	8.70	239.564	11.30	652.490	13.90	1620.14	16.50	3808.87
3.55	17.5473	6.15	76.2537	8.75	244.528	11.35	664.476	13.95	1647.69	16.55	3870.56
3.60	18.2093	6.20	78.1688	8.80	249.580	11.40	676.661	14.00	1675.69	16.60	3933.20

A.4 Oblique Shock Tables and Charts and Conical Shock Charts

TABLE A.4.1

Oblique Shock Table, k = 1.4

M=	1.1	1.2	1.3	1.4	1.5	1.6	1.7	1.8	1.9	2.0	2.1
δ=1°	69.8033	58.5479	51.8115	46.8424	42.9130	39.6844	36.9641	34.6300	32.5988	30.8114	29.2239
δ=2°		61.0501	53.4736	48.1732	44.0646	40.7240	37.9274	35.5382	33.4657	31.6463	30.0334
δ=3°		64.3390	55.3170	49.5914	45.2718	41.8046	38.9240	36.4753	34.3587	32.5055	30.8661
δ=4°			57.4228	51.1174	46.5429	42.9308	39.9568	37.4431	35.2792	33.3902	31.7230
δ=5°			59.9613	52.7815	47.8893	44.1082	41.0290	38.4440	36.2288	34.3016	32.6050
δ=6°			63.4584	54.6330	49.3262	45.3438	42.1447	39.4804	37.2093	35.2409	33.5131
δ=7°				56.7615	50.8757	46.6470	43.3090	40.5556	38.2229	36.2098	34.4484
δ=8°				59.3672	52.5715	48.0302	44.5282	41.6734	39.2722	37.2101	35.4125
δ=9°				63.1866	54.4699	49.5111	45.8105	42.8384	40.3601	38.2440	36.4068
δ=10°					56.6787	51.1153	47.1669	44.0567	41.4904	39.3139	37.4332
δ=11°					59.4651	52.8839	48.6122	45.3358	42.6678	40.4231	38.4939
δ=12°					64.3587	54.8890	50.1685	46.6860	43.8981	41.5752	39.5917
δ=13°						57.2828	51.8691	48.1211	45.1887	42.7750	40.7297
δ=14°						60.5370	53.7707	49.6611	46.5498	44.0286	41.9119
δ=15°							55.9840	51.3365	47.9951	45.3436	43.1434
δ=16°							58.7942	53.1976	49.5443	46.7306	44.4307
δ=17°							64.6319	55.3400	51.2276	48.2041	45.7822
δ=18°								57.9947	53.0952	49.7851	47.2095
δ=19°								62.3071	55.2424	51.5063	48.7289
δ=20°									57.9008	53.4229	50.3645
δ=21°									62.2527	55.6443	52.1549
δ=22°										58.4566	54.1686
δ=23°											56.5515
δ=24°											59.7678
δ=25°											
δ=26°											
δ=27°											
δ=28°											
δ=29°											
δ=30°											
δ=31°											
δ=32°											
δ=33°											
δ=34°											
δ=35°											
δ=36°											
δ=37°											
δ=38°											
δ=39°											
δ=40°											
δ=41°											
δ=42°											
δ=43°											
δ=44°											
δ=45°											

(Continued)

TABLE A.4.1 (*Continued*)

Oblique Shock Table, k = 1.4

M=	2.2	2.3	2.4	2.5	2.6	2.7	2.8	2.9	3.0	3.2	3.4
δ=1°	27.8027	26.5220	25.3610	24.3031	23.3348	22.4447	21.6235	20.8634	20.1576	18.8867	17.7737
δ=2°	28.5917	27.2942	26.1192	25.0496	24.0714	23.1728	22.3444	21.5780	20.8667	19.5870	18.4672
δ=3°	29.4033	28.0886	26.8995	25.8181	24.8300	23.9232	23.0877	22.3153	21.5990	20.3111	19.1855
δ=4°	30.2382	28.9057	27.7022	26.6090	25.6110	24.6960	23.8537	23.0756	22.3544	21.0592	19.9284
δ=5°	31.0971	29.7463	28.5279	27.4227	26.4148	25.4917	24.6427	23.8590	23.1333	21.8312	20.6961
δ=6°	31.9809	30.6109	29.3772	28.2596	27.2417	26.3104	25.4548	24.6657	23.9356	22.6273	21.4884
δ=7°	32.8905	31.5003	30.2507	29.1203	28.0922	27.1526	26.2903	25.4959	24.7616	23.4474	22.3053
δ=8°	33.8269	32.4154	31.1489	30.0053	28.9666	28.0186	27.1496	26.3498	25.6114	24.2916	23.1466
δ=9°	34.7915	33.3571	32.0728	30.9151	29.8654	28.9086	28.0327	27.2277	26.4850	25.1598	24.0123
δ=10°	35.7855	34.3264	33.0231	31.8506	30.7892	29.8233	28.9402	28.1296	27.3827	26.0521	24.9022
δ=11°	36.8107	35.3248	34.0009	32.8124	31.7386	30.7630	29.8724	29.0559	28.3046	26.9684	25.8162
δ=12°	37.8690	36.3536	35.0073	33.8016	32.7144	31.7283	30.8297	30.0070	29.2510	27.9089	26.7542
δ=13°	38.9628	37.4147	36.0438	34.8191	33.7173	32.7200	31.8126	30.9832	30.2221	28.8737	27.7162
δ=14°	40.0948	38.5102	37.1119	35.8664	34.7485	33.7388	32.8218	31.9851	31.2184	29.8628	28.7021
δ=15°	41.2688	39.6427	38.2137	36.9449	35.8092	34.7858	33.8582	33.0133	32.2404	30.8767	29.7121
δ=16°	42.4890	40.8155	39.3515	38.0566	36.9009	35.8621	34.9226	34.0686	33.2886	31.9156	30.7463
δ=17°	43.7612	42.0324	40.5284	39.2037	38.0254	36.9692	36.0163	35.1519	34.3639	32.9802	31.8050
δ=18°	45.0925	43.2986	41.7482	40.3891	39.1850	38.1090	37.1408	36.2645	35.4673	34.0710	32.8887
δ=19°	46.4927	44.6206	43.0154	41.6164	40.3825	39.2837	38.2979	37.4079	36.6000	35.1889	33.9980
δ=20°	47.9755	46.0071	44.3362	42.8902	41.6213	40.4960	39.4898	38.5839	37.7636	36.3351	35.1337
δ=21°	49.5604	47.4700	45.7187	44.2162	42.9059	41.7495	40.7195	39.7950	38.9601	37.5111	36.2968
δ=22°	51.2773	49.0263	47.1738	45.6021	44.2421	43.0487	41.9904	41.0440	40.1920	38.7185	37.4887
δ=23°	53.1768	50.7012	48.7168	47.0584	45.6374	44.3993	43.3073	42.3348	41.4624	39.9598	38.7112
δ=24°	55.3558	52.5361	50.3708	48.5998	47.1023	45.8092	44.6761	43.6723	42.7753	41.2378	39.9666
δ=25°	58.0559	54.6058	52.1719	50.2478	48.6510	47.2886	46.1048	45.0626	44.1359	42.5562	41.2575
δ=26°	62.6951	57.0770	54.1842	52.0365	50.3051	48.8526	47.6045	46.5145	45.5512	43.9199	42.5877
δ=27°		60.5476	56.5423	54.0252	52.0980	50.5228	49.1906	48.0396	47.0303	45.3352	43.9619
δ=28°			59.6559	56.3348	54.0881	52.3340	50.8864	49.6548	48.5862	46.8106	45.3863
δ=29°				59.3112	56.3941	54.3468	52.7292	51.3856	50.2377	48.3578	46.8690
δ=30°					59.3524	56.6870	54.7855	53.2738	52.0138	49.9937	48.4215
δ=31°						59.7249	57.1987	55.3968	53.9636	51.7441	50.0603
δ=32°							60.4328	57.9310	56.1821	53.6509	51.8099
δ=33°								61.5687	58.9089	55.7927	53.7110
δ=34°									63.6732	58.3503	55.8378
δ=35°										62.0613	58.3588
δ=36°											61.9147
δ=37°											
δ=38°											
δ=39°											
δ=40°											
δ=41°											
δ=42°											
δ=43°											
δ=44°											
δ=45°											

(*Continued*)

TABLE A.4.1 (*Continued*)

Oblique Shock Table, k = 1.4

M=	3.6	3.8	4.0	4.5	5.0	5.5	6.0	7.0	8.0	10.0	15.0	20.0
δ=1°	16.7905	15.9154	15.1313	13.4859	12.1785	11.1142	10.2307	8.8483	7.8159	6.3770	4.4723	3.5298
δ=2°	17.4789	16.5998	15.8126	14.1625	12.8532	11.7888	10.9064	9.5281	8.5017	7.0773	5.2131	4.3110
δ=3°	18.1928	17.3107	16.5215	14.8693	13.5608	12.4990	11.6204	10.2520	9.2370	7.8374	6.0362	5.1912
δ=4°	18.9324	18.0481	17.2578	15.6060	14.3009	13.2442	12.3719	11.0182	10.0193	8.6531	6.9298	6.1486
δ=5°	19.6974	18.8119	18.0213	16.3721	15.0727	14.0235	13.1598	11.8250	10.8460	9.5193	7.8818	7.1639
δ=6°	20.4878	19.6018	18.8117	17.1673	15.8755	14.8357	13.9824	12.6700	11.7138	10.4306	8.8809	8.2220
δ=7°	21.3034	20.4175	19.6287	17.9907	16.7083	15.6796	14.8382	13.5507	12.6192	11.3817	9.9179	9.3121
δ=8°	22.1439	21.2587	20.4717	18.8416	17.5700	16.5537	15.7254	14.4645	13.5588	12.3676	10.9853	10.4264
δ=9°	23.0092	22.1251	21.3404	19.7193	18.4597	17.4566	16.6422	15.4091	14.5296	13.3838	12.0775	11.5595
δ=10°	23.8989	23.0162	22.2341	20.6230	19.3760	18.3869	17.5869	16.3818	15.5284	14.4266	13.1899	12.7076
δ=11°	24.8129	23.9318	23.1525	21.5518	20.3180	19.3431	18.5576	17.3806	16.5526	15.4926	14.3193	13.8681
δ=12°	25.7508	24.8714	24.0950	22.5050	21.2845	20.3240	19.5529	18.4033	17.5998	16.5791	15.4632	15.0392
δ=13°	26.7125	25.8346	25.0611	23.4818	22.2746	21.3282	20.5712	19.4482	18.6679	17.6839	16.6198	16.2195
δ=14°	27.6979	26.8214	26.0505	24.4815	23.2872	22.3546	21.6112	20.5136	19.7553	18.8052	17.7877	17.4082
δ=15°	28.7069	27.8313	27.0629	25.5036	24.3217	23.4022	22.6719	21.5983	20.8605	19.9416	18.9658	18.6046
δ=16°	29.7394	28.8644	28.0979	26.5475	25.3773	24.4702	23.7521	22.7010	21.9822	21.0918	20.1535	19.8083
δ=17°	30.7957	29.9205	29.1555	27.6129	26.4534	25.5578	24.8511	23.8208	23.1195	22.2551	21.3502	21.0192
δ=18°	31.8759	30.9998	30.2356	28.6995	27.5495	26.6645	25.9683	24.9570	24.2717	23.4308	22.5558	22.2373
δ=19°	32.9806	32.1025	31.3383	29.8071	28.6654	27.7898	27.1031	26.1091	25.4382	24.6184	23.7700	23.4625
δ=20°	34.1102	33.2291	32.4639	30.9357	29.8009	28.9335	28.2551	27.2767	26.6187	25.8178	24.9930	24.6951
δ=21°	35.2655	34.3800	33.6128	32.0856	30.9560	30.0955	29.4244	28.4595	27.8130	27.0288	26.2248	25.9354
δ=22°	36.4477	35.5563	34.7857	33.2570	32.1310	31.2759	30.6108	29.6577	29.0211	28.2515	27.4657	27.1838
δ=23°	37.6580	36.7589	35.9835	34.4506	33.3261	32.4719	31.8147	30.8712	30.2432	29.4862	28.7161	28.4407
δ=24°	38.8984	37.9892	37.2074	35.6671	34.5419	33.6931	33.0363	32.1006	31.4796	30.7332	29.9766	29.7067
δ=25°	40.1709	39.2493	38.4589	36.9077	35.7794	34.9311	34.2763	33.3462	32.7307	31.9931	31.2478	30.9825
δ=26°	41.4786	40.5413	39.7400	38.1737	37.0397	36.1898	35.5356	34.6089	33.9974	33.2666	32.5303	32.2688
δ=27°	42.8250	41.8684	41.0533	39.4671	38.3242	37.4706	36.8152	35.8896	35.2805	34.5545	33.8252	33.5667
δ=28°	44.2150	43.2344	42.4021	40.7902	39.6347	38.7749	38.1167	37.1895	36.5813	35.8579	35.1335	34.8773
δ=29°	45.6548	44.6443	43.7905	42.1459	40.9737	40.1049	39.4417	38.5103	37.9010	37.1783	36.4565	36.2018
δ=30°	47.1530	46.1049	45.2241	43.5382	42.3443	41.4631	40.7925	39.8538	39.2415	38.5171	37.7958	37.5417
δ=31°	48.7211	47.6251	46.7101	44.9720	43.7501	42.8528	42.1721	41.2224	40.6050	39.8765	39.1532	38.8988
δ=32°	50.3762	49.2175	48.2585	46.4538	45.1964	44.2779	43.5839	42.6193	41.9942	41.2588	40.5308	40.2753
δ=33°	52.1437	50.9004	49.8833	47.9925	46.6896	45.7440	45.0327	44.0481	43.4125	42.6670	41.9313	41.6737
δ=34°	54.0659	52.7018	51.6052	49.6004	48.2385	47.2581	46.5244	45.5137	44.8641	44.1048	43.3581	43.0972
δ=35°	56.2214	54.6692	53.4569	51.2953	49.8554	48.8297	48.0671	47.0225	46.3546	45.5690	44.8151	44.5497
δ=36°	58.7932	56.8941	55.4956	53.1046	51.5582	50.4722	49.6715	48.5830	47.8911	47.0893	46.3076	46.0360
δ=37°	62.5461	59.6060	57.8389	55.0741	53.3741	52.2052	51.3535	50.2066	49.4834	48.6502	47.8422	47.5625
δ=38°		64.1923	60.8271	57.2921	55.3490	54.0595	53.1363	51.9102	51.1451	50.2703	49.4279	49.1373
δ=39°				59.9766	57.5711	56.0881	55.0581	53.7189	52.8963	51.9652	51.0768	50.7721
δ=40°				64.3401	60.2593	58.3979	57.1882	55.6749	54.7684	53.7582	52.8069	52.4830
δ=41°					64.6552	61.2847	59.6820	57.8564	56.8161	55.6870	54.6454	54.2947
δ=42°							63.1049	60.4497	59.1517	57.8207	56.6384	56.2474
δ=43°								64.2844	62.0971	60.3138	58.8748	58.4167
δ=44°										63.7354	61.5801	60.9784
δ=45°											66.2215	64.6796

TABLE A.4.2

Oblique Shock Table, k = 1.3

M=	1.1	1.2	1.3	1.4	1.5	1.6	1.7	1.8	1.9	2.0	2.1
δ=1°	69.5437	58.4511	51.7441	46.7877	42.8654	39.6414	36.9242	34.5923	32.5629	30.7768	29.1903
δ=2°		60.8067	53.3226	48.0551	43.9635	40.6334	37.8438	35.4597	33.3909	31.5744	29.9638
δ=3°			63.8069	55.0576	49.3984	45.1101	41.6613	38.7927	36.3523	34.2419	32.3935	30.7578
δ=4°			69.0210	57.0115	50.8339	46.3118	42.7286	39.7729	37.2718	35.1170	33.2349	31.5730
δ=5°				59.3057	52.3847	47.5773	43.8397	40.7870	38.2198	36.0174	34.0996	32.4102
δ=6°				62.2495	54.0867	48.9182	44.9999	41.8382	39.1984	36.9443	34.9885	33.2701
δ=7°				68.0443	56.0003	50.3505	46.2163	42.9303	40.2098	37.8996	35.9028	34.1536
δ=8°					58.2438	51.8976	47.4978	44.0678	41.2568	38.8850	36.8439	35.0617
δ=9°					61.1236	53.5956	48.8565	45.2566	42.3430	39.9027	37.8133	35.9955
δ=10°					67.5576	55.5066	50.3095	46.5043	43.4723	40.9554	38.8127	36.9564
δ=11°						57.7544	51.8819	47.8206	44.6501	42.0463	39.8444	37.9458
δ=12°						60.6754	53.6133	49.2194	45.8830	43.1792	40.9110	38.9656
δ=13°							55.5743	50.7201	47.1798	44.3591	42.0154	40.0178
δ=14°							57.9135	52.3522	48.5522	45.5922	43.1616	41.1052
δ=15°							61.0951	54.1648	50.0168	46.8866	44.3543	42.2308
δ=16°								56.2503	51.5984	48.2533	45.5997	43.3985
δ=17°								58.8303	53.3363	49.7077	46.9058	44.6133
δ=18°								62.9657	55.3008	51.2722	48.2835	45.8815
δ=19°									57.6432	52.9824	49.7479	47.2114
δ=20°									60.8574	54.9003	51.3213	48.6145
δ=21°										57.1539	53.0392	50.1067
δ=22°										60.1183	54.9633	51.7120
δ=23°											57.2216	53.4690
δ=24°											60.1932	55.4468
δ=25°												57.7956
δ=26°												61.0202
δ=27°												
δ=28°												
δ=29°												
δ=30°												
δ=31°												
δ=32°												
δ=33°												
δ=34°												
δ=35°												
δ=36°												
δ=37°												
δ=38°												
δ=39°												
δ=40°												
δ=41°												
δ=42°												
δ=43°												
δ=44°												
δ=45°												

(Continued)

TABLE A.4.2 (*Continued*)

Oblique Shock Table, k = 1.3

M=	2.2	2.3	2.4	2.5	2.6	2.7	2.8	2.9	3.0	3.2	3.4
δ=1°	27.7701	26.4900	25.3296	24.2723	23.3043	22.4146	21.5938	20.8339	20.1283	18.8578	17.7452
δ=2°	28.5240	27.2279	26.0542	24.9857	24.0083	23.1105	22.2827	21.5168	20.8061	19.5271	18.4079
δ=3°	29.2979	27.9855	26.7984	25.7187	24.7319	23.8262	22.9917	22.2202	21.5045	20.2178	19.0929
δ=4°	30.0924	28.7633	27.5625	26.4716	25.4755	24.5621	23.7211	22.9441	22.2238	20.9300	19.8002
δ=5°	30.9080	29.5617	28.3470	27.2447	26.2394	25.3182	24.4709	23.6886	22.9641	21.6638	20.5298
δ=6°	31.7454	30.3812	29.1522	28.0384	27.0237	26.0949	25.2413	24.4540	23.7253	22.4191	21.2816
δ=7°	32.6051	31.2223	29.9786	28.8529	27.8287	26.8923	26.0326	25.2403	24.5076	23.1960	22.0554
δ=8°	33.4880	32.0857	30.8266	29.6887	28.6547	27.7106	26.8447	26.0475	25.3110	23.9943	22.8511
δ=9°	34.3950	32.9720	31.6967	30.5461	29.5021	28.5500	27.6779	26.8758	26.1356	24.8139	23.6686
δ=10°	35.3269	33.8819	32.5895	31.4256	30.3712	29.4108	28.5323	27.7253	26.9813	25.6549	24.5075
δ=11°	36.2851	34.8163	33.5057	32.3277	31.2623	30.2934	29.4081	28.5960	27.8482	26.5170	25.3678
δ=12°	37.2706	35.7762	34.4461	33.2530	32.1759	31.1979	30.3057	29.4882	28.7364	27.4003	26.2493
δ=13°	38.2852	36.7628	35.4115	34.2022	33.1125	32.1248	31.2252	30.4020	29.6460	28.3048	27.1518
δ=14°	39.3307	37.7775	36.4031	35.1761	34.0728	33.0747	32.1670	31.3377	30.5771	29.2303	28.0751
δ=15°	40.4093	38.8219	37.4219	36.1755	35.0574	34.0480	33.1315	32.2956	31.5300	30.1770	29.0192
δ=16°	41.5236	39.8978	38.4695	37.2018	36.0674	35.0454	34.1194	33.2761	32.5050	31.1449	29.9840
δ=17°	42.6770	41.0079	39.5477	38.2561	37.1036	36.0678	35.1311	34.2797	33.5023	32.1343	30.9695
δ=18°	43.8737	42.1548	40.6586	39.3403	38.1675	37.1161	36.1675	35.3070	34.5226	33.1453	31.9759
δ=19°	45.1189	43.3423	41.8049	40.4562	39.2605	38.1917	37.2297	36.3588	35.5664	34.1785	33.0033
δ=20°	46.4196	44.5750	42.9899	41.6063	40.3846	39.2960	38.3187	37.4360	36.6346	35.2342	34.0520
δ=21°	47.7847	45.8588	44.2177	42.7938	41.5421	40.4308	39.4361	38.5400	37.7280	36.3131	35.1224
δ=22°	49.2270	47.2014	45.4936	44.0224	42.7360	41.5985	40.5838	39.6721	38.8480	37.4162	36.2153
δ=23°	50.7646	48.6136	46.8245	45.2971	43.9699	42.8019	41.7639	40.8342	39.9960	38.5444	37.3312
δ=24°	52.4253	50.1100	48.2199	46.6243	45.2485	44.0446	42.9794	42.0287	41.1741	39.6992	38.4713
δ=25°	54.2554	51.7130	49.6926	48.0125	46.5780	45.3313	44.2339	43.2584	42.3845	40.8822	39.6367
δ=26°	56.3423	53.4574	51.2615	49.4737	47.9665	46.6678	45.5319	44.5271	43.6303	42.0956	40.8292
δ=27°	58.8985	55.4052	52.9557	51.0247	49.4252	48.0624	46.8796	45.8394	44.9154	43.3421	42.0507
δ=28°	62.9824	57.6864	54.8244	52.6914	50.9701	49.5258	48.2850	47.2017	46.2447	44.6251	43.3038
δ=29°		60.6976	56.9633	54.5163	52.6254	51.0739	49.7593	48.6223	47.6248	45.9491	44.5916
δ=30°			59.6173	56.579	54.4303	52.7301	51.3182	50.1127	49.0645	47.3198	45.9185
δ=31°			64.5505	59.0656	56.4568	54.5329	52.9858	51.6895	50.5762	48.7449	47.2897
δ=32°				62.7765	58.8667	56.5518	54.8013	53.3780	52.1776	50.2350	48.7125
δ=33°					62.2620	58.9424	56.8363	55.2199	53.8964	51.8051	50.1966
δ=34°						62.2641	59.2534	57.2936	55.7792	53.4780	51.7560
δ=35°							62.6660	59.7834	57.9164	55.2903	53.4111
δ=36°								63.4943	60.5364	57.3086	55.1947
δ=37°									65.1201	59.6801	57.1646
δ=38°										62.9217	59.4419
δ=39°											62.3949
δ=40°											
δ=41°											
δ=42°											
δ=43°											
δ=44°											
δ=45°											

(*Continued*)

TABLE A.4.2 (*Continued*)

Oblique Shock Table, k = 1.3

M=	3.6	3.8	4.0	4.5	5.0	5.5	6.0	7.0	8.0	10.0	15.0	20.0
δ=1°	16.7622	15.8873	15.1034	13.4583	12.1510	11.0868	10.2034	8.8208	7.7883	6.3491	4.4434	3.4997
δ=2°	17.4200	16.5412	15.7542	14.1045	12.7954	11.7309	10.8482	9.4695	8.4424	7.0164	5.1479	4.2418
δ=3°	18.1009	17.2191	16.4302	14.7783	13.4697	12.4075	11.5283	10.1584	9.1418	7.7387	5.9288	5.0762
δ=4°	18.8049	17.9211	17.1310	15.4792	14.1736	13.1161	12.2427	10.8864	9.8847	8.5125	6.7758	5.9839
δ=5°	19.5319	18.6468	17.8564	16.2070	14.9065	13.8558	12.9903	11.6516	10.6684	9.3333	7.6779	6.9473
δ=6°	20.2818	19.3962	18.6063	16.9611	15.6676	14.6257	13.7699	12.4520	11.4902	10.1963	8.6252	7.9523
δ=7°	21.0544	20.1689	19.3801	17.7409	16.4562	15.4246	14.5800	13.2855	12.3471	11.0967	9.6089	8.9885
δ=8°	21.8494	20.9646	20.1775	18.5457	17.2712	16.2513	15.4190	14.1498	13.2360	12.0301	10.6220	10.0484
δ=9°	22.6665	21.7828	20.9980	19.3748	18.1117	17.1044	16.2853	15.0426	14.1540	12.9922	11.6590	11.1267
δ=10°	23.5056	22.6233	21.8411	20.2275	18.9765	17.9825	17.1773	15.9617	15.0983	13.9794	12.7156	12.2195
δ=11°	24.3661	23.4856	22.7062	21.1029	19.8647	18.8845	18.0933	16.9049	16.0663	14.9887	13.7885	13.3241
δ=12°	25.2479	24.3693	23.5929	22.0003	20.7751	19.8090	19.0319	17.8704	17.0559	16.0174	14.8751	14.4386
δ=13°	26.1507	25.2740	24.5007	22.9189	21.7069	20.7547	19.9915	18.8563	18.0650	17.0632	15.9735	15.5614
δ=14°	27.0742	26.1993	25.4290	23.8580	22.6591	21.7206	20.9711	19.8612	19.0918	18.1243	17.0822	16.6917
δ=15°	28.0182	27.1450	26.3775	24.8170	23.6308	22.7056	21.9692	20.8835	20.1349	19.1991	18.2001	17.8286
δ=16°	28.9825	28.1106	27.3458	25.7952	24.6212	23.7088	22.9849	21.9222	21.1930	20.2865	19.3263	18.9716
δ=17°	29.9670	29.0961	28.3335	26.7921	25.6296	24.7294	24.0173	22.9761	22.2650	21.3854	20.4602	20.1204
δ=18°	30.9717	30.1011	29.3404	27.8072	26.6555	25.7666	25.0656	24.0444	23.3500	22.4950	21.6013	21.2748
δ=19°	31.9966	31.1258	30.3663	28.8403	27.6983	26.8198	26.1291	25.1264	24.4473	23.6147	22.7493	22.4346
δ=20°	33.0418	32.1700	31.4112	29.8909	28.7576	27.8886	27.2073	26.2215	25.5564	24.7441	23.9041	23.5998
δ=21°	34.1077	33.2339	32.4750	30.9590	29.8331	28.9726	28.2997	27.3292	26.6768	25.8828	25.0655	24.7705
δ=22°	35.1946	34.3178	33.5579	32.0444	30.9247	30.0716	29.4061	28.4493	27.8082	27.0307	26.2335	25.9467
δ=23°	36.3031	35.4221	34.6601	33.1474	32.0323	31.1853	30.5263	29.5816	28.9505	28.1875	27.4083	27.1288
δ=24°	37.4339	36.5473	35.7822	34.2679	33.1560	32.3139	31.6602	30.7259	30.1037	29.3535	28.5901	28.3169
δ=25°	38.5881	37.6942	36.9247	35.4065	34.2960	33.4573	32.8080	31.8824	31.2677	30.5286	29.7790	29.5114
δ=26°	39.7668	38.8639	38.0884	36.5636	35.4525	34.6160	33.9699	33.0512	32.4428	31.7132	30.9753	30.7126
δ=27°	40.9716	40.0574	39.2743	37.7399	36.6263	35.7903	35.1462	34.2326	33.6292	32.9074	32.1796	31.9209
δ=28°	42.2045	41.2766	40.4838	38.9364	37.8179	36.9808	36.3374	35.4271	34.8275	34.1118	33.3922	33.1370
δ=29°	43.4681	42.5233	41.7187	40.1541	39.0282	38.1883	37.5441	36.6353	36.0381	35.3270	34.6138	34.3614
δ=30°	44.7656	43.8002	42.9809	41.3948	40.2586	39.4137	38.7674	37.8578	37.2617	36.5535	35.8450	35.5947
δ=31°	46.1009	45.1106	44.2733	42.6601	41.5103	40.6584	40.0082	39.0957	38.4992	37.7921	37.0866	36.8378
δ=32°	47.4795	46.4585	45.5992	43.9526	42.7854	41.9237	41.2680	40.3502	39.7517	39.0440	38.3396	38.0916
δ=33°	48.9083	47.8495	46.9632	45.2751	44.0860	43.2118	42.5484	41.6225	41.0205	40.3102	39.6050	39.3571
δ=34°	50.3971	49.2909	48.3709	46.6316	45.4152	44.5249	43.8515	42.9145	42.3071	41.5921	40.8842	40.6357
δ=35°	51.9595	50.7924	49.8300	48.0270	46.7766	45.8662	45.1799	44.2284	43.6134	42.8915	42.1786	41.9288
δ=36°	53.6154	52.3682	51.3512	49.4678	48.1749	47.2393	46.5369	45.5668	44.9418	44.2103	43.4900	43.2382
δ=37°	55.3969	54.0392	52.9498	50.9627	49.6164	48.6493	47.9267	46.9329	46.2953	45.5511	44.8207	44.5658
δ=38°	57.3601	55.8389	54.6485	52.5241	51.1095	50.1026	49.3545	48.3310	47.6773	46.9169	46.1732	45.9143
δ=39°	59.6216	57.8269	56.4849	54.1700	52.6657	51.6078	50.8275	49.7666	49.0925	48.3116	47.5509	47.2867
δ=40°	62.5292	60.1304	58.5278	55.9282	54.3019	53.1771	52.3553	51.2467	50.5469	49.7401	48.9578	48.6868
δ=41°		63.1544	60.9332	57.8468	56.0441	54.8282	53.9515	52.7811	52.0484	51.2088	50.3991	50.1196
δ=42°			64.2945	60.0207	57.9359	56.5886	55.6362	54.3836	53.6081	52.7262	51.8816	51.5914
δ=43°				62.7036	60.0618	58.5054	57.4420	56.0748	55.2418	54.3044	53.4147	53.1106
δ=44°					62.6349	60.6720	59.4264	57.8875	56.9736	55.9606	55.0113	54.6890
δ=45°					66.9841	63.3378	61.7126	59.8811	58.8437	57.7219	56.6905	56.3438
δ=46°							64.6948	62.1839	60.9286	59.6346	58.4827	58.1016
δ=47°								65.2219	63.4138	61.7910	60.4417	60.0078
δ=48°									67.2287	64.4388	62.6803	62.1527
δ=49°											65.5445	64.7771

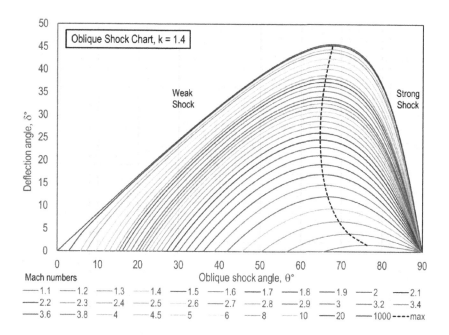

FIGURE A.4.1
Oblique Shock Chart, k = 1.4.

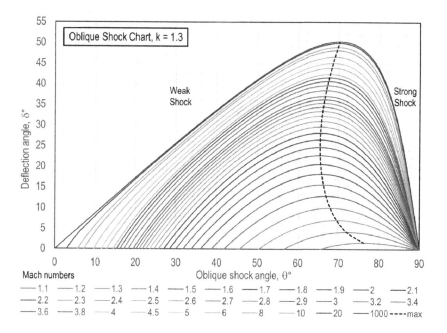

FIGURE A.4.2
Oblique Shock Chart, k = 1.3.

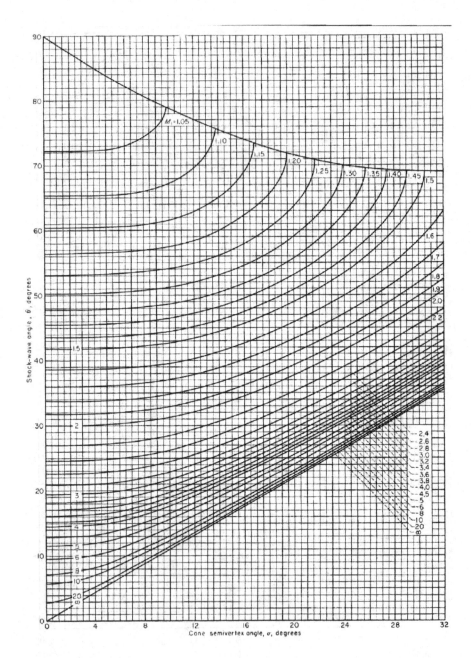

FIGURE A.4.3A
Conical shock, shock angle versus semivertex angle, k = 1.405 [NACA 1135].

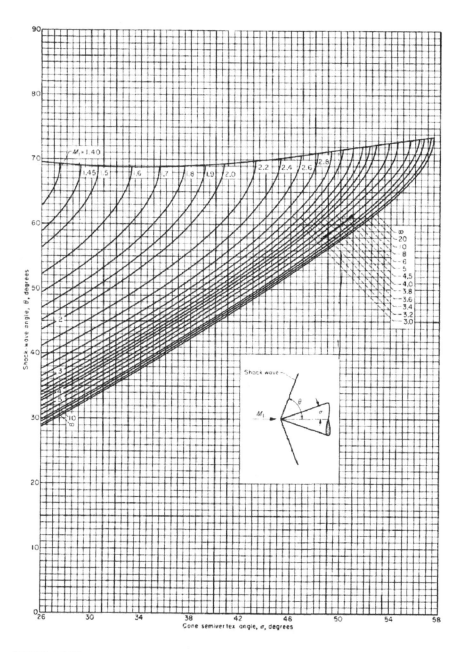

FIGURE A.4.3B
Conical shock, shock angle versus semivertex angle, k = 1.405 [NACA 1135].

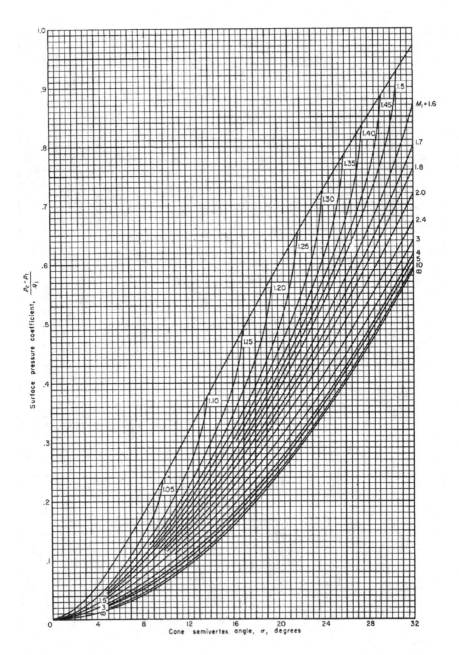

FIGURE A.4.4A
Conical shock chart, surface pressure coef. vs. Mach No. and semivertex angle.

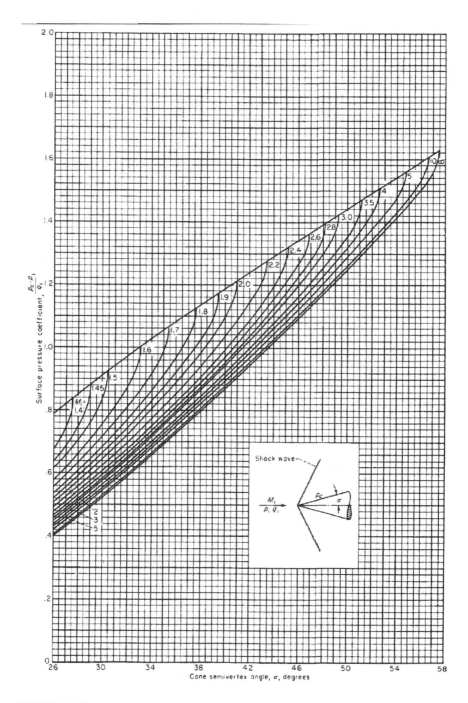

FIGURE A.4.4B
Conical shock chart, surface pressure coef. vs. Mach No. and semivertex angle.

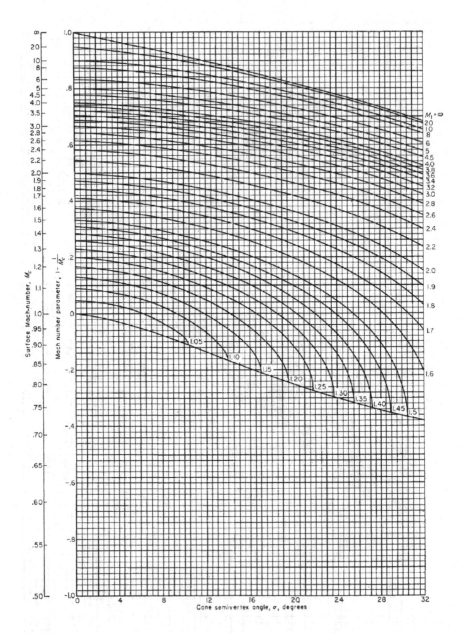

FIGURE A.4.5A
Conical shock surface Mach No. vs. approach Mach number and semivertex angle.

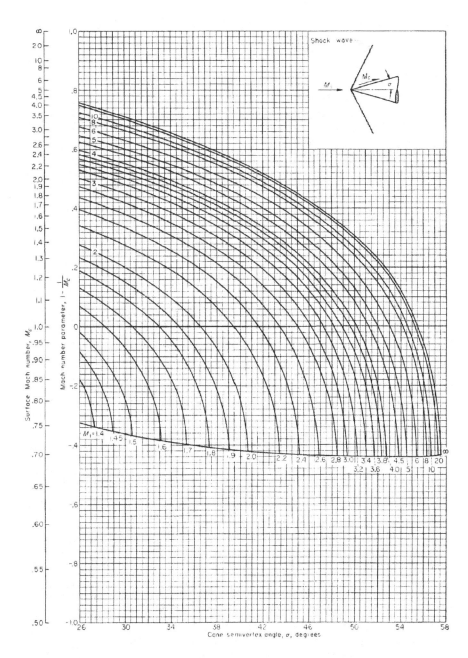

FIGURE A.4.5B

Conical shock surface Mach No. vs approach Mach number and semivertex angle.

A.5 Prandtl Meyer Flow Tables

TABLE A.5.1

Prandtl Meyer flow table, k = 1.4

M	ν	μ	M	ν	μ	M	ν	μ	M	ν	μ	M	ν	μ
1.00	0.0000	90.0000	1.50	11.9052	41.8103	2.00	26.3798	30.0000	2.50	39.1236	23.5782	3.00	49.7573	19.4712
1.01	0.0447	81.9307	1.51	12.1999	41.4718	2.01	26.6550	29.8356	2.51	39.3565	23.4786	3.01	49.9499	19.4039
1.02	0.1257	78.6351	1.52	12.4949	41.1395	2.02	26.9295	29.6730	2.52	39.5886	23.3799	3.02	50.1417	19.3371
1.03	0.2294	76.1376	1.53	12.7901	40.8132	2.03	27.2033	29.5123	2.53	39.8199	23.2820	3.03	50.3328	19.2708
1.04	0.3510	74.0576	1.54	13.0856	40.4927	2.04	27.4762	29.3535	2.54	40.0503	23.1850	3.04	50.5231	19.2049
1.05	0.4874	72.2472	1.55	13.3812	40.1778	2.05	27.7484	29.1964	2.55	40.2798	23.0888	3.05	50.7127	19.1395
1.06	0.6367	70.6300	1.56	13.6770	39.8683	2.06	28.0197	29.0411	2.56	40.5085	22.9934	3.06	50.9016	19.0745
1.07	0.7973	69.1603	1.57	13.9728	39.5642	2.07	28.2903	28.8875	2.57	40.7363	22.8988	3.07	51.0897	19.0100
1.08	0.9680	67.8084	1.58	14.2686	39.2652	2.08	28.5600	28.7357	2.58	40.9633	22.8051	3.08	51.2771	18.9459
1.09	1.1479	66.5534	1.59	14.5645	38.9713	2.09	28.8290	28.5855	2.59	41.1894	22.7121	3.09	51.4638	18.8823
1.10	1.3362	65.3800	1.60	14.8604	38.6822	2.10	29.0971	28.4369	2.60	41.4147	22.6199	3.10	51.6497	18.8191
1.11	1.5321	64.2767	1.61	15.1561	38.3978	2.11	29.3644	28.2899	2.61	41.6392	22.5284	3.11	51.8350	18.7563
1.12	1.7350	63.2345	1.62	15.4518	38.1181	2.12	29.6308	28.1446	2.62	41.8628	22.4377	3.12	52.0195	18.6939
1.13	1.9445	62.2461	1.63	15.7473	37.8428	2.13	29.8965	28.0008	2.63	42.0855	22.3478	3.13	52.2033	18.6320
1.14	2.1600	61.3056	1.64	16.0427	37.5719	2.14	30.1613	27.8585	2.64	42.3074	22.2586	3.14	52.3864	18.5705
1.15	2.3810	60.4082	1.65	16.3379	37.3052	2.15	30.4253	27.7177	2.65	42.5285	22.1702	3.15	52.5688	18.5094
1.16	2.6073	59.5497	1.66	16.6328	37.0427	2.16	30.6884	27.5785	2.66	42.7488	22.0824	3.16	52.7505	18.4487
1.17	2.8385	58.7267	1.67	16.9276	36.7842	2.17	30.9507	27.4406	2.67	42.9682	21.9954	3.17	52.9315	18.3884
1.18	3.0743	57.9362	1.68	17.2220	36.5296	2.18	31.2121	27.3043	2.68	43.1868	21.9090	3.18	53.1118	18.3285
1.19	3.3142	57.1756	1.69	17.5161	36.2789	2.19	31.4727	27.1693	2.69	43.4045	21.8234	3.19	53.2914	18.2691
1.20	3.5582	56.4427	1.70	17.8099	36.0319	2.20	31.7325	27.0357	2.70	43.6215	21.7385	3.20	53.4703	18.2100
1.21	3.8060	55.7354	1.71	18.1034	35.7885	2.21	31.9914	26.9035	2.71	43.8376	21.6542	3.21	53.6486	18.1512
1.22	4.0572	55.0520	1.72	18.3964	35.5487	2.22	32.2494	26.7726	2.72	44.0529	21.5706	3.22	53.8261	18.0929
1.23	4.3117	54.3909	1.73	18.6891	35.3124	2.23	32.5066	26.6430	2.73	44.2673	21.4876	3.23	54.0029	18.0350
1.24	4.5694	53.7507	1.74	18.9814	35.0795	2.24	32.7629	26.5148	2.74	44.4810	21.4053	3.24	54.1791	17.9774
1.25	4.8299	53.1301	1.75	19.2732	34.8499	2.25	33.0184	26.3878	2.75	44.6938	21.3237	3.25	54.3546	17.9202
1.26	5.0931	52.5280	1.76	19.5646	34.6235	2.26	33.2730	26.2621	2.76	44.9059	21.2427	3.26	54.5294	17.8634
1.27	5.3590	51.9433	1.77	19.8554	34.4003	2.27	33.5268	26.1376	2.77	45.1171	21.1623	3.27	54.7036	17.8069
1.28	5.6272	51.3752	1.78	20.1458	34.1802	2.28	33.7796	26.0144	2.78	45.3275	21.0825	3.28	54.8770	17.7508
1.29	5.8977	50.8226	1.79	20.4357	33.9631	2.29	34.0316	25.8923	2.79	45.5371	21.0034	3.29	55.0498	17.6951
1.30	6.1703	50.2849	1.80	20.7251	33.7490	2.30	34.2828	25.7715	2.80	45.7459	20.9248	3.30	55.2220	17.6397
1.31	6.4449	49.7612	1.81	21.0139	33.5377	2.31	34.5331	25.6518	2.81	45.9539	20.8469	3.31	55.3935	17.5847
1.32	6.7213	49.2509	1.82	21.3021	33.3293	2.32	34.7825	25.5332	2.82	46.1611	20.7695	3.32	55.5643	17.5300
1.33	6.9995	48.7535	1.83	21.5898	33.1237	2.33	35.0310	25.4158	2.83	46.3675	20.6928	3.33	55.7344	17.4756
1.34	7.2794	48.2682	1.84	21.8768	32.9207	2.34	35.2787	25.2995	2.84	46.5731	20.6166	3.34	55.9040	17.4216
1.35	7.5607	47.7946	1.85	22.1633	32.7204	2.35	35.5255	25.1843	2.85	46.7779	20.5410	3.35	56.0728	17.3680
1.36	7.8435	47.3321	1.86	22.4492	32.5227	2.36	35.7715	25.0702	2.86	46.9820	20.4659	3.36	56.2411	17.3147
1.37	8.1276	46.8803	1.87	22.7344	32.3276	2.37	36.0165	24.9572	2.87	47.1852	20.3914	3.37	56.4086	17.2617
1.38	8.4130	46.4387	1.88	23.0190	32.1349	2.38	36.2607	24.8452	2.88	47.3877	20.3175	3.38	56.5756	17.2090
1.39	8.6995	46.0070	1.89	23.3029	31.9447	2.39	36.5041	24.7342	2.89	47.5894	20.2441	3.39	56.7419	17.1567
1.40	8.9870	45.5847	1.90	23.5861	31.7569	2.40	36.7465	24.6243	2.90	47.7903	20.1713	3.40	56.9075	17.1046
1.41	9.2756	45.1715	1.91	23.8687	31.5714	2.41	36.9881	24.5154	2.91	47.9905	20.0990	3.41	57.0725	17.0529
1.42	9.5650	44.7670	1.92	24.1506	31.3882	2.42	37.2289	24.4075	2.92	48.1898	20.0272	3.42	57.2369	17.0016
1.43	9.8553	44.3709	1.93	24.4318	31.2072	2.43	37.4687	24.3005	2.93	48.3884	19.9559	3.43	57.4007	16.9505
1.44	10.1464	43.9830	1.94	24.7123	31.0285	2.44	37.7077	24.1945	2.94	48.5863	19.8852	3.44	57.5639	16.8997
1.45	10.4381	43.6028	1.95	24.9920	30.8519	2.45	37.9459	24.0895	2.95	48.7833	19.8149	3.45	57.7264	16.8493
1.46	10.7305	43.2302	1.96	25.2711	30.6774	2.46	38.1831	23.9854	2.96	48.9796	19.7452	3.46	57.8883	16.7991
1.47	11.0235	42.8649	1.97	25.5494	30.5050	2.47	38.4195	23.8822	2.97	49.1752	19.6760	3.47	58.0496	16.7493
1.48	11.3169	42.5066	1.98	25.8269	30.3347	2.48	38.6551	23.7800	2.98	49.3700	19.6072	3.48	58.2102	16.6997
1.49	11.6109	42.1552	1.99	26.1037	30.1664	2.49	38.8897	23.6786	2.99	49.5640	19.5390	3.49	58.3703	16.6505

(Continued)

TABLE A.5.1 (*Continued*)

Prandtl Meyer flow table, k = 1.4

M	v (°)	μ (°)	M	v (°)	μ (°)	M	v (°)	μ (°)	M	v (°)	μ (°)	M	v (°)	μ (°)
3.50	58.5298	16.6015	4.00	65.7848	14.4775	4.50	71.8317	12.8396	5.00	76.9202	11.5370	5.50	81.2448	10.4757
3.51	58.6886	16.5529	4.01	65.9167	14.4406	4.51	71.9422	12.8106	5.01	77.0136	11.5136	5.51	81.3246	10.4565
3.52	58.8469	16.5045	4.02	66.0480	14.4039	4.52	72.0522	12.7818	5.02	77.1067	11.4904	5.52	81.4041	10.4373
3.53	59.0045	16.4564	4.03	66.1789	14.3674	4.53	72.1619	12.7531	5.03	77.1995	11.4672	5.53	81.4834	10.4182
3.54	59.1616	16.4086	4.04	66.3093	14.3311	4.54	72.2712	12.7246	5.04	77.2921	11.4442	5.54	81.5625	10.3992
3.55	59.3180	16.3611	4.05	66.4393	14.2949	4.55	72.3801	12.6961	5.05	77.3843	11.4212	5.55	81.6413	10.3803
3.56	59.4739	16.3139	4.06	66.5688	14.2590	4.56	72.4887	12.6678	5.06	77.4762	11.3983	5.56	81.7199	10.3614
3.57	59.6291	16.2669	4.07	66.6978	14.2232	4.57	72.5968	12.6396	5.07	77.5678	11.3755	5.57	81.7983	10.3426
3.58	59.7838	16.2202	4.08	66.8263	14.1876	4.58	72.7046	12.6116	5.08	77.6591	11.3528	5.58	81.8764	10.3238
3.59	59.9379	16.1738	4.09	66.9544	14.1522	4.59	72.8121	12.5837	5.09	77.7501	11.3302	5.59	81.9543	10.3052
3.60	60.0915	16.1276	4.10	67.0820	14.1170	4.60	72.9192	12.5559	5.10	77.8409	11.3077	5.60	82.0319	10.2866
3.61	60.2444	16.0817	4.11	67.2092	14.0819	4.61	73.0259	12.5282	5.11	77.9313	11.2853	5.61	82.1093	10.2680
3.62	60.3968	16.0361	4.12	67.3359	14.0470	4.62	73.1322	12.5006	5.12	78.0215	11.2630	5.62	82.1865	10.2496
3.63	60.5486	15.9907	4.13	67.4621	14.0123	4.63	73.2382	12.4732	5.13	78.1113	11.2407	5.63	82.2634	10.2312
3.64	60.6998	15.9456	4.14	67.5879	13.9778	4.64	73.3438	12.4459	5.14	78.2009	11.2186	5.64	82.3401	10.2128
3.65	60.8504	15.9008	4.15	67.7132	13.9434	4.65	73.4491	12.4187	5.15	78.2902	11.1965	5.65	82.4166	10.1946
3.66	61.0005	15.8562	4.16	67.8381	13.9092	4.66	73.5540	12.3916	5.16	78.3792	11.1745	5.66	82.4929	10.1763
3.67	61.1501	15.8119	4.17	67.9626	13.8752	4.67	73.6586	12.3647	5.17	78.4679	11.1526	5.67	82.5689	10.1582
3.68	61.2990	15.7678	4.18	68.0866	13.8414	4.68	73.7628	12.3378	5.18	78.5564	11.1308	5.68	82.6447	10.1401
3.69	61.4474	15.7239	4.19	68.2101	13.8077	4.69	73.8666	12.3111	5.19	78.6445	11.1091	5.69	82.7202	10.1221
3.70	61.5953	15.6804	4.20	68.3332	13.7741	4.70	73.9701	12.2845	5.20	78.7324	11.0875	5.70	82.7956	10.1042
3.71	61.7426	15.6370	4.21	68.4559	13.7408	4.71	74.0733	12.2580	5.21	78.8200	11.0659	5.71	82.8707	10.0863
3.72	61.8893	15.5939	4.22	68.5782	13.7076	4.72	74.1761	12.2316	5.22	78.9074	11.0445	5.72	82.9456	10.0685
3.73	62.0355	15.5510	4.23	68.7000	13.6746	4.73	74.2786	12.2054	5.23	78.9944	11.0231	5.73	83.0203	10.0507
3.74	62.1812	15.5084	4.24	68.8213	13.6417	4.74	74.3807	12.1792	5.24	79.0812	11.0018	5.74	83.0947	10.0330
3.75	62.3263	15.4660	4.25	68.9423	13.6090	4.75	74.4824	12.1532	5.25	79.1677	10.9806	5.75	83.1689	10.0154
3.76	62.4709	15.4239	4.26	69.0628	13.5764	4.76	74.5839	12.1273	5.26	79.2539	10.9594	5.76	83.2430	9.9978
3.77	62.6149	15.3819	4.27	69.1829	13.5440	4.77	74.6850	12.1015	5.27	79.3399	10.9384	5.77	83.3167	9.9803
3.78	62.7584	15.3402	4.28	69.3026	13.5118	4.78	74.7858	12.0758	5.28	79.4256	10.9174	5.78	83.3903	9.9629
3.79	62.9014	15.2988	4.29	69.4218	13.4797	4.79	74.8862	12.0502	5.29	79.5110	10.8965	5.79	83.4637	9.9455
3.80	63.0438	15.2575	4.30	69.5406	13.4477	4.80	74.9863	12.0247	5.30	79.5962	10.8757	5.80	83.5368	9.9282
3.81	63.1857	15.2165	4.31	69.6590	13.4159	4.81	75.0860	11.9993	5.31	79.6810	10.8550	5.81	83.6097	9.9109
3.82	63.3271	15.1757	4.32	69.7770	13.3843	4.82	75.1855	11.9741	5.32	79.7657	10.8343	5.82	83.6824	9.8937
3.83	63.4679	15.1351	4.33	69.8946	13.3528	4.83	75.2846	11.9489	5.33	79.8500	10.8138	5.83	83.7549	9.8766
3.84	63.6083	15.0948	4.34	70.0118	13.3215	4.84	75.3833	11.9239	5.34	79.9341	10.7933	5.84	83.8272	9.8595
3.85	63.7481	15.0547	4.35	70.1285	13.2903	4.85	75.4818	11.8989	5.35	80.0180	10.7729	5.85	83.8993	9.8425
3.86	63.8874	15.0147	4.36	70.2449	13.2593	4.86	75.5799	11.8741	5.36	80.1015	10.7525	5.86	83.9711	9.8255
3.87	64.0262	14.9750	4.37	70.3608	13.2284	4.87	75.6777	11.8493	5.37	80.1848	10.7323	5.87	84.0428	9.8086
3.88	64.1645	14.9355	4.38	70.4763	13.1976	4.88	75.7752	11.8247	5.38	80.2679	10.7121	5.88	84.1142	9.7918
3.89	64.3023	14.8962	4.39	70.5914	13.1670	4.89	75.8723	11.8002	5.39	80.3507	10.6920	5.89	84.1855	9.7750
3.90	64.4395	14.8572	4.40	70.7062	13.1366	4.90	75.9691	11.7757	5.40	80.4332	10.6719	5.90	84.2565	9.7583
3.91	64.5763	14.8183	4.41	70.8205	13.1062	4.91	76.0657	11.7514	5.41	80.5155	10.6520	5.91	84.3273	9.7416
3.92	64.7125	14.7796	4.42	70.9344	13.0761	4.92	76.1619	11.7272	5.42	80.5976	10.6321	5.92	84.3979	9.7250
3.93	64.8483	14.7412	4.43	71.0479	13.0460	4.93	76.2577	11.7031	5.43	80.6793	10.6123	5.93	84.4683	9.7084
3.94	64.9836	14.7029	4.44	71.1611	13.0161	4.94	76.3533	11.6790	5.44	80.7609	10.5925	5.94	84.5385	9.6919
3.95	65.1183	14.6649	4.45	71.2738	12.9864	4.95	76.4486	11.6551	5.45	80.8421	10.5729	5.95	84.6085	9.6755
3.96	65.2526	14.6270	4.46	71.3862	12.9567	4.96	76.5435	11.6313	5.46	80.9232	10.5533	5.96	84.6783	9.6591
3.97	65.3864	14.5893	4.47	71.4982	12.9272	4.97	76.6382	11.6076	5.47	81.0039	10.5338	5.97	84.7479	9.6427
3.98	65.5197	14.5519	4.48	71.6097	12.8979	4.98	76.7325	11.5839	5.48	81.0845	10.5143	5.98	84.8173	9.6265
3.99	65.6525	14.5146	4.49	71.7209	12.8687	4.99	76.8265	11.5604	5.49	81.1648	10.4950	5.99	84.8865	9.6102

APPENDIX A.5.2

Prandtl Meyer flow table, k = 1.3

M	v (°)	μ (°)	M	v (°)	μ (°)	M	v (°)	μ (°)	M	v (°)	μ (°)	M	v (°)	μ (°)
1.00	0.0000	90.0000	1.50	12.6928	41.8103	2.00	28.6809	30.0000	2.50	43.2486	23.5782	3.00	55.7584	19.4712
1.01	0.0467	81.9307	1.51	13.0125	41.4718	2.01	28.9907	29.8356	2.51	43.5192	23.4786	3.01	55.9879	19.4039
1.02	0.1313	78.6351	1.52	13.3326	41.1395	2.02	29.2998	29.6730	2.52	43.7890	23.3799	3.02	56.2166	19.3371
1.03	0.2397	76.1376	1.53	13.6533	40.8132	2.03	29.6083	29.5123	2.53	44.0579	23.2820	3.03	56.4445	19.2708
1.04	0.3669	74.0576	1.54	13.9745	40.4927	2.04	29.9161	29.3535	2.54	44.3260	23.1850	3.04	56.6716	19.2049
1.05	0.5097	72.2472	1.55	14.2961	40.1778	2.05	30.2232	29.1964	2.55	44.5932	23.0888	3.05	56.8980	19.1395
1.06	0.6661	70.6300	1.56	14.6181	39.8683	2.06	30.5296	29.0411	2.56	44.8596	22.9934	3.06	57.1236	19.0745
1.07	0.8345	69.1603	1.57	14.9404	39.5642	2.07	30.8354	28.8875	2.57	45.1252	22.8988	3.07	57.3484	19.0100
1.08	1.0137	67.8084	1.58	15.2630	39.2652	2.08	31.1404	28.7357	2.58	45.3899	22.8051	3.08	57.5724	18.9459
1.09	1.2026	66.5534	1.59	15.5859	38.9713	2.09	31.4447	28.5855	2.59	45.6537	22.7121	3.09	57.7957	18.8823
1.10	1.4004	65.3800	1.60	15.9089	38.6822	2.10	31.7483	28.4369	2.60	45.9168	22.6199	3.10	58.0182	18.8191
1.11	1.6064	64.2767	1.61	16.2322	38.3978	2.11	32.0512	28.2899	2.61	46.1789	22.5284	3.11	58.2400	18.7563
1.12	1.8200	63.2345	1.62	16.5556	38.1181	2.12	32.3533	28.1446	2.62	46.4403	22.4377	3.12	58.4610	18.6939
1.13	2.0405	62.2461	1.63	16.8791	37.8428	2.13	32.6547	28.0008	2.63	46.7008	22.3478	3.13	58.6812	18.6320
1.14	2.2677	61.3056	1.64	17.2026	37.5719	2.14	32.9554	27.8585	2.64	46.9604	22.2586	3.14	58.9007	18.5705
1.15	2.5008	60.4082	1.65	17.5262	37.3052	2.15	33.2553	27.7177	2.65	47.2193	22.1702	3.15	59.1194	18.5094
1.16	2.7397	59.5497	1.66	17.8498	37.0427	2.16	33.5544	27.5785	2.66	47.4773	22.0824	3.16	59.3373	18.4487
1.17	2.9839	58.7267	1.67	18.1734	36.7842	2.17	33.8528	27.4406	2.67	47.7344	21.9954	3.17	59.5546	18.3884
1.18	3.2332	57.9362	1.68	18.4969	36.5296	2.18	34.1504	27.3043	2.68	47.9907	21.9090	3.18	59.7710	18.3285
1.19	3.4871	57.1756	1.69	18.8203	36.2789	2.19	34.4473	27.1693	2.69	48.2462	21.8234	3.19	59.9868	18.2691
1.20	3.7454	56.4427	1.70	19.1436	36.0319	2.20	34.7433	27.0357	2.70	48.5008	21.7385	3.20	60.2017	18.2100
1.21	4.0079	55.7354	1.71	19.4667	35.7885	2.21	35.0386	26.9035	2.71	48.7546	21.6542	3.21	60.4160	18.1512
1.22	4.2743	55.0520	1.72	19.7897	35.5487	2.22	35.3331	26.7726	2.72	49.0076	21.5706	3.22	60.6295	18.0929
1.23	4.5444	54.3909	1.73	20.1125	35.3124	2.23	35.6269	26.6430	2.73	49.2597	21.4876	3.23	60.8423	18.0350
1.24	4.8181	53.7507	1.74	20.4350	35.0795	2.24	35.9198	26.5148	2.74	49.5110	21.4053	3.24	61.0543	17.9774
1.25	5.0950	53.1301	1.75	20.7574	34.8499	2.25	36.2119	26.3878	2.75	49.7615	21.3237	3.25	61.2656	17.9202
1.26	5.3750	52.5280	1.76	21.0794	34.6235	2.26	36.5032	26.2621	2.76	50.0112	21.2427	3.26	61.4762	17.8634
1.27	5.6580	51.9433	1.77	21.4012	34.4003	2.27	36.7938	26.1376	2.77	50.2600	21.1623	3.27	61.6861	17.8069
1.28	5.9437	51.3752	1.78	21.7226	34.1802	2.28	37.0835	26.0144	2.78	50.5080	21.0825	3.28	61.8952	17.7508
1.29	6.2321	50.8226	1.79	22.0438	33.9631	2.29	37.3724	25.8923	2.79	50.7552	21.0034	3.29	62.1036	17.6951
1.30	6.5230	50.2849	1.80	22.3645	33.7490	2.30	37.6605	25.7715	2.80	51.0015	20.9248	3.30	62.3113	17.6397
1.31	6.8162	49.7612	1.81	22.6850	33.5377	2.31	37.9478	25.6518	2.81	51.2471	20.8469	3.31	62.5183	17.5847
1.32	7.1116	49.2509	1.82	23.0050	33.3293	2.32	38.2342	25.5332	2.82	51.4918	20.7695	3.32	62.7245	17.5300
1.33	7.4092	48.7535	1.83	23.3246	33.1237	2.33	38.5199	25.4158	2.83	51.7357	20.6928	3.33	62.9301	17.4756
1.34	7.7087	48.2682	1.84	23.6438	32.9207	2.34	38.8047	25.2995	2.84	51.9787	20.6166	3.34	63.1349	17.4216
1.35	8.0100	47.7946	1.85	23.9625	32.7204	2.35	39.0887	25.1843	2.85	52.2210	20.5410	3.35	63.3391	17.3680
1.36	8.3132	47.3321	1.86	24.2808	32.5227	2.36	39.3719	25.0702	2.86	52.4625	20.4659	3.36	63.5425	17.3147
1.37	8.6180	46.8803	1.87	24.5986	32.3276	2.37	39.6542	24.9572	2.87	52.7031	20.3914	3.37	63.7452	17.2617
1.38	8.9243	46.4387	1.88	24.9159	32.1349	2.38	39.9357	24.8452	2.88	52.9429	20.3175	3.38	63.9473	17.2090
1.39	9.2322	46.0070	1.89	25.2328	31.9447	2.39	40.2164	24.7342	2.89	53.1820	20.2441	3.39	64.1486	17.1567
1.40	9.5414	45.5847	1.90	25.5491	31.7569	2.40	40.4962	24.6243	2.90	53.4202	20.1713	3.40	64.3493	17.1046
1.41	9.8519	45.1715	1.91	25.8648	31.5714	2.41	40.7752	24.5154	2.91	53.6576	20.0990	3.41	64.5492	17.0529
1.42	10.1636	44.7670	1.92	26.1801	31.3882	2.42	41.0534	24.4075	2.92	53.8942	20.0272	3.42	64.7485	17.0016
1.43	10.4765	44.3709	1.93	26.4947	31.2072	2.43	41.3307	24.3005	2.93	54.1300	19.9559	3.43	64.9471	16.9505
1.44	10.7905	43.9830	1.94	26.8089	31.0285	2.44	41.6072	24.1945	2.94	54.3650	19.8852	3.44	65.1450	16.8997
1.45	11.1054	43.6028	1.95	27.1224	30.8519	2.45	41.8829	24.0895	2.95	54.5992	19.8149	3.45	65.3422	16.8493
1.46	11.4213	43.2302	1.96	27.4353	30.6774	2.46	42.1577	23.9854	2.96	54.8326	19.7452	3.46	65.5387	16.7991
1.47	11.7380	42.8649	1.97	27.7476	30.5050	2.47	42.4317	23.8822	2.97	55.0653	19.6760	3.47	65.7346	16.7493
1.48	12.0556	42.5066	1.98	28.0593	30.3347	2.48	42.7048	23.7800	2.98	55.2971	19.6072	3.48	65.9297	16.6997
1.49	12.3739	42.1552	1.99	28.3704	30.1664	2.49	42.9771	23.6786	2.99	55.5282	19.5390	3.49	66.1242	16.6505

(Continued)

TABLE A.5.2 (*Continued*)

Prandtl Meyer flow table, k = 1.3

M	ν (°)	μ (°)	M	ν (°)	μ (°)	M	ν (°)	μ (°)	M	ν (°)	μ (°)	M	ν (°)	μ (°)
3.50	66.3181	16.6015	4.00	75.2102	14.4775	4.50	82.7271	12.8396	5.00	89.1234	11.5370	5.50	94.6078	10.4757
3.51	66.5113	16.5529	4.01	75.3731	14.4406	4.51	82.8652	12.8106	5.01	89.2414	11.5136	5.51	94.7094	10.4565
3.52	66.7038	16.5045	4.02	75.5354	14.4039	4.52	83.0029	12.7818	5.02	89.3591	11.4904	5.52	94.8107	10.4373
3.53	66.8956	16.4564	4.03	75.6972	14.3674	4.53	83.1402	12.7531	5.03	89.4764	11.4672	5.53	94.9117	10.4182
3.54	67.0868	16.4086	4.04	75.8585	14.3311	4.54	83.2770	12.7246	5.04	89.5933	11.4442	5.54	95.0124	10.3992
3.55	67.2773	16.3611	4.05	76.0192	14.2949	4.55	83.4134	12.6961	5.05	89.7098	11.4212	5.55	95.1129	10.3803
3.56	67.4672	16.3139	4.06	76.1794	14.2590	4.56	83.5493	12.6678	5.06	89.8261	11.3983	5.56	95.2130	10.3614
3.57	67.6564	16.2669	4.07	76.3391	14.2232	4.57	83.6848	12.6396	5.07	89.9419	11.3755	5.57	95.3128	10.3426
3.58	67.8450	16.2202	4.08	76.4982	14.1876	4.58	83.8199	12.6116	5.08	90.0574	11.3528	5.58	95.4124	10.3238
3.59	68.0329	16.1738	4.09	76.6567	14.1522	4.59	83.9545	12.5837	5.09	90.1726	11.3302	5.59	95.5116	10.3052
3.60	68.2202	16.1276	4.10	76.8148	14.1170	4.60	84.0887	12.5559	5.10	90.2874	11.3077	5.60	95.6106	10.2866
3.61	68.4068	16.0817	4.11	76.9723	14.0819	4.61	84.2225	12.5282	5.11	90.4018	11.2853	5.61	95.7093	10.2680
3.62	68.5928	16.0361	4.12	77.1293	14.0470	4.62	84.3558	12.5006	5.12	90.5159	11.2630	5.62	95.8077	10.2496
3.63	68.7782	15.9907	4.13	77.2858	14.0123	4.63	84.4887	12.4732	5.13	90.6297	11.2407	5.63	95.9058	10.2312
3.64	68.9629	15.9456	4.14	77.4418	13.9778	4.64	84.6212	12.4459	5.14	90.7431	11.2186	5.64	96.0037	10.2128
3.65	69.1470	15.9008	4.15	77.5972	13.9434	4.65	84.7533	12.4187	5.15	90.8562	11.1965	5.65	96.1012	10.1946
3.66	69.3305	15.8562	4.16	77.7521	13.9092	4.66	84.8850	12.3916	5.16	90.9689	11.1745	5.66	96.1985	10.1763
3.67	69.5133	15.8119	4.17	77.9065	13.8752	4.67	85.0162	12.3647	5.17	91.0813	11.1526	5.67	96.2955	10.1582
3.68	69.6955	15.7678	4.18	78.0604	13.8414	4.68	85.1470	12.3378	5.18	91.1933	11.1308	5.68	96.3923	10.1401
3.69	69.8771	15.7239	4.19	78.2138	13.8077	4.69	85.2774	12.3111	5.19	91.3050	11.1091	5.69	96.4887	10.1221
3.70	70.0581	15.6804	4.20	78.3667	13.7741	4.70	85.4074	12.2845	5.20	91.4164	11.0875	5.70	96.5849	10.1042
3.71	70.2384	15.6370	4.21	78.5191	13.7408	4.71	85.5370	12.2580	5.21	91.5274	11.0659	5.71	96.6808	10.0863
3.72	70.4182	15.5939	4.22	78.6710	13.7076	4.72	85.6662	12.2316	5.22	91.6382	11.0445	5.72	96.7764	10.0685
3.73	70.5973	15.5510	4.23	78.8223	13.6746	4.73	85.7949	12.2054	5.23	91.7485	11.0231	5.73	96.8718	10.0507
3.74	70.7758	15.5084	4.24	78.9732	13.6417	4.74	85.9233	12.1792	5.24	91.8586	11.0018	5.74	96.9668	10.0330
3.75	70.9537	15.4660	4.25	79.1236	13.6090	4.75	86.0512	12.1532	5.25	91.9683	10.9806	5.75	97.0617	10.0154
3.76	71.1310	15.4239	4.26	79.2734	13.5764	4.76	86.1788	12.1273	5.26	92.0777	10.9594	5.76	97.1562	9.9978
3.77	71.3077	15.3819	4.27	79.4228	13.5440	4.77	86.3059	12.1015	5.27	92.1867	10.9384	5.77	97.2505	9.9803
3.78	71.4838	15.3402	4.28	79.5717	13.5118	4.78	86.4327	12.0758	5.28	92.2955	10.9174	5.78	97.3445	9.9629
3.79	71.6592	15.2988	4.29	79.7201	13.4797	4.79	86.5591	12.0502	5.29	92.4039	10.8965	5.79	97.4382	9.9455
3.80	71.8341	15.2575	4.30	79.8680	13.4477	4.80	86.6850	12.0247	5.30	92.5120	10.8757	5.80	97.5317	9.9282
3.81	72.0084	15.2165	4.31	80.0155	13.4159	4.81	86.8106	11.9993	5.31	92.6197	10.8550	5.81	97.6249	9.9109
3.82	72.1821	15.1757	4.32	80.1624	13.3843	4.82	86.9357	11.9741	5.32	92.7272	10.8343	5.82	97.7179	9.8937
3.83	72.3553	15.1351	4.33	80.3089	13.3528	4.83	87.0605	11.9489	5.33	92.8343	10.8138	5.83	97.8106	9.8766
3.84	72.5278	15.0948	4.34	80.4549	13.3215	4.84	87.1849	11.9239	5.34	92.9411	10.7933	5.84	97.9030	9.8595
3.85	72.6997	15.0547	4.35	80.6004	13.2903	4.85	87.3089	11.8989	5.35	93.0476	10.7729	5.85	97.9952	9.8425
3.86	72.8711	15.0147	4.36	80.7454	13.2593	4.86	87.4325	11.8741	5.36	93.1538	10.7525	5.86	98.0871	9.8255
3.87	73.0419	14.9750	4.37	80.8900	13.2284	4.87	87.5558	11.8493	5.37	93.2596	10.7323	5.87	98.1787	9.8086
3.88	73.2121	14.9355	4.38	81.0341	13.1976	4.88	87.6786	11.8247	5.38	93.3652	10.7121	5.88	98.2701	9.7918
3.89	73.3817	14.8962	4.39	81.1777	13.1670	4.89	87.8011	11.8002	5.39	93.4704	10.6920	5.89	98.3613	9.7750
3.90	73.5507	14.8572	4.40	81.3209	13.1366	4.90	87.9232	11.7757	5.40	93.5753	10.6719	5.90	98.4522	9.7583
3.91	73.7192	14.8183	4.41	81.4635	13.1062	4.91	88.0449	11.7514	5.41	93.6800	10.6520	5.91	98.5428	9.7416
3.92	73.8871	14.7796	4.42	81.6058	13.0761	4.92	88.1662	11.7272	5.42	93.7843	10.6321	5.92	98.6332	9.7250
3.93	74.0545	14.7412	4.43	81.7475	13.0460	4.93	88.2871	11.7031	5.43	93.8883	10.6123	5.93	98.7233	9.7084
3.94	74.2213	14.7029	4.44	81.8888	13.0161	4.94	88.4077	11.6790	5.44	93.9920	10.5925	5.94	98.8132	9.6919
3.95	74.3875	14.6649	4.45	82.0297	12.9864	4.95	88.5279	11.6551	5.45	94.0954	10.5729	5.95	98.9028	9.6755
3.96	74.5531	14.6270	4.46	82.1701	12.9567	4.96	88.6478	11.6313	5.46	94.1985	10.5533	5.96	98.9922	9.6591
3.97	74.7182	14.5893	4.47	82.3100	12.9272	4.97	88.7672	11.6076	5.47	94.3012	10.5338	5.97	99.0813	9.6427
3.98	74.8828	14.5519	4.48	82.4495	12.8979	4.98	88.8863	11.5839	5.48	94.4037	10.5143	5.98	99.1702	9.6265
3.99	75.0468	14.5146	4.49	82.5885	12.8687	4.99	89.0051	11.5604	5.49	94.5059	10.4950	5.99	99.2589	9.6102

A.6 Fanno Line Flow Tables

TABLE A.6.1

Fanno line flow table, k = 1.4

M	fL_{MAX}/D	P/P*	T/T*	V/V*	Pt/Pt*	M	fL_{MAX}/D	P/P*	T/T*	V/V*	Pt/Pt*
0.000	∞	∞	1.2000	0.00000	∞	0.26	7.6876	4.1851	1.1840	0.2829	2.3173
0.005	28562	219.09	1.2000	0.00548	115.74	0.27	6.9832	4.0279	1.1828	0.2936	2.2385
0.010	7134.4	109.54	1.2000	0.01095	57.874	0.28	6.3572	3.8820	1.1815	0.3043	2.1656
0.015	3166.9	73.028	1.1999	0.01643	38.586	0.29	5.7989	3.7460	1.1801	0.3150	2.0979
0.020	1778.5	54.770	1.1999	0.02191	28.942	0.30	5.2993	3.6191	1.1788	0.3257	2.0351
0.025	1136.0	43.815	1.1999	0.02738	23.157	0.31	4.8507	3.5002	1.1774	0.3364	1.9765
0.030	787.08	36.512	1.1998	0.03286	19.301	0.32	4.4467	3.3887	1.1759	0.3470	1.9219
0.035	576.79	31.295	1.1997	0.03834	16.547	0.33	4.0821	3.2840	1.1744	0.3576	1.8707
0.040	440.35	27.382	1.1996	0.04381	14.482	0.34	3.7520	3.1853	1.1729	0.3682	1.8229
0.045	346.86	24.338	1.1995	0.04928	12.876	0.35	3.4525	3.0922	1.1713	0.3788	1.7780
0.050	280.02	21.903	1.1994	0.05476	11.591	0.36	3.1801	3.0042	1.1697	0.3893	1.7358
0.055	230.60	19.911	1.1993	0.06023	10.541	0.37	2.9320	2.9209	1.1680	0.3999	1.6961
0.060	193.03	18.251	1.1991	0.06570	9.6659	0.38	2.7054	2.8420	1.1663	0.4104	1.6587
0.065	163.82	16.846	1.1990	0.07117	8.9257	0.39	2.4983	2.7671	1.1646	0.4209	1.6234
0.070	140.66	15.642	1.1988	0.07664	8.2915	0.40	2.3085	2.6958	1.1628	0.4313	1.5901
0.075	121.99	14.598	1.1987	0.08211	7.7421	0.41	2.1344	2.6280	1.1610	0.4418	1.5587
0.080	106.72	13.684	1.1985	0.08758	7.2616	0.42	1.9744	2.5634	1.1591	0.4522	1.5289
0.085	94.078	12.878	1.1983	0.09305	6.8378	0.43	1.8272	2.5017	1.1572	0.4626	1.5007
0.090	83.496	12.162	1.1981	0.09851	6.4613	0.44	1.6915	2.4428	1.1553	0.4729	1.4740
0.095	74.551	11.521	1.1978	0.10397	6.1247	0.45	1.5664	2.3865	1.1533	0.4833	1.4487
0.100	66.922	10.944	1.1976	0.10944	5.8218	0.46	1.4509	2.3326	1.1513	0.4936	1.4246
0.105	60.364	10.421	1.1974	0.11490	5.5480	0.47	1.3441	2.2809	1.1492	0.5038	1.4018
0.110	54.688	9.9466	1.1971	0.12035	5.2992	0.48	1.2453	2.2313	1.1471	0.5141	1.3801
0.115	49.742	9.5130	1.1968	0.12581	5.0722	0.49	1.1539	2.1838	1.1450	0.5243	1.3595
0.120	45.408	9.1156	1.1966	0.13126	4.8643	0.50	1.0691	2.1381	1.1429	0.5345	1.3398
0.125	41.589	8.7499	1.1963	0.13672	4.6732	0.51	0.9904	2.0942	1.1407	0.5447	1.3212
0.130	38.207	8.4123	1.1960	0.14217	4.4969	0.52	0.9174	2.0519	1.1384	0.5548	1.3034
0.135	35.199	8.0997	1.1956	0.14762	4.3337	0.53	0.8496	2.0112	1.1362	0.5649	1.2865
0.140	32.511	7.8093	1.1953	0.15306	4.1824	0.54	0.7866	1.9719	1.1339	0.5750	1.2703
0.145	30.101	7.5390	1.1950	0.15851	4.0416	0.55	0.7281	1.9341	1.1315	0.5851	1.2549
0.150	27.932	7.2866	1.1946	0.16395	3.9103	0.56	0.6736	1.8975	1.1292	0.5951	1.2403
0.155	25.973	7.0505	1.1943	0.16939	3.7877	0.57	0.6229	1.8623	1.1268	0.6051	1.2263
0.160	24.198	6.8291	1.1939	0.17482	3.6727	0.58	0.5757	1.8282	1.1244	0.6150	1.2130
0.165	22.585	6.6211	1.1935	0.18026	3.5649	0.59	0.5317	1.7952	1.1219	0.6249	1.2003
0.170	21.115	6.4253	1.1931	0.18569	3.4635	0.60	0.4908	1.7634	1.1194	0.6348	1.1882
0.175	19.772	6.2406	1.1927	0.19112	3.3680	0.61	0.4527	1.7325	1.1169	0.6447	1.1767
0.180	18.543	6.0662	1.1923	0.19654	3.2779	0.62	0.4172	1.7026	1.1143	0.6545	1.1656
0.185	17.414	5.9012	1.1918	0.20197	3.1928	0.63	0.3841	1.6736	1.1117	0.6643	1.1552
0.190	16.375	5.7448	1.1914	0.20739	3.1123	0.64	0.3533	1.6456	1.1091	0.6740	1.1451
0.195	15.418	5.5964	1.1909	0.21280	3.0359	0.65	0.3246	1.6183	1.1065	0.6837	1.1356
0.200	14.533	5.4554	1.1905	0.21822	2.9635	0.66	0.2979	1.5919	1.1038	0.6934	1.1265
0.205	13.715	5.3213	1.1900	0.22363	2.8947	0.67	0.2730	1.5662	1.1011	0.7031	1.1179
0.210	12.956	5.1936	1.1895	0.22904	2.8293	0.68	0.2498	1.5413	1.0984	0.7127	1.1097
0.215	12.251	5.0717	1.1890	0.23444	2.7670	0.69	0.2282	1.5170	1.0957	0.7223	1.1018
0.220	11.596	4.9554	1.1885	0.23984	2.7076	0.70	0.2081	1.4935	1.0929	0.7318	1.0944
0.225	10.986	4.8442	1.1880	0.24524	2.6509	0.71	0.1895	1.4705	1.0901	0.7413	1.0873
0.230	10.416	4.7378	1.1874	0.25063	2.5968	0.72	0.1721	1.4482	1.0873	0.7508	1.0806
0.235	9.8841	4.6359	1.1869	0.25602	2.5451	0.73	0.1561	1.4265	1.0844	0.7602	1.0742
0.240	9.3865	4.5383	1.1863	0.26141	2.4956	0.74	0.1411	1.4054	1.0815	0.7696	1.0681
0.245	8.9204	4.4446	1.1858	0.26679	2.4482	0.75	0.1273	1.3848	1.0787	0.7789	1.0624
0.250	8.4834	4.3546	1.1852	0.27217	2.4027	0.76	0.1145	1.3647	1.0757	0.7883	1.0570

(Continued)

TABLE A.6.1 (*Continued*)

Fanno line flow table, k = 1.4

M	f L_{MAX}/D	P/P*	T/T*	V/V*	Pt/Pt*	M	f L_{MAX}/D	P/P*	T/T*	V/V*	Pt/Pt*
0.77	0.10262	1.34514	1.0728	0.79753	1.0519	1.27	0.0549	0.7500	0.9073	1.2097	1.0542
0.78	0.09167	1.32606	1.0698	0.80677	1.0471	1.28	0.0582	0.7427	0.9038	1.2169	1.0581
0.79	0.08158	1.30744	1.0668	0.81597	1.0425	1.29	0.0615	0.7356	0.9003	1.2240	1.0621
0.80	0.07229	1.28928	1.0638	0.82514	1.0382	1.30	0.0648	0.7285	0.8969	1.2311	1.0663
0.81	0.06376	1.27155	1.0608	0.83426	1.0342	1.31	0.0682	0.7215	0.8934	1.2382	1.0706
0.82	0.05593	1.25423	1.0578	0.84335	1.0305	1.32	0.0716	0.7147	0.8899	1.2452	1.0750
0.83	0.04878	1.23732	1.0547	0.85239	1.0270	1.33	0.0750	0.7079	0.8864	1.2522	1.0796
0.84	0.04226	1.22080	1.0516	0.86140	1.0237	1.34	0.0785	0.7012	0.8829	1.2591	1.0842
0.85	0.03633	1.20466	1.0485	0.87037	1.0207	1.35	0.0820	0.6947	0.8794	1.2660	1.0890
0.86	0.03097	1.18888	1.0454	0.87929	1.0179	1.36	0.0855	0.6882	0.8760	1.2729	1.0940
0.87	0.02613	1.17344	1.0422	0.88818	1.0153	1.37	0.0890	0.6818	0.8725	1.2797	1.0990
0.88	0.02179	1.15835	1.0391	0.89703	1.0129	1.38	0.0926	0.6755	0.8690	1.2864	1.1042
0.89	0.01793	1.14358	1.0359	0.90583	1.0108	1.39	0.0962	0.6693	0.8655	1.2932	1.1095
0.90	0.01451	1.12913	1.0327	0.91460	1.0089	1.40	0.0997	0.6632	0.8621	1.2999	1.1149
0.91	0.01151	1.11499	1.0295	0.92332	1.0071	1.41	0.1033	0.6572	0.8586	1.3065	1.1205
0.92	0.00891	1.10114	1.0263	0.93201	1.0056	1.42	0.1069	0.6512	0.8551	1.3131	1.1262
0.93	0.00669	1.08758	1.0230	0.94065	1.0043	1.43	0.1106	0.6454	0.8517	1.3197	1.1320
0.94	0.00482	1.07430	1.0198	0.94925	1.0031	1.44	0.1142	0.6396	0.8482	1.3262	1.1379
0.95	0.00328	1.06129	1.0165	0.95781	1.0021	1.45	0.1178	0.6339	0.8448	1.3327	1.1440
0.96	0.00206	1.04854	1.0132	0.96633	1.0014	1.46	0.1215	0.6282	0.8413	1.3392	1.1501
0.97	0.00113	1.03604	1.0099	0.97481	1.0008	1.47	0.1251	0.6227	0.8379	1.3456	1.1565
0.98	0.00049	1.02379	1.0066	0.98325	1.0003	1.48	0.1288	0.6172	0.8344	1.3520	1.1629
0.99	0.00012	1.01178	1.0033	0.99165	1.0001	1.49	0.1324	0.6118	0.8310	1.3583	1.1695
1.00	0.00000	1.00000	1.0000	1.00000	1.0000	1.50	0.1361	0.6065	0.8276	1.3646	1.1762
1.01	0.00012	0.98844	0.9967	1.00831	1.0001	1.51	0.1397	0.6012	0.8242	1.3708	1.1830
1.02	0.00046	0.97711	0.9933	1.01658	1.0003	1.52	0.1433	0.5960	0.8207	1.3770	1.1899
1.03	0.00101	0.96598	0.9900	1.02481	1.0007	1.53	0.1470	0.5909	0.8173	1.3832	1.1970
1.04	0.00177	0.95507	0.9866	1.03300	1.0013	1.54	0.1506	0.5858	0.8139	1.3894	1.2042
1.05	0.00271	0.94435	0.9832	1.04114	1.0020	1.55	0.1543	0.5808	0.8105	1.3955	1.2116
1.06	0.00384	0.93383	0.9798	1.04925	1.0029	1.56	0.1579	0.5759	0.8071	1.4015	1.2190
1.07	0.00513	0.92349	0.9764	1.05731	1.0039	1.57	0.1615	0.5710	0.8038	1.4075	1.2266
1.08	0.00658	0.91335	0.9730	1.06533	1.0051	1.58	0.1651	0.5662	0.8004	1.4135	1.2344
1.09	0.00819	0.90338	0.9696	1.07331	1.0064	1.59	0.1688	0.5615	0.7970	1.4195	1.2422
1.10	0.00993	0.89359	0.9662	1.08124	1.0079	1.60	0.1724	0.5568	0.7937	1.4254	1.2502
1.11	0.01182	0.88397	0.9628	1.08913	1.0095	1.61	0.1760	0.5522	0.7903	1.4313	1.2584
1.12	0.01382	0.87451	0.9593	1.09699	1.0113	1.62	0.1795	0.5476	0.7869	1.4371	1.2666
1.13	0.01595	0.86522	0.9559	1.10479	1.0132	1.63	0.1831	0.5431	0.7836	1.4429	1.2750
1.14	0.01819	0.85608	0.9524	1.11256	1.0153	1.64	0.1867	0.5386	0.7803	1.4487	1.2836
1.15	0.02053	0.84710	0.9490	1.12029	1.0175	1.65	0.1902	0.5342	0.7770	1.4544	1.2922
1.16	0.02298	0.83826	0.9455	1.12797	1.0198	1.66	0.1938	0.5299	0.7736	1.4601	1.3010
1.17	0.02552	0.82958	0.9421	1.13561	1.0222	1.67	0.1973	0.5256	0.7703	1.4657	1.3100
1.18	0.02814	0.82103	0.9386	1.14321	1.0248	1.68	0.2008	0.5213	0.7670	1.4713	1.3190
1.19	0.03085	0.81263	0.9351	1.15077	1.0276	1.69	0.2043	0.5171	0.7637	1.4769	1.3283
1.20	0.03364	0.80436	0.9317	1.15828	1.0304	1.70	0.2078	0.5130	0.7605	1.4825	1.3376
1.21	0.03650	0.79623	0.9282	1.16575	1.0334	1.71	0.2113	0.5089	0.7572	1.4880	1.3471
1.22	0.03943	0.78822	0.9247	1.17319	1.0366	1.72	0.2147	0.5048	0.7539	1.4935	1.3567
1.23	0.04242	0.78034	0.9212	1.18057	1.0398	1.73	0.2182	0.5008	0.7507	1.4989	1.3665
1.24	0.04547	0.77258	0.9178	1.18792	1.0432	1.74	0.2216	0.4969	0.7474	1.5043	1.3764
1.25	0.04858	0.76495	0.9143	1.19523	1.0468	1.75	0.2250	0.4929	0.7442	1.5097	1.3865
1.26	0.05174	0.75743	0.9108	1.20249	1.0504	1.76	0.2284	0.4891	0.7410	1.5150	1.3967

(*Continued*)

TABLE A.6.1 (*Continued*)

Fanno line flow table, k = 1.4

M	fL_{MAX}/D	P/P*	T/T*	V/V*	Pt/Pt*	M	fL_{MAX}/D	P/P*	T/T*	V/V*	Pt/Pt*
1.77	0.23182	0.48527	0.7377	1.52029	1.4070	2.27	0.3788	0.3387	0.5910	1.7450	2.1345
1.78	0.23519	0.48149	0.7345	1.52555	1.4175	2.28	0.3813	0.3364	0.5883	1.7488	2.1538
1.79	0.23855	0.47776	0.7313	1.53078	1.4282	2.29	0.3838	0.3342	0.5857	1.7526	2.1734
1.80	0.24189	0.47407	0.7282	1.53598	1.4390	2.30	0.3862	0.3320	0.5831	1.7563	2.1931
1.81	0.24521	0.47042	0.7250	1.54114	1.4499	2.31	0.3887	0.3298	0.5805	1.7600	2.2131
1.82	0.24851	0.46681	0.7218	1.54626	1.4610	2.32	0.3911	0.3277	0.5779	1.7637	2.2333
1.83	0.25180	0.46324	0.7187	1.55136	1.4723	2.33	0.3935	0.3255	0.5753	1.7673	2.2538
1.84	0.25507	0.45972	0.7155	1.55642	1.4836	2.34	0.3959	0.3234	0.5728	1.7709	2.2744
1.85	0.25832	0.45623	0.7124	1.56145	1.4952	2.35	0.3983	0.3213	0.5702	1.7745	2.2953
1.86	0.26156	0.45278	0.7093	1.56644	1.5069	2.36	0.4006	0.3193	0.5677	1.7781	2.3164
1.87	0.26478	0.44937	0.7061	1.57140	1.5187	2.37	0.4030	0.3172	0.5651	1.7817	2.3377
1.88	0.26798	0.44600	0.7030	1.57633	1.5308	2.38	0.4053	0.3152	0.5626	1.7852	2.3593
1.89	0.27116	0.44266	0.6999	1.58123	1.5429	2.39	0.4076	0.3131	0.5601	1.7887	2.3811
1.90	0.27433	0.43936	0.6969	1.58609	1.5553	2.40	0.4099	0.3111	0.5576	1.7922	2.4031
1.91	0.27748	0.43610	0.6938	1.59092	1.5677	2.41	0.4122	0.3092	0.5551	1.7956	2.4254
1.92	0.28061	0.43287	0.6907	1.59572	1.5804	2.42	0.4144	0.3072	0.5527	1.7991	2.4479
1.93	0.28372	0.42967	0.6877	1.60049	1.5932	2.43	0.4167	0.3053	0.5502	1.8025	2.4706
1.94	0.28681	0.42651	0.6847	1.60523	1.6062	2.44	0.4189	0.3033	0.5478	1.8059	2.4936
1.95	0.28989	0.42339	0.6816	1.60993	1.6193	2.45	0.4211	0.3014	0.5453	1.8092	2.5168
1.96	0.29295	0.42029	0.6786	1.61460	1.6326	2.46	0.4233	0.2995	0.5429	1.8126	2.5403
1.97	0.29599	0.41724	0.6756	1.61925	1.6461	2.47	0.4255	0.2976	0.5405	1.8159	2.5640
1.98	0.29931	0.41391	0.6723	1.62432	1.6611	2.48	0.4277	0.2958	0.5381	1.8192	2.5880
1.99	0.30201	0.41121	0.6696	1.62844	1.6735	2.49	0.4298	0.2939	0.5357	1.8225	2.6122
2.00	0.30500	0.40825	0.6667	1.63299	1.6875	2.50	0.4320	0.2921	0.5333	1.8257	2.6367
2.01	0.30796	0.40532	0.6637	1.63751	1.7016	2.51	0.4341	0.2903	0.5310	1.8290	2.6615
2.02	0.31091	0.40241	0.6608	1.64201	1.7160	2.52	0.4362	0.2885	0.5286	1.8322	2.6865
2.03	0.31384	0.39954	0.6578	1.64647	1.7305	2.53	0.4383	0.2867	0.5263	1.8354	2.7117
2.04	0.31676	0.39670	0.6549	1.65090	1.7451	2.54	0.4404	0.2850	0.5239	1.8386	2.7372
2.05	0.31965	0.39388	0.6520	1.65530	1.7600	2.55	0.4425	0.2832	0.5216	1.8417	2.7630
2.06	0.32253	0.39110	0.6491	1.65967	1.7750	2.56	0.4445	0.2815	0.5193	1.8448	2.7891
2.07	0.32538	0.38834	0.6462	1.66402	1.7902	2.57	0.4466	0.2798	0.5170	1.8479	2.8154
2.08	0.32822	0.38562	0.6433	1.66833	1.8056	2.58	0.4486	0.2781	0.5147	1.8510	2.8420
2.09	0.33105	0.38292	0.6405	1.67262	1.8212	2.59	0.4506	0.2764	0.5125	1.8541	2.8688
2.10	0.33385	0.38024	0.6376	1.67687	1.8369	2.60	0.4526	0.2747	0.5102	1.8571	2.8960
2.11	0.33664	0.37760	0.6348	1.68110	1.8529	2.61	0.4546	0.2731	0.5080	1.8602	2.9234
2.12	0.33940	0.37498	0.6320	1.68530	1.8690	2.62	0.4565	0.2714	0.5057	1.8632	2.9511
2.13	0.34215	0.37239	0.6291	1.68947	1.8853	2.63	0.4585	0.2698	0.5035	1.8662	2.9791
2.14	0.34489	0.36982	0.6263	1.69362	1.9018	2.64	0.4604	0.2682	0.5013	1.8691	3.0073
2.15	0.34760	0.36728	0.6235	1.69774	1.9185	2.65	0.4624	0.2666	0.4991	1.8721	3.0359
2.16	0.35030	0.36476	0.6208	1.70183	1.9354	2.66	0.4643	0.2650	0.4969	1.8750	3.0647
2.17	0.35298	0.36227	0.6180	1.70589	1.9525	2.67	0.4662	0.2634	0.4947	1.8779	3.0938
2.18	0.35564	0.35980	0.6152	1.70992	1.9698	2.68	0.4681	0.2619	0.4925	1.8808	3.1233
2.19	0.35828	0.35736	0.6125	1.71393	1.9873	2.69	0.4700	0.2603	0.4904	1.8837	3.1530
2.20	0.36091	0.35494	0.6098	1.71791	2.0050	2.70	0.4718	0.2588	0.4882	1.8865	3.1830
2.21	0.36352	0.35255	0.6070	1.72187	2.0229	2.71	0.4737	0.2573	0.4861	1.8894	3.2133
2.22	0.36611	0.35017	0.6043	1.72579	2.0409	2.72	0.4755	0.2558	0.4839	1.8922	3.2440
2.23	0.36869	0.34782	0.6016	1.72970	2.0592	2.73	0.4773	0.2543	0.4818	1.8950	3.2749
2.24	0.37124	0.34550	0.5989	1.73357	2.0777	2.74	0.4791	0.2528	0.4797	1.8978	3.3061
2.25	0.37378	0.34319	0.5963	1.73742	2.0964	2.75	0.4809	0.2513	0.4776	1.9005	3.3377
2.26	0.37631	0.34091	0.5936	1.74125	2.1153	2.76	0.4827	0.2498	0.4755	1.9033	3.3695

(Continued)

TABLE A.6.1 (*Continued*)

Fanno line flow table, k = 1.4

M	f L_{MAX}/D	P/P*	T/T*	V/V*	Pt/Pt*	M	f L_{MAX}/D	P/P*	T/T*	V/V*	Pt/Pt*
2.77	0.48451	0.24840	0.4735	1.90598	3.4017	5.70	0.7208	0.0702	0.1600	2.2803	42.797
2.78	0.48627	0.24697	0.4714	1.90868	3.4342	5.80	0.7240	0.0679	0.1553	2.2855	46.050
2.79	0.48803	0.24555	0.4693	1.91137	3.4670	5.90	0.7270	0.0658	0.1507	2.2905	49.507
2.80	0.48976	0.24414	0.4673	1.91404	3.5001	6.00	0.7299	0.0638	0.1463	2.2953	53.180
2.81	0.49149	0.24274	0.4653	1.91669	3.5336	6.10	0.7326	0.0618	0.1421	2.2998	57.077
2.82	0.49321	0.24135	0.4632	1.91933	3.5674	6.20	0.7353	0.0599	0.1381	2.3042	61.210
2.83	0.49491	0.23998	0.4612	1.92195	3.6015	6.30	0.7378	0.0582	0.1343	2.3084	65.590
2.84	0.49660	0.23861	0.4592	1.92455	3.6359	6.40	0.7402	0.0565	0.1305	2.3124	70.227
2.85	0.49828	0.23726	0.4572	1.92714	3.6707	6.50	0.7425	0.0548	0.1270	2.3163	75.134
2.86	0.49995	0.23592	0.4552	1.92970	3.7058	6.60	0.7448	0.0533	0.1236	2.3200	80.323
2.87	0.50161	0.23459	0.4533	1.93225	3.7413	6.70	0.7469	0.0518	0.1203	2.3235	85.805
2.88	0.50326	0.23326	0.4513	1.93479	3.7771	6.80	0.7489	0.0503	0.1171	2.3269	91.593
2.89	0.50489	0.23195	0.4494	1.93731	3.8133	6.90	0.7509	0.0489	0.1140	2.3302	97.702
2.90	0.50652	0.23066	0.4474	1.93981	3.8498	7.00	0.7528	0.0476	0.1111	2.3333	104.14
2.91	0.50813	0.22937	0.4455	1.94230	3.8866	7.10	0.7546	0.0463	0.1083	2.3364	110.93
2.92	0.50973	0.22809	0.4436	1.94477	3.9238	7.20	0.7564	0.0451	0.1056	2.3393	118.08
2.93	0.51132	0.22682	0.4417	1.94722	3.9614	7.30	0.7580	0.0439	0.1029	2.3421	125.60
2.94	0.51290	0.22556	0.4398	1.94966	3.9993	7.40	0.7597	0.0428	0.1004	2.3448	133.52
2.95	0.51447	0.22431	0.4379	1.95208	4.0376	7.50	0.7612	0.0417	0.0980	2.3474	141.84
2.96	0.51603	0.22307	0.4360	1.95449	4.0763	7.60	0.7627	0.0407	0.0956	2.3499	150.58
2.97	0.51758	0.22185	0.4341	1.95688	4.1153	7.70	0.7642	0.0397	0.0933	2.3523	159.77
2.98	0.51912	0.22063	0.4323	1.95925	4.1547	7.80	0.7656	0.0387	0.0911	2.3546	169.40
2.99	0.52064	0.21942	0.4304	1.96162	4.1944	7.90	0.7669	0.0378	0.0890	2.3569	179.51
3.00	0.52216	0.21822	0.4286	1.96396	4.2346	8.00	0.7682	0.0369	0.0870	2.3591	190.11
3.10	0.53678	0.20672	0.4107	1.98661	4.6573	8.10	0.7694	0.0360	0.0850	2.3612	201.21
3.20	0.55044	0.19608	0.3937	2.00786	5.1210	8.20	0.7707	0.0351	0.0831	2.3632	212.85
3.30	0.56323	0.18621	0.3776	2.02781	5.6286	8.30	0.7718	0.0343	0.0812	2.3652	225.02
3.40	0.57521	0.17704	0.3623	2.04656	6.1837	8.40	0.7729	0.0335	0.0794	2.3671	237.76
3.50	0.58643	0.16851	0.3478	2.06419	6.7896	8.50	0.7740	0.0328	0.0777	2.3689	251.09
3.60	0.59695	0.16055	0.3341	2.08077	7.4501	8.60	0.7751	0.0321	0.0760	2.3707	265.01
3.70	0.60684	0.15313	0.3210	2.09639	8.1691	8.70	0.7761	0.0313	0.0744	2.3724	279.57
3.80	0.61612	0.14620	0.3086	2.11111	8.9506	8.80	0.7771	0.0307	0.0728	2.3740	294.77
3.90	0.62485	0.13971	0.2969	2.12499	9.7990	8.90	0.7781	0.0300	0.0713	2.3757	310.63
4.00	0.63307	0.13363	0.2857	2.13809	10.719	9.00	0.7790	0.0293	0.0698	2.3772	327.19
4.10	0.64080	0.12793	0.2751	2.15046	11.715	9.10	0.7799	0.0287	0.0683	2.3787	344.46
4.20	0.64810	0.12257	0.2650	2.16216	12.792	9.20	0.7808	0.0281	0.0669	2.3802	362.46
4.30	0.65499	0.11753	0.2554	2.17322	13.955	9.30	0.7816	0.0275	0.0656	2.3816	381.23
4.40	0.66149	0.11279	0.2463	2.18368	15.210	9.40	0.7824	0.0270	0.0643	2.3830	400.77
4.50	0.66763	0.10833	0.2376	2.19360	16.562	9.50	0.7832	0.0264	0.0630	2.3843	421.13
4.60	0.67345	0.10411	0.2294	2.20300	18.018	9.60	0.7840	0.0259	0.0618	2.3856	442.32
4.70	0.67895	0.10013	0.2215	2.21192	19.583	9.70	0.7847	0.0254	0.0606	2.3869	464.37
4.80	0.68417	0.09637	0.2140	2.22038	21.264	9.80	0.7854	0.0249	0.0594	2.3881	487.30
4.90	0.68911	0.09281	0.2068	2.22843	23.067	9.90	0.7861	0.0244	0.0582	2.3893	511.15
5.00	0.69380	0.08944	0.2000	2.23607	25.000	10.0	0.7868	0.0239	0.0571	2.3905	535.94
5.10	0.69826	0.08625	0.1935	2.24334	27.070						
5.20	0.70249	0.08322	0.1873	2.25026	29.283						
5.30	0.70652	0.08034	0.1813	2.25685	31.649						
5.40	0.71035	0.07761	0.1756	2.26314	34.175						
5.50	0.71400	0.07501	0.1702	2.26913	36.869						
5.60	0.71748	0.07254	0.1650	2.27485	39.740						

TABLE A.6.2

Fanno line flow table, k = 1.31

M	f L$_{MAX}$/D	P/P*	T/T*	V/V*	Pt/Pt*	M	f L$_{MAX}$/D	P/P*	T/T*	V/V*	Pt/Pt*
0.000	∞	∞	1.1550	0.00000	∞	0.26	8.2714	4.1120	1.1430	0.2780	2.3374
0.005	30525	214.94	1.1550	0.00537	116.91	0.27	7.5163	3.9581	1.1421	0.2885	2.2576
0.010	7624.8	107.47	1.1550	0.01075	58.460	0.28	6.8451	3.8151	1.1411	0.2991	2.1838
0.015	3384.7	71.646	1.1550	0.01612	38.976	0.29	6.2463	3.6820	1.1401	0.3097	2.1154
0.020	1900.9	53.734	1.1549	0.02149	29.235	0.30	5.7102	3.5576	1.1391	0.3202	2.0518
0.025	1214.2	42.986	1.1549	0.02687	23.391	0.31	5.2288	3.4413	1.1380	0.3307	1.9925
0.030	841.36	35.821	1.1548	0.03224	19.496	0.32	4.7953	3.3321	1.1370	0.3412	1.9371
0.035	616.60	30.703	1.1548	0.03761	16.714	0.33	4.4037	3.2296	1.1358	0.3517	1.8854
0.040	470.79	26.864	1.1547	0.04298	14.628	0.34	4.0492	3.1330	1.1347	0.3622	1.8369
0.045	370.86	23.879	1.1546	0.04835	13.006	0.35	3.7274	3.0419	1.1335	0.3726	1.7914
0.050	299.43	21.490	1.1546	0.05372	11.708	0.36	3.4347	2.9558	1.1323	0.3831	1.7487
0.055	246.60	19.536	1.1545	0.05910	10.647	0.37	3.1680	2.8743	1.1310	0.3935	1.7085
0.060	206.45	17.907	1.1544	0.06446	9.7630	0.38	2.9244	2.7971	1.1297	0.4039	1.6706
0.065	175.22	16.529	1.1542	0.06983	9.0153	0.39	2.7016	2.7237	1.1284	0.4143	1.6348
0.070	150.46	15.347	1.1541	0.07520	8.3746	0.40	2.4973	2.6541	1.1270	0.4247	1.6011
0.075	130.50	14.323	1.1540	0.08057	7.8195	0.41	2.3099	2.5877	1.1257	0.4350	1.5692
0.080	114.18	13.427	1.1539	0.08593	7.3341	0.42	2.1376	2.5246	1.1243	0.4453	1.5390
0.085	100.671	12.637	1.1537	0.09130	6.9060	0.43	1.9791	2.4643	1.1228	0.4556	1.5104
0.090	89.359	11.934	1.1536	0.09666	6.5256	0.44	1.8329	2.4067	1.1214	0.4659	1.4833
0.095	79.795	11.305	1.1534	0.10203	6.1854	0.45	1.6981	2.3516	1.1199	0.4762	1.4576
0.100	71.638	10.739	1.1532	0.10739	5.8795	0.46	1.5735	2.2989	1.1183	0.4865	1.4332
0.105	64.627	10.227	1.1530	0.11275	5.6028	0.47	1.4583	2.2484	1.1168	0.4967	1.4100
0.110	58.557	9.7609	1.1528	0.11811	5.3514	0.48	1.3517	2.2000	1.1152	0.5069	1.3879
0.115	53.269	9.3357	1.1526	0.12347	5.1221	0.49	1.2529	2.1536	1.1136	0.5171	1.3670
0.120	48.634	8.9459	1.1524	0.12882	4.9120	0.50	1.1613	2.1089	1.1119	0.5272	1.3470
0.125	44.550	8.5873	1.1522	0.13418	4.7188	0.51	1.0764	2.0660	1.1102	0.5374	1.3281
0.130	40.933	8.2562	1.1520	0.13953	4.5407	0.52	0.9975	2.0248	1.1085	0.5475	1.3100
0.135	37.715	7.9496	1.1517	0.14488	4.3759	0.53	0.9241	1.9850	1.1068	0.5576	1.2928
0.140	34.841	7.6649	1.1515	0.15023	4.2229	0.54	0.8560	1.9467	1.1051	0.5677	1.2763
0.145	32.263	7.3997	1.1512	0.15558	4.0806	0.55	0.7926	1.9098	1.1033	0.5777	1.2607
0.150	29.942	7.1523	1.1510	0.16093	3.9480	0.56	0.7336	1.8741	1.1015	0.5877	1.2458
0.155	27.846	6.9207	1.1507	0.16627	3.8240	0.57	0.6787	1.8397	1.0996	0.5977	1.2316
0.160	25.947	6.7036	1.1504	0.17161	3.7078	0.58	0.6275	1.8064	1.0978	0.6077	1.2180
0.165	24.222	6.4997	1.1501	0.17695	3.5988	0.59	0.5799	1.7743	1.0959	0.6176	1.2051
0.170	22.649	6.3077	1.1498	0.18229	3.4963	0.60	0.5355	1.7432	1.0940	0.6276	1.1927
0.175	21.212	6.1267	1.1495	0.18763	3.3998	0.61	0.4941	1.7131	1.0920	0.6374	1.1810
0.180	19.896	5.9557	1.1492	0.19296	3.3088	0.62	0.4556	1.6840	1.0901	0.6473	1.1697
0.185	18.688	5.7939	1.1489	0.19830	3.2227	0.63	0.4196	1.6557	1.0881	0.6572	1.1590
0.190	17.576	5.6406	1.1486	0.20363	3.1413	0.64	0.3861	1.6283	1.0860	0.6670	1.1488
0.195	16.551	5.4952	1.1482	0.20895	3.0641	0.65	0.3549	1.6018	1.0840	0.6768	1.1391
0.200	15.604	5.3570	1.1479	0.21428	2.9909	0.66	0.3258	1.5760	1.0819	0.6865	1.1298
0.205	14.728	5.2255	1.1475	0.21960	2.9214	0.67	0.2987	1.5510	1.0799	0.6962	1.1210
0.210	13.915	5.1003	1.1472	0.22492	2.8552	0.68	0.2735	1.5267	1.0778	0.7059	1.1126
0.215	13.161	4.9808	1.1468	0.23024	2.7922	0.69	0.2500	1.5031	1.0756	0.7156	1.1046
0.220	12.459	4.8668	1.1464	0.23555	2.7321	0.70	0.2281	1.4801	1.0735	0.7253	1.0969
0.225	11.805	4.7579	1.1460	0.24087	2.6748	0.71	0.2077	1.4578	1.0713	0.7349	1.0897
0.230	11.195	4.6536	1.1456	0.24618	2.6201	0.72	0.1888	1.4361	1.0691	0.7445	1.0828
0.235	10.6252	4.5538	1.1452	0.25148	2.5678	0.73	0.1712	1.4149	1.0669	0.7540	1.0763
0.240	10.0921	4.4581	1.1448	0.25679	2.5177	0.74	0.1549	1.3943	1.0646	0.7635	1.0701
0.245	9.5927	4.3663	1.1444	0.26209	2.4697	0.75	0.1398	1.3743	1.0624	0.7730	1.0642
0.250	9.1244	4.2782	1.1439	0.26739	2.4238	0.76	0.1258	1.3547	1.0601	0.7825	1.0587

(Continued)

TABLE A.6.2 (*Continued*)

Fanno line flow table, k = 1.31

M	fL_{MAX}/D	P/P*	T/T*	V/V*	Pt/Pt*	M	fL_{MAX}/D	P/P*	T/T*	V/V*	Pt/Pt*
0.77	0.11280	1.33570	1.0578	0.79194	1.0534	1.27	0.0617	0.7569	0.9240	1.2208	1.0570
0.78	0.10081	1.31713	1.0555	0.80134	1.0485	1.28	0.0654	0.7498	0.9211	1.2285	1.0612
0.79	0.08975	1.29901	1.0531	0.81071	1.0438	1.29	0.0691	0.7428	0.9182	1.2361	1.0655
0.80	0.07957	1.28133	1.0508	0.82005	1.0394	1.30	0.0729	0.7359	0.9153	1.2437	1.0699
0.81	0.07020	1.26408	1.0484	0.82936	1.0353	1.31	0.0767	0.7291	0.9123	1.2513	1.0745
0.82	0.06162	1.24724	1.0460	0.83864	1.0314	1.32	0.0806	0.7224	0.9094	1.2588	1.0792
0.83	0.05376	1.23079	1.0436	0.84789	1.0278	1.33	0.0845	0.7159	0.9065	1.2663	1.0840
0.84	0.04659	1.21471	1.0411	0.85710	1.0245	1.34	0.0884	0.7094	0.9035	1.2737	1.0890
0.85	0.04008	1.19901	1.0387	0.86628	1.0213	1.35	0.0924	0.7030	0.9006	1.2811	1.0942
0.86	0.03417	1.18366	1.0362	0.87543	1.0185	1.36	0.0964	0.6967	0.8977	1.2885	1.0994
0.87	0.02885	1.16865	1.0337	0.88455	1.0158	1.37	0.1004	0.6904	0.8947	1.2959	1.1048
0.88	0.02407	1.15397	1.0312	0.89363	1.0134	1.38	0.1044	0.6843	0.8918	1.3032	1.1104
0.89	0.01981	1.13961	1.0287	0.90268	1.0112	1.39	0.1085	0.6783	0.8888	1.3105	1.1161
0.90	0.01604	1.12555	1.0262	0.91170	1.0092	1.40	0.1126	0.6723	0.8859	1.3177	1.1219
0.91	0.01273	1.11180	1.0236	0.92068	1.0074	1.41	0.1167	0.6664	0.8829	1.3249	1.1279
0.92	0.00986	1.09834	1.0210	0.92963	1.0058	1.42	0.1208	0.6606	0.8800	1.3321	1.1340
0.93	0.00740	1.08515	1.0185	0.93855	1.0044	1.43	0.1249	0.6549	0.8770	1.3392	1.1402
0.94	0.00533	1.07224	1.0159	0.94743	1.0032	1.44	0.1291	0.6492	0.8741	1.3463	1.1466
0.95	0.00363	1.05959	1.0133	0.95628	1.0022	1.45	0.1333	0.6437	0.8711	1.3533	1.1532
0.96	0.00228	1.04719	1.0106	0.96509	1.0014	1.46	0.1374	0.6382	0.8682	1.3604	1.1599
0.97	0.00126	1.03504	1.0080	0.97387	1.0008	1.47	0.1416	0.6328	0.8652	1.3673	1.1667
0.98	0.00055	1.02313	1.0053	0.98261	1.0003	1.48	0.1458	0.6274	0.8623	1.3743	1.1737
0.99	0.00013	1.01145	1.0027	0.99132	1.0001	1.49	0.1500	0.6221	0.8593	1.3812	1.1808
1.00	0.00000	1.00000	1.0000	1.00000	1.0000	1.50	0.1542	0.6169	0.8563	1.3881	1.1881
1.01	0.00013	0.98877	0.9973	1.00864	1.0001	1.51	0.1584	0.6118	0.8534	1.3949	1.1955
1.02	0.00051	0.97775	0.9946	1.01725	1.0003	1.52	0.1626	0.6067	0.8504	1.4017	1.2030
1.03	0.00113	0.96693	0.9919	1.02582	1.0008	1.53	0.1667	0.6017	0.8475	1.4085	1.2108
1.04	0.00197	0.95632	0.9892	1.03435	1.0014	1.54	0.1709	0.5967	0.8445	1.4152	1.2186
1.05	0.00302	0.94590	0.9864	1.04285	1.0021	1.55	0.1751	0.5919	0.8416	1.4220	1.2266
1.06	0.00427	0.93567	0.9837	1.05132	1.0030	1.56	0.1793	0.5870	0.8387	1.4286	1.2348
1.07	0.00571	0.92562	0.9809	1.05975	1.0041	1.57	0.1835	0.5823	0.8357	1.4352	1.2431
1.08	0.00734	0.91576	0.9782	1.06814	1.0053	1.58	0.1877	0.5776	0.8328	1.4418	1.2516
1.09	0.00913	0.90607	0.9754	1.07650	1.0067	1.59	0.1919	0.5729	0.8298	1.4484	1.2602
1.10	0.01108	0.89655	0.9726	1.08482	1.0083	1.60	0.1960	0.5683	0.8269	1.4549	1.2690
1.11	0.01318	0.88719	0.9698	1.09311	1.0100	1.61	0.2002	0.5638	0.8240	1.4614	1.2779
1.12	0.01543	0.87800	0.9670	1.10136	1.0118	1.62	0.2043	0.5593	0.8210	1.4679	1.2870
1.13	0.01781	0.86896	0.9642	1.10957	1.0138	1.63	0.2085	0.5549	0.8181	1.4743	1.2963
1.14	0.02032	0.86007	0.9613	1.11775	1.0160	1.64	0.2126	0.5505	0.8152	1.4807	1.3057
1.15	0.02294	0.85134	0.9585	1.12589	1.0183	1.65	0.2167	0.5462	0.8122	1.4871	1.3152
1.16	0.02569	0.84275	0.9557	1.13400	1.0207	1.66	0.2208	0.5419	0.8093	1.4934	1.3250
1.17	0.02854	0.83430	0.9528	1.14207	1.0233	1.67	0.2249	0.5377	0.8064	1.4997	1.3349
1.18	0.03149	0.82599	0.9500	1.15011	1.0260	1.68	0.2290	0.5336	0.8035	1.5059	1.3450
1.19	0.03453	0.81781	0.9471	1.15810	1.0289	1.69	0.2331	0.5294	0.8006	1.5121	1.3552
1.20	0.03767	0.80977	0.9442	1.16607	1.0319	1.70	0.2372	0.5254	0.7977	1.5183	1.3656
1.21	0.04089	0.80185	0.9414	1.17399	1.0351	1.71	0.2412	0.5213	0.7948	1.5245	1.3762
1.22	0.04419	0.79406	0.9385	1.18188	1.0384	1.72	0.2452	0.5174	0.7919	1.5306	1.3869
1.23	0.04756	0.78639	0.9356	1.18974	1.0418	1.73	0.2493	0.5134	0.7890	1.5367	1.3978
1.24	0.05100	0.77885	0.9327	1.19755	1.0454	1.74	0.2533	0.5096	0.7861	1.5427	1.4089
1.25	0.05451	0.77141	0.9298	1.20533	1.0492	1.75	0.2573	0.5057	0.7832	1.5487	1.4202
1.26	0.05808	0.76410	0.9269	1.21308	1.0530	1.76	0.2612	0.5019	0.7803	1.5547	1.4316

(Continued)

TABLE A.6.2 (*Continued*)

Fanno line flow table, k = 1.31

M	f L$_{MAX}$/D	P/P*	T/T*	V/V*	Pt/Pt*	M	f L$_{MAX}$/D	P/P*	T/T*	V/V*	Pt/Pt*
1.77	0.26521	0.49816	0.7775	1.5607	1.4432	2.27	0.4398	0.3530	0.6421	1.8190	2.2947
1.78	0.26915	0.49444	0.7746	1.5666	1.4550	2.28	0.4428	0.3508	0.6396	1.8235	2.3182
1.79	0.27308	0.49077	0.7717	1.5725	1.4670	2.29	0.4458	0.3486	0.6371	1.8279	2.3420
1.80	0.27699	0.48714	0.7689	1.5783	1.4792	2.30	0.4488	0.3464	0.6346	1.8323	2.3661
1.81	0.28089	0.48355	0.7660	1.5842	1.4915	2.31	0.4517	0.3442	0.6322	1.8366	2.3905
1.82	0.28477	0.48000	0.7632	1.5899	1.5040	2.32	0.4546	0.3420	0.6297	1.8410	2.4152
1.83	0.28863	0.47649	0.7603	1.5957	1.5168	2.33	0.4575	0.3399	0.6272	1.8453	2.4403
1.84	0.29247	0.47301	0.7575	1.6014	1.5297	2.34	0.4604	0.3378	0.6248	1.8496	2.4656
1.85	0.29629	0.46957	0.7547	1.6071	1.5428	2.35	0.4633	0.3357	0.6223	1.8538	2.4913
1.86	0.30010	0.46617	0.7518	1.6128	1.5561	2.36	0.4662	0.3336	0.6199	1.8581	2.5173
1.87	0.30389	0.46281	0.7490	1.6184	1.5696	2.37	0.4690	0.3316	0.6174	1.8623	2.5436
1.88	0.30766	0.45948	0.7462	1.6240	1.5832	2.38	0.4718	0.3295	0.6150	1.8665	2.5703
1.89	0.31142	0.45619	0.7434	1.6296	1.5971	2.39	0.4746	0.3275	0.6126	1.8706	2.5972
1.90	0.31515	0.45294	0.7406	1.6351	1.6112	2.40	0.4774	0.3255	0.6102	1.8748	2.6246
1.91	0.31887	0.44971	0.7378	1.6406	1.6255	2.41	0.4802	0.3235	0.6078	1.8789	2.6522
1.92	0.32256	0.44653	0.7350	1.6461	1.6400	2.42	0.4829	0.3215	0.6054	1.8830	2.6803
1.93	0.32624	0.44337	0.7322	1.6515	1.6547	2.43	0.4857	0.3196	0.6031	1.8871	2.7086
1.94	0.32990	0.44025	0.7295	1.6569	1.6696	2.44	0.4884	0.3176	0.6007	1.8911	2.7374
1.95	0.33354	0.43716	0.7267	1.6623	1.6848	2.45	0.4911	0.3157	0.5983	1.8951	2.7664
1.96	0.33716	0.43410	0.7239	1.6677	1.7001	2.46	0.4938	0.3138	0.5960	1.8991	2.7959
1.97	0.34076	0.43108	0.7212	1.6730	1.7157	2.47	0.4964	0.3119	0.5936	1.9031	2.8257
1.98	0.34470	0.42779	0.7182	1.6788	1.7330	2.48	0.4991	0.3101	0.5913	1.9070	2.8559
1.99	0.34791	0.42512	0.7157	1.6835	1.7474	2.49	0.5017	0.3082	0.5890	1.9110	2.8864
2.00	0.35145	0.42219	0.7130	1.6887	1.7637	2.50	0.5043	0.3064	0.5867	1.9149	2.9174
2.01	0.35498	0.41928	0.7102	1.6939	1.7801	2.51	0.5069	0.3046	0.5844	1.9187	2.9487
2.02	0.35848	0.41641	0.7075	1.6991	1.7968	2.52	0.5095	0.3028	0.5821	1.9226	2.9804
2.03	0.36197	0.41356	0.7048	1.7042	1.8137	2.53	0.5121	0.3010	0.5798	1.9264	3.0124
2.04	0.36543	0.41074	0.7021	1.7094	1.8308	2.54	0.5146	0.2992	0.5775	1.9302	3.0449
2.05	0.36888	0.40796	0.6994	1.7144	1.8482	2.55	0.5172	0.2974	0.5752	1.9340	3.0778
2.06	0.37231	0.40519	0.6967	1.7195	1.8658	2.56	0.5197	0.2957	0.5730	1.9378	3.1111
2.07	0.37571	0.40246	0.6940	1.7245	1.8836	2.57	0.5222	0.2940	0.5707	1.9415	3.1448
2.08	0.37910	0.39975	0.6914	1.7295	1.9017	2.58	0.5246	0.2922	0.5685	1.9453	3.1789
2.09	0.38247	0.39707	0.6887	1.7345	1.9200	2.59	0.5271	0.2905	0.5662	1.9490	3.2134
2.10	0.38582	0.39442	0.6861	1.7394	1.9386	2.60	0.5296	0.2889	0.5640	1.9526	3.2483
2.11	0.38915	0.39179	0.6834	1.7443	1.9574	2.61	0.5320	0.2872	0.5618	1.9563	3.2836
2.12	0.39246	0.38919	0.6808	1.7492	1.9765	2.62	0.5344	0.2855	0.5596	1.9599	3.3194
2.13	0.39575	0.38661	0.6781	1.7540	1.9958	2.63	0.5368	0.2839	0.5574	1.9635	3.3556
2.14	0.39902	0.38406	0.6755	1.7588	2.0154	2.64	0.5392	0.2822	0.5552	1.9671	3.3923
2.15	0.40227	0.38153	0.6729	1.7636	2.0353	2.65	0.5416	0.2806	0.5530	1.9707	3.4294
2.16	0.40550	0.37903	0.6703	1.7684	2.0554	2.66	0.5439	0.2790	0.5509	1.9743	3.4669
2.17	0.40872	0.37655	0.6677	1.7731	2.0757	2.67	0.5463	0.2774	0.5487	1.9778	3.5049
2.18	0.41191	0.37409	0.6651	1.7778	2.0964	2.68	0.5486	0.2759	0.5465	1.9813	3.5434
2.19	0.41509	0.37166	0.6625	1.7825	2.1173	2.69	0.5509	0.2743	0.5444	1.9848	3.5823
2.20	0.41824	0.36925	0.6599	1.7872	2.1385	2.70	0.5532	0.2727	0.5423	1.9882	3.6217
2.21	0.42138	0.36687	0.6574	1.7918	2.1600	2.71	0.5555	0.2712	0.5401	1.9917	3.6615
2.22	0.42449	0.36450	0.6548	1.7964	2.1817	2.72	0.5578	0.2697	0.5380	1.9951	3.7018
2.23	0.42759	0.36216	0.6522	1.8010	2.2037	2.73	0.5600	0.2682	0.5359	1.9985	3.7426
2.24	0.43067	0.35984	0.6497	1.8055	2.2260	2.74	0.5623	0.2667	0.5338	2.0019	3.7839
2.25	0.43373	0.35754	0.6472	1.8101	2.2486	2.75	0.5645	0.2652	0.5317	2.0053	3.8257
2.26	0.43677	0.35527	0.6446	1.8146	2.2715	2.76	0.5667	0.2637	0.5296	2.0086	3.8680

(*Continued*)

TABLE A.6.2 (*Continued*)

Fanno line flow table, k = 1.31

M	f L$_{MAX}$/D	P/P*	T/T*	V/V*	Pt/Pt*	M	f L$_{MAX}$/D	P/P*	T/T*	V/V*	Pt/Pt*
2.77	0.56890	0.26222	0.5276	2.0120	3.9108	5.70	0.8712	0.0767	0.1914	2.4934	83.150
2.78	0.57109	0.26076	0.5255	2.0153	3.9541	5.80	0.8754	0.0743	0.1859	2.5005	91.076
2.79	0.57326	0.25932	0.5234	2.0186	3.9979	5.90	0.8795	0.0720	0.1806	2.5073	99.661
2.80	0.57541	0.25789	0.5214	2.0218	4.0422	6.00	0.8833	0.0698	0.1755	2.5138	108.95
2.81	0.57755	0.25646	0.5194	2.0251	4.0870	6.10	0.8870	0.0677	0.1707	2.5200	118.99
2.82	0.57968	0.25506	0.5173	2.0283	4.1324	6.20	0.8905	0.0657	0.1660	2.5260	129.84
2.83	0.58179	0.25366	0.5153	2.0315	4.1783	6.30	0.8939	0.0638	0.1615	2.5317	141.55
2.84	0.58389	0.25227	0.5133	2.0347	4.2247	6.40	0.8971	0.0619	0.1572	2.5372	154.17
2.85	0.58598	0.25089	0.5113	2.0379	4.2717	6.50	0.9002	0.0602	0.1530	2.5425	167.77
2.86	0.58805	0.24953	0.5093	2.0410	4.3192	6.60	0.9032	0.0585	0.1490	2.5476	182.40
2.87	0.59011	0.24817	0.5073	2.0442	4.3673	6.70	0.9060	0.0569	0.1451	2.5525	198.14
2.88	0.59215	0.24683	0.5053	2.0473	4.4160	6.80	0.9088	0.0553	0.1414	2.5572	215.05
2.89	0.59419	0.24549	0.5034	2.0504	4.4652	6.90	0.9114	0.0538	0.1378	2.5617	233.20
2.90	0.59621	0.24417	0.5014	2.0535	4.5150	7.00	0.9140	0.0524	0.1344	2.5661	252.67
2.91	0.59821	0.24286	0.4994	2.0565	4.5654	7.10	0.9164	0.0510	0.1310	2.5702	273.54
2.92	0.60021	0.24155	0.4975	2.0596	4.6163	7.20	0.9187	0.0497	0.1278	2.5743	295.90
2.93	0.60219	0.24026	0.4956	2.0626	4.6679	7.30	0.9210	0.0484	0.1247	2.5782	319.82
2.94	0.60416	0.23898	0.4936	2.0656	4.7200	7.40	0.9232	0.0471	0.1217	2.5819	345.41
2.95	0.60612	0.23770	0.4917	2.0686	4.7728	7.50	0.9253	0.0460	0.1188	2.5855	372.75
2.96	0.60806	0.23644	0.4898	2.0716	4.8261	7.60	0.9273	0.0448	0.1160	2.5890	401.95
2.97	0.60999	0.23519	0.4879	2.0746	4.8801	7.70	0.9292	0.0437	0.1133	2.5924	433.11
2.98	0.61210	0.23382	0.4858	2.0778	4.9402	7.80	0.9311	0.0427	0.1107	2.5956	466.33
2.99	0.61382	0.23271	0.4841	2.0804	4.9899	7.90	0.9329	0.0416	0.1082	2.5987	501.75
3.00	0.61571	0.23148	0.4823	2.0833	5.0458	8.00	0.9347	0.0407	0.1058	2.6018	539.46
3.10	0.63401	0.21972	0.4639	2.1115	5.6408	8.10	0.9364	0.0397	0.1034	2.6047	579.59
3.20	0.65119	0.20880	0.4464	2.1381	6.3068	8.20	0.9380	0.0388	0.1011	2.6075	622.28
3.30	0.66732	0.19864	0.4297	2.1632	7.0511	8.30	0.9396	0.0379	0.0989	2.6103	667.66
3.40	0.68247	0.18918	0.4137	2.1869	7.8819	8.40	0.9411	0.0370	0.0968	2.6129	715.86
3.50	0.69672	0.18035	0.3984	2.2093	8.8079	8.50	0.9426	0.0362	0.0947	2.6155	767.03
3.60	0.71013	0.17210	0.3839	2.2305	9.8385	8.60	0.9440	0.0354	0.0927	2.6180	821.32
3.70	0.72276	0.16439	0.3700	2.2505	######	8.70	0.9454	0.0346	0.0907	2.6204	878.89
3.80	0.73465	0.15716	0.3567	2.2695	######	8.80	0.9467	0.0339	0.0888	2.6227	939.91
3.90	0.74587	0.15039	0.3440	2.2874	######	8.90	0.9480	0.0331	0.0870	2.6250	1004.5
4.00	0.75645	0.14403	0.3319	2.3044	15.226	9.00	0.9493	0.0324	0.0852	2.6271	1072.9
4.10	0.76645	0.13805	0.3203	2.3205	16.952	9.10	0.9505	0.0318	0.0835	2.6293	1145.3
4.20	0.77589	0.13242	0.3093	2.3358	18.857	9.20	0.9517	0.0311	0.0818	2.6313	1221.8
4.30	0.78483	0.12711	0.2988	2.3503	20.959	9.30	0.9528	0.0304	0.0802	2.6333	1302.7
4.40	0.79328	0.12211	0.2887	2.3641	23.273	9.40	0.9539	0.0298	0.0786	2.6353	1388.2
4.50	0.80128	0.11739	0.2791	2.3772	25.820	9.50	0.9550	0.0292	0.0771	2.6371	1478.4
4.60	0.80887	0.11293	0.2699	2.3897	28.618	9.60	0.9561	0.0286	0.0756	2.6390	1573.6
4.70	0.81607	0.10871	0.2611	2.4015	31.688	9.70	0.9571	0.0281	0.0741	2.6407	1674.0
4.80	0.82290	0.10472	0.2527	2.4128	35.054	9.80	0.9581	0.0275	0.0727	2.6425	1779.8
4.90	0.82939	0.10094	0.2446	2.4235	38.739	9.90	0.9590	0.0270	0.0713	2.6441	1891.3
5.00	0.83556	0.09735	0.2369	2.4337	42.768	10.0	0.9599	0.0265	0.0700	2.6458	2008.8
5.10	0.84142	0.09394	0.2296	2.4435	47.170						
5.20	0.84700	0.09071	0.2225	2.4528	51.973						
5.30	0.85232	0.08764	0.2157	2.4617	57.208						
5.40	0.85739	0.08471	0.2092	2.4702	62.907						
5.50	0.86222	0.08193	0.2030	2.4783	69.106						
5.60	0.86683	0.07927	0.1971	2.4860	75.841						

A.7 Rayleigh Line Flow Tables

TABLE A.7.1

Rayleigh line flow table, k = 1.4

M	P/P*	T/T*	V/V*	T_t/T_t^*	P_t/P_t^*	M	P/P*	T/T*	V/V*	T_t/T_t^*	P_t/P_t^*
0.01	2.39966	0.00058	0.00024	0.00048	1.26779	0.51	1.75935	0.80509	0.45761	0.70581	1.10995
0.02	2.39866	0.00230	0.00096	0.00192	1.26752	0.52	1.74095	0.81955	0.47075	0.71990	1.10588
0.03	2.39698	0.00517	0.00216	0.00431	1.26708	0.53	1.72258	0.83351	0.48387	0.73361	1.10186
0.04	2.39464	0.00917	0.00383	0.00765	1.26646	0.54	1.70425	0.84695	0.49696	0.74695	1.09789
0.05	2.39163	0.01430	0.00598	0.01192	1.26567	0.55	1.68599	0.85987	0.51001	0.75991	1.09397
0.06	2.38796	0.02053	0.00860	0.01712	1.26470	0.56	1.66778	0.87227	0.52302	0.77249	1.09011
0.07	2.38365	0.02784	0.01168	0.02322	1.26356	0.57	1.64964	0.88416	0.53597	0.78468	1.08630
0.07	2.38196	0.03070	0.01289	0.02561	1.26312	0.58	1.63159	0.89552	0.54887	0.79648	1.08256
0.09	2.37309	0.04562	0.01922	0.03807	1.26078	0.59	1.61362	0.90637	0.56170	0.80789	1.07887
0.10	2.36686	0.05602	0.02367	0.04678	1.25915	0.60	1.59574	0.91670	0.57447	0.81892	1.07525
0.10	2.36553	0.05824	0.02462	0.04864	1.25880	0.61	1.57797	0.92653	0.58716	0.82957	1.07170
0.12	2.35257	0.07970	0.03388	0.06661	1.25539	0.62	1.56031	0.93584	0.59978	0.83983	1.06822
0.13	2.34453	0.09290	0.03962	0.07768	1.25329	0.63	1.54275	0.94466	0.61232	0.84970	1.06481
0.14	2.33590	0.10695	0.04578	0.08947	1.25103	0.64	1.52532	0.95298	0.62477	0.85920	1.06147
0.15	2.32671	0.12181	0.05235	0.10196	1.24863	0.65	1.50801	0.96081	0.63713	0.86833	1.05821
0.16	2.31696	0.13743	0.05931	0.11511	1.24608	0.66	1.49083	0.96816	0.64941	0.87708	1.05503
0.17	2.30667	0.15377	0.06666	0.12888	1.24340	0.67	1.47379	0.97503	0.66158	0.88547	1.05193
0.18	2.29586	0.17078	0.07439	0.14324	1.24059	0.68	1.45688	0.98144	0.67366	0.89350	1.04890
0.19	2.28454	0.18841	0.08247	0.15814	1.23765	0.69	1.44011	0.98739	0.68564	0.90118	1.04596
0.20	2.27273	0.20661	0.09091	0.17355	1.23460	0.70	1.42349	0.99290	0.69751	0.90850	1.04310
0.21	2.26044	0.22533	0.09969	0.18943	1.23142	0.71	1.40701	0.99796	0.70928	0.91548	1.04033
0.22	2.24770	0.24452	0.10879	0.20574	1.22814	0.72	1.39069	1.00260	0.72093	0.92212	1.03764
0.23	2.23451	0.26413	0.11821	0.22244	1.22475	0.73	1.37452	1.00682	0.73248	0.92843	1.03504
0.24	2.22091	0.28411	0.12792	0.23948	1.22126	0.74	1.35851	1.01062	0.74392	0.93442	1.03253
0.25	2.20690	0.30440	0.13793	0.25684	1.21767	0.75	1.34266	1.01403	0.75524	0.94009	1.03010
0.26	2.19250	0.32496	0.14821	0.27446	1.21400	0.76	1.32696	1.01706	0.76645	0.94546	1.02777
0.27	2.17774	0.34573	0.15876	0.29231	1.21025	0.77	1.31143	1.01970	0.77755	0.95052	1.02552
0.28	2.16263	0.36667	0.16955	0.31035	1.20642	0.78	1.29606	1.02198	0.78853	0.95528	1.02337
0.29	2.14719	0.38774	0.18058	0.32855	1.20251	0.79	1.28086	1.02390	0.79939	0.95975	1.02131
0.30	2.13144	0.40887	0.19183	0.34686	1.19855	0.80	1.26582	1.02548	0.81013	0.96395	1.01934
0.31	2.11539	0.43004	0.20329	0.36525	1.19452	0.81	1.25095	1.02672	0.82075	0.96787	1.01747
0.32	2.09908	0.45119	0.21495	0.38369	1.19045	0.82	1.23625	1.02763	0.83125	0.97152	1.01569
0.33	2.08250	0.47228	0.22678	0.40214	1.18632	0.83	1.22171	1.02823	0.84164	0.97492	1.01400
0.34	2.06569	0.49327	0.23879	0.42056	1.18215	0.84	1.20734	1.02853	0.85190	0.97807	1.01241
0.35	2.04866	0.51413	0.25096	0.43894	1.17795	0.85	1.19314	1.02854	0.86204	0.98097	1.01091
0.36	2.03142	0.53482	0.26327	0.45723	1.17371	0.86	1.17911	1.02826	0.87207	0.98363	1.00951
0.37	2.01400	0.55529	0.27572	0.47541	1.16945	0.87	1.16524	1.02771	0.88197	0.98607	1.00820
0.38	1.99641	0.57553	0.28828	0.49346	1.16517	0.88	1.15154	1.02689	0.89175	0.98828	1.00699
0.39	1.97866	0.59549	0.30095	0.51134	1.16088	0.89	1.13801	1.02583	0.90142	0.99028	1.00587
0.40	1.96078	0.61515	0.31373	0.52903	1.15658	0.90	1.12465	1.02452	0.91097	0.99207	1.00486
0.41	1.94278	0.63448	0.32658	0.54651	1.15227	0.91	1.11145	1.02297	0.92039	0.99366	1.00393
0.42	1.92468	0.65346	0.33951	0.56376	1.14796	0.92	1.09842	1.02120	0.92970	0.99506	1.00311
0.43	1.90649	0.67205	0.35251	0.58076	1.14366	0.93	1.08555	1.01922	0.93889	0.99627	1.00238
0.44	1.88822	0.69025	0.36556	0.59748	1.13936	0.94	1.07285	1.01702	0.94797	0.99729	1.00175
0.45	1.86989	0.70804	0.37865	0.61393	1.13508	0.95	1.06030	1.01463	0.95693	0.99814	1.00122
0.46	1.85151	0.72538	0.39178	0.63007	1.13082	0.96	1.04793	1.01205	0.96577	0.99883	1.00078
0.47	1.83310	0.74228	0.40493	0.64589	1.12659	0.97	1.03571	1.00929	0.97450	0.99935	1.00044
0.48	1.81466	0.75871	0.41810	0.66139	1.12238	0.98	1.02365	1.00636	0.98311	0.99971	1.00019
0.49	1.79622	0.77466	0.43127	0.67655	1.11820	0.99	1.01174	1.00326	0.99161	0.99993	1.00005
0.50	1.77778	0.79012	0.44444	0.69136	1.11405	1.00	1.00000	1.00000	1.00000	1.00000	1.00000

(Continued)

TABLE A.7.1 (*Continued*)

Rayleigh line flow table, k = 1.4

M	P/P*	T/T*	V/V*	T$_t$/T$_t$*	P$_t$/P$_t$*	M	P/P*	T/T*	V/V*	T$_t$/T$_t$*	P$_t$/P$_t$*
1.01	0.98841	0.99659	1.00828	0.99993	1.00005	1.51	0.57250	0.74732	1.30536	0.90676	1.12649
1.02	0.97698	0.99304	1.01645	0.99973	1.00019	1.52	0.56676	0.74215	1.30945	0.90424	1.13153
1.03	0.96569	0.98936	1.02450	0.99940	1.00044	1.53	0.56111	0.73701	1.31350	0.90172	1.13668
1.04	0.95456	0.98554	1.03246	0.99895	1.00078	1.54	0.55552	0.73189	1.31748	0.89920	1.14193
1.05	0.94358	0.98161	1.04030	0.99838	1.00122	1.55	0.55002	0.72680	1.32142	0.89669	1.14729
1.06	0.93275	0.97755	1.04804	0.99769	1.00175	1.56	0.54458	0.72173	1.32530	0.89418	1.15274
1.07	0.92206	0.97339	1.05567	0.99690	1.00238	1.57	0.53922	0.71669	1.32913	0.89168	1.15830
1.08	0.91152	0.96913	1.06320	0.99601	1.00311	1.58	0.53393	0.71168	1.33291	0.88917	1.16397
1.09	0.90112	0.96477	1.07063	0.99501	1.00394	1.59	0.52871	0.70669	1.33663	0.88668	1.16974
1.10	0.89087	0.96031	1.07795	0.99392	1.00486	1.60	0.52356	0.70174	1.34031	0.88419	1.17561
1.11	0.88075	0.95577	1.08518	0.99275	1.00588	1.61	0.51848	0.69680	1.34394	0.88170	1.18159
1.12	0.87078	0.95115	1.09230	0.99148	1.00699	1.62	0.51346	0.69190	1.34753	0.87922	1.18768
1.13	0.86094	0.94645	1.09933	0.99013	1.00821	1.63	0.50851	0.68703	1.35106	0.87675	1.19387
1.14	0.85123	0.94169	1.10626	0.98871	1.00952	1.64	0.50363	0.68219	1.35455	0.87429	1.20017
1.15	0.84166	0.93685	1.11310	0.98721	1.01093	1.65	0.49880	0.67738	1.35800	0.87184	1.20657
1.16	0.83222	0.93196	1.11984	0.98564	1.01243	1.66	0.49405	0.67259	1.36140	0.86939	1.21309
1.17	0.82292	0.92701	1.12649	0.98400	1.01403	1.67	0.48935	0.66784	1.36475	0.86696	1.21971
1.18	0.81374	0.92200	1.13305	0.98230	1.01573	1.68	0.48472	0.66312	1.36806	0.86453	1.22644
1.19	0.80468	0.91695	1.13951	0.98054	1.01752	1.69	0.48014	0.65843	1.37133	0.86212	1.23328
1.20	0.79576	0.91185	1.14589	0.97872	1.01942	1.70	0.47562	0.65377	1.37455	0.85971	1.24024
1.21	0.78695	0.90671	1.15218	0.97684	1.02140	1.71	0.47117	0.64914	1.37774	0.85731	1.24730
1.22	0.77827	0.90153	1.15838	0.97492	1.02349	1.72	0.46677	0.64455	1.38088	0.85493	1.25447
1.23	0.76971	0.89632	1.16449	0.97294	1.02567	1.73	0.46242	0.63999	1.38398	0.85256	1.26175
1.24	0.76127	0.89108	1.17052	0.97092	1.02795	1.74	0.45813	0.63545	1.38705	0.85019	1.26915
1.25	0.75294	0.88581	1.17647	0.96886	1.03033	1.75	0.45390	0.63095	1.39007	0.84784	1.27666
1.26	0.74473	0.88052	1.18233	0.96675	1.03280	1.76	0.44972	0.62649	1.39306	0.84551	1.28428
1.27	0.73663	0.87521	1.18812	0.96461	1.03537	1.77	0.44559	0.62205	1.39600	0.84318	1.29202
1.28	0.72865	0.86988	1.19382	0.96243	1.03803	1.78	0.44152	0.61765	1.39891	0.84087	1.29987
1.29	0.72078	0.86453	1.19945	0.96022	1.04080	1.79	0.43750	0.61328	1.40179	0.83857	1.30784
1.30	0.71301	0.85917	1.20499	0.95798	1.04366	1.80	0.43353	0.60894	1.40462	0.83628	1.31592
1.31	0.70536	0.85380	1.21046	0.95571	1.04662	1.81	0.42960	0.60464	1.40743	0.83400	1.32413
1.32	0.69780	0.84843	1.21585	0.95341	1.04968	1.82	0.42573	0.60036	1.41019	0.83174	1.33244
1.33	0.69036	0.84305	1.22117	0.95108	1.05283	1.83	0.42191	0.59612	1.41292	0.82949	1.34088
1.34	0.68301	0.83766	1.22642	0.94873	1.05608	1.84	0.41813	0.59191	1.41562	0.82726	1.34943
1.35	0.67577	0.83227	1.23159	0.94637	1.05943	1.85	0.41440	0.58774	1.41829	0.82504	1.35811
1.36	0.66863	0.82689	1.23669	0.94398	1.06288	1.86	0.41072	0.58359	1.42092	0.82283	1.36690
1.37	0.66158	0.82151	1.24173	0.94157	1.06642	1.87	0.40708	0.57948	1.42351	0.82064	1.37582
1.38	0.65464	0.81613	1.24669	0.93914	1.07007	1.88	0.40349	0.57540	1.42608	0.81845	1.38486
1.39	0.64778	0.81076	1.25158	0.93671	1.07381	1.89	0.39994	0.57136	1.42862	0.81629	1.39402
1.40	0.64103	0.80539	1.25641	0.93425	1.07765	1.90	0.39643	0.56734	1.43112	0.81414	1.40330
1.41	0.63436	0.80004	1.26117	0.93179	1.08159	1.91	0.39297	0.56336	1.43359	0.81200	1.41271
1.42	0.62779	0.79469	1.26587	0.92931	1.08563	1.92	0.38955	0.55941	1.43604	0.80987	1.42224
1.43	0.62130	0.78936	1.27050	0.92683	1.08977	1.93	0.38617	0.55549	1.43845	0.80776	1.43190
1.44	0.61491	0.78405	1.27507	0.92434	1.09401	1.94	0.38283	0.55160	1.44083	0.80567	1.44168
1.45	0.60860	0.77874	1.27957	0.92184	1.09835	1.95	0.37954	0.54774	1.44319	0.80358	1.45159
1.46	0.60237	0.77346	1.28402	0.91933	1.10278	1.96	0.37628	0.54392	1.44551	0.80152	1.46164
1.47	0.59623	0.76819	1.28840	0.91682	1.10732	1.97	0.37306	0.54012	1.44781	0.79946	1.47180
1.48	0.59018	0.76294	1.29273	0.91431	1.11196	1.98	0.36988	0.53636	1.45008	0.79742	1.48210
1.49	0.58421	0.75771	1.29700	0.91179	1.11670	1.99	0.36674	0.53263	1.45233	0.79540	1.49253
1.50	0.57831	0.75250	1.30120	0.90928	1.12155	2.00	0.36364	0.52893	1.45455	0.79339	1.50310

(*Continued*)

TABLE A.7.1 (*Continued*)

Rayleigh line flow table, k = 1.4

M	P/P*	T/T*	V/V*	T_t/T_t^*	P_t/P_t^*	M	P/P*	T/T*	V/V*	T_t/T_t^*	P_t/P_t^*
2.01	0.36057	0.52525	1.45674	0.79139	1.51379	2.51	0.24440	0.37630	1.53972	0.70871	2.24054
2.02	0.35754	0.52161	1.45890	0.78941	1.52462	2.52	0.24266	0.37392	1.54096	0.70736	2.25944
2.03	0.35454	0.51800	1.46104	0.78744	1.53558	2.53	0.24093	0.37157	1.54219	0.70603	2.27853
2.04	0.35158	0.51442	1.46315	0.78549	1.54668	2.54	0.23923	0.36923	1.54341	0.70471	2.29782
2.05	0.34866	0.51087	1.46524	0.78355	1.55791	2.55	0.23754	0.36691	1.54461	0.70340	2.31730
2.06	0.34577	0.50735	1.46731	0.78162	1.56928	2.56	0.23587	0.36461	1.54581	0.70210	2.33699
2.07	0.34291	0.50386	1.46935	0.77971	1.58079	2.57	0.23422	0.36233	1.54699	0.70081	2.35687
2.08	0.34009	0.50040	1.47136	0.77782	1.59244	2.58	0.23258	0.36007	1.54816	0.69952	2.37696
2.09	0.33730	0.49696	1.47336	0.77593	1.60423	2.59	0.23096	0.35783	1.54931	0.69826	2.39725
2.10	0.33454	0.49356	1.47533	0.77406	1.61616	2.60	0.22936	0.35561	1.55046	0.69700	2.41774
2.11	0.33182	0.49018	1.47727	0.77221	1.62823	2.61	0.22777	0.35341	1.55159	0.69575	2.43844
2.12	0.32912	0.48684	1.47920	0.77037	1.64045	2.62	0.22620	0.35122	1.55272	0.69451	2.45935
2.13	0.32646	0.48352	1.48110	0.76854	1.65281	2.63	0.22464	0.34906	1.55383	0.69328	2.48047
2.14	0.32382	0.48023	1.48298	0.76673	1.66531	2.64	0.22310	0.34691	1.55493	0.69206	2.50179
2.15	0.32122	0.47696	1.48484	0.76493	1.67796	2.65	0.22158	0.34478	1.55602	0.69084	2.52334
2.16	0.31865	0.47373	1.48668	0.76314	1.69076	2.66	0.22007	0.34266	1.55710	0.68964	2.54509
2.17	0.31610	0.47052	1.48850	0.76137	1.70371	2.67	0.21857	0.34057	1.55816	0.68845	2.56706
2.18	0.31359	0.46734	1.49029	0.75961	1.71680	2.68	0.21709	0.33849	1.55922	0.68727	2.58925
2.19	0.31110	0.46418	1.49207	0.75787	1.73005	2.69	0.21562	0.33643	1.56027	0.68610	2.61166
2.20	0.30864	0.46106	1.49383	0.75613	1.74345	2.70	0.21417	0.33439	1.56131	0.68494	2.63429
2.21	0.30621	0.45796	1.49556	0.75442	1.75700	2.71	0.21273	0.33236	1.56233	0.68378	2.65714
2.22	0.30381	0.45488	1.49728	0.75271	1.77070	2.72	0.21131	0.33035	1.56335	0.68264	2.68021
2.23	0.30143	0.45184	1.49898	0.75102	1.78456	2.73	0.20990	0.32836	1.56436	0.68150	2.70351
2.24	0.29908	0.44882	1.50066	0.74934	1.79858	2.74	0.20850	0.32638	1.56536	0.68037	2.72704
2.25	0.29675	0.44582	1.50232	0.74768	1.81275	2.75	0.20712	0.32442	1.56634	0.67926	2.75080
2.26	0.29446	0.44285	1.50396	0.74602	1.82708	2.76	0.20575	0.32248	1.56732	0.67815	2.77478
2.27	0.29218	0.43990	1.50558	0.74438	1.84157	2.77	0.20439	0.32055	1.56829	0.67705	2.79900
2.28	0.28993	0.43698	1.50719	0.74276	1.85623	2.78	0.20305	0.31864	1.56925	0.67595	2.82346
2.29	0.28771	0.43409	1.50878	0.74114	1.87104	2.79	0.20172	0.31674	1.57020	0.67487	2.84815
2.30	0.28551	0.43122	1.51035	0.73954	1.88602	2.80	0.20040	0.31486	1.57114	0.67380	2.87308
2.31	0.28333	0.42838	1.51190	0.73795	1.90116	2.81	0.19910	0.31299	1.57207	0.67273	2.89825
2.32	0.28118	0.42555	1.51344	0.73638	1.91647	2.82	0.19780	0.31114	1.57300	0.67167	2.92366
2.33	0.27905	0.42276	1.51496	0.73482	1.93195	2.83	0.19652	0.30931	1.57391	0.67062	2.94931
2.34	0.27695	0.41998	1.51646	0.73326	1.94759	2.84	0.19525	0.30749	1.57482	0.66958	2.97521
2.35	0.27487	0.41723	1.51795	0.73173	1.96340	2.85	0.19399	0.30568	1.57572	0.66855	3.00136
2.36	0.27281	0.41451	1.51942	0.73020	1.97939	2.86	0.19275	0.30389	1.57661	0.66752	3.02775
2.37	0.27077	0.41181	1.52088	0.72868	1.99554	2.87	0.19151	0.30211	1.57749	0.66651	3.05440
2.38	0.26875	0.40913	1.52232	0.72718	2.01187	2.88	0.19029	0.30035	1.57836	0.66550	3.08129
2.39	0.26676	0.40647	1.52374	0.72569	2.02837	2.89	0.18908	0.29860	1.57923	0.66450	3.10844
2.40	0.26478	0.40384	1.52515	0.72421	2.04505	2.90	0.18788	0.29687	1.58008	0.66350	3.13585
2.41	0.26283	0.40122	1.52655	0.72275	2.06191	2.91	0.18669	0.29515	1.58093	0.66252	3.16352
2.42	0.26090	0.39864	1.52793	0.72129	2.07895	2.92	0.18551	0.29344	1.58178	0.66154	3.19145
2.43	0.25899	0.39607	1.52929	0.71985	2.09616	2.93	0.18435	0.29175	1.58261	0.66057	3.21963
2.44	0.25710	0.39352	1.53065	0.71842	2.11356	2.94	0.18319	0.29007	1.58343	0.65960	3.24809
2.45	0.25522	0.39100	1.53198	0.71699	2.13114	2.95	0.18205	0.28841	1.58425	0.65865	3.27680
2.46	0.25337	0.38850	1.53331	0.71558	2.14891	2.96	0.18091	0.28675	1.58506	0.65770	3.30579
2.47	0.25154	0.38602	1.53461	0.71419	2.16685	2.97	0.17979	0.28512	1.58587	0.65676	3.33505
2.48	0.24973	0.38356	1.53591	0.71280	2.18499	2.98	0.17867	0.28349	1.58666	0.65583	3.36457
2.49	0.24793	0.38112	1.53719	0.71142	2.20332	2.99	0.17757	0.28188	1.58745	0.65490	3.39437
2.50	0.24615	0.37870	1.53846	0.71006	2.22183	3.00	0.17647	0.28028	1.58824	0.65398	3.42445

(*Continued*)

TABLE A.7.1 (*Continued*)

Rayleigh line flow table, k = 1.4

M	P/P*	T/T*	V/V*	T_t/T_t^*	P_t/P_t^*
3.10	0.16604	0.26495	1.59568	0.64516	3.74084
3.20	0.15649	0.25078	1.60250	0.63699	4.08712
3.30	0.14773	0.23766	1.60877	0.62940	4.46549
3.40	0.13966	0.22549	1.61453	0.62236	4.87830
3.50	0.13223	0.21419	1.61983	0.61580	5.32804
3.60	0.12537	0.20369	1.62474	0.60970	5.81730
3.70	0.11901	0.19390	1.62928	0.60401	6.34884
3.80	0.11312	0.18478	1.63348	0.59870	6.92557
3.90	0.10765	0.17627	1.63739	0.59373	7.55050
4.00	0.10256	0.16831	1.64103	0.58909	8.22685
4.10	0.09782	0.16086	1.64441	0.58473	8.95794
4.20	0.09340	0.15388	1.64757	0.58065	9.74729
4.30	0.08927	0.14734	1.65052	0.57682	10.5985
4.40	0.08540	0.14119	1.65329	0.57322	11.5155
4.50	0.08177	0.13540	1.65588	0.56982	12.5023
4.60	0.07837	0.12996	1.65831	0.56663	13.5629
4.70	0.07517	0.12483	1.66059	0.56362	14.7017
4.80	0.07217	0.12000	1.66274	0.56078	15.9234
4.90	0.06934	0.11543	1.66476	0.55809	17.2325
5.00	0.06667	0.11111	1.66667	0.55556	18.6339
5.10	0.06415	0.10703	1.66847	0.55315	20.1328
5.20	0.06177	0.10316	1.67017	0.55088	21.7344
5.30	0.05951	0.09950	1.67178	0.54872	23.4442
5.40	0.05738	0.09602	1.67330	0.54667	25.2679
5.50	0.05536	0.09272	1.67474	0.54473	27.2113
5.60	0.05345	0.08958	1.67611	0.54288	29.2806
5.70	0.05163	0.08660	1.67741	0.54112	31.4821
5.80	0.04990	0.08376	1.67864	0.53944	33.8223
5.90	0.04826	0.08106	1.67982	0.53785	36.3079
6.00	0.04669	0.07849	1.68093	0.53633	38.9459
6.20	0.04378	0.07369	1.68301	0.53349	44.7084
6.40	0.04114	0.06931	1.68490	0.53091	51.1700
6.60	0.03872	0.06531	1.68663	0.52854	58.3953
6.80	0.03651	0.06164	1.68821	0.52637	66.4524
7.00	0.03448	0.05826	1.68966	0.52438	75.4138
7.20	0.03262	0.05516	1.69099	0.52254	85.3562
7.40	0.03090	0.05229	1.69221	0.52084	96.3605
7.60	0.02932	0.04964	1.69335	0.51927	108.5124
7.80	0.02785	0.04719	1.69439	0.51782	121.9017
8.00	0.02649	0.04491	1.69536	0.51647	136.6235
8.20	0.02523	0.04279	1.69627	0.51521	152.7774
8.40	0.02405	0.04082	1.69711	0.51404	170.4680
8.60	0.02296	0.03898	1.69789	0.51295	189.8050
8.80	0.02193	0.03726	1.69862	0.51193	210.9036
9.00	0.02098	0.03565	1.69930	0.51098	233.8840
9.20	0.02008	0.03414	1.69994	0.51008	258.8719
9.40	0.01925	0.03273	1.70054	0.50925	285.9989
9.60	0.01846	0.03140	1.70110	0.50846	315.4021
9.80	0.01772	0.03015	1.70163	0.50772	347.2245
10.0	0.01702	0.02897	1.70213	0.50702	381.6149

TABLE A.7.2

Rayleigh line flow table, k = 1.3

M	P/P*	T/T*	V/V*	T_t/T_t^*	P_t/P_t^*	M	P/P*	T/T*	V/V*	T_t/T_t^*	P_t/P_t^*
0.01	2.29970	0.00053	0.00023	0.00046	1.25509	0.51	1.71882	0.76842	0.44706	0.69426	1.10722
0.02	2.29880	0.00211	0.00092	0.00184	1.25485	0.52	1.70179	0.78310	0.46016	0.70858	1.10333
0.03	2.29731	0.00475	0.00207	0.00413	1.25444	0.53	1.68477	0.79732	0.47325	0.72254	1.09948
0.04	2.29523	0.00843	0.00367	0.00733	1.25387	0.54	1.66778	0.81108	0.48632	0.73614	1.09567
0.05	2.29255	0.01314	0.00573	0.01143	1.25314	0.55	1.65082	0.82437	0.49937	0.74937	1.09191
0.06	2.28929	0.01887	0.00824	0.01641	1.25225	0.56	1.63389	0.83719	0.51239	0.76224	1.08820
0.07	2.28544	0.02559	0.01120	0.02227	1.25121	0.57	1.61702	0.84953	0.52537	0.77473	1.08453
0.07	2.28102	0.03330	0.01460	0.02898	1.25000	0.58	1.60020	0.86140	0.53831	0.78684	1.08092
0.09	2.27603	0.04196	0.01844	0.03653	1.24865	0.59	1.58344	0.87279	0.55120	0.79858	1.07737
0.10	2.27048	0.05155	0.02270	0.04489	1.24714	0.60	1.56676	0.88370	0.56403	0.80993	1.07388
0.10	2.26438	0.06204	0.02740	0.05405	1.24548	0.61	1.55015	0.89414	0.57681	0.82091	1.07044
0.12	2.25774	0.07340	0.03251	0.06397	1.24368	0.62	1.53362	0.90410	0.58952	0.83151	1.06707
0.13	2.25056	0.08560	0.03803	0.07462	1.24174	0.63	1.51718	0.91360	0.60217	0.84173	1.06377
0.14	2.24285	0.09860	0.04396	0.08599	1.23966	0.64	1.50084	0.92263	0.61474	0.85158	1.06053
0.15	2.23464	0.11236	0.05028	0.09803	1.23744	0.65	1.48459	0.93119	0.62724	0.86105	1.05736
0.16	2.22592	0.12684	0.05698	0.11072	1.23509	0.66	1.46845	0.93930	0.63966	0.87015	1.05427
0.17	2.21672	0.14201	0.06406	0.12402	1.23261	0.67	1.45241	0.94696	0.65199	0.87889	1.05124
0.18	2.20704	0.15782	0.07151	0.13790	1.23001	0.68	1.43649	0.95417	0.66424	0.88726	1.04830
0.19	2.19690	0.17423	0.07931	0.15233	1.22730	0.69	1.42069	0.96094	0.67639	0.89528	1.04542
0.20	2.18631	0.19120	0.08745	0.16726	1.22446	0.70	1.40501	0.96728	0.68845	0.90294	1.04263
0.21	2.17529	0.20868	0.09593	0.18266	1.22152	0.71	1.38945	0.97320	0.70042	0.91025	1.03992
0.22	2.16385	0.22662	0.10473	0.19849	1.21848	0.72	1.37402	0.97870	0.71229	0.91722	1.03728
0.23	2.15201	0.24499	0.11384	0.21472	1.21533	0.73	1.35872	0.98380	0.72406	0.92386	1.03473
0.24	2.13977	0.26373	0.12325	0.23131	1.21209	0.74	1.34355	0.98849	0.73573	0.93016	1.03227
0.25	2.12717	0.28280	0.13295	0.24822	1.20876	0.75	1.32852	0.99279	0.74729	0.93614	1.02988
0.26	2.11420	0.30216	0.14292	0.26541	1.20534	0.76	1.31363	0.99671	0.75875	0.94180	1.02759
0.27	2.10090	0.32176	0.15316	0.28285	1.20184	0.77	1.29887	1.00026	0.77010	0.94715	1.02538
0.28	2.08727	0.34156	0.16364	0.30050	1.19827	0.78	1.28426	1.00344	0.78134	0.95219	1.02325
0.29	2.07332	0.36152	0.17437	0.31833	1.19464	0.79	1.26979	1.00627	0.79247	0.95693	1.02122
0.30	2.05909	0.38159	0.18532	0.33629	1.19093	0.80	1.25546	1.00875	0.80349	0.96139	1.01927
0.31	2.04457	0.40172	0.19648	0.35436	1.18717	0.81	1.24128	1.01090	0.81440	0.96555	1.01742
0.32	2.02979	0.42189	0.20785	0.37250	1.18335	0.82	1.22724	1.01272	0.82520	0.96944	1.01565
0.33	2.01477	0.44206	0.21941	0.39068	1.17949	0.83	1.21336	1.01422	0.83588	0.97306	1.01398
0.34	1.99951	0.46217	0.23114	0.40886	1.17558	0.84	1.19962	1.01541	0.84645	0.97642	1.01240
0.35	1.98404	0.48221	0.24305	0.42702	1.17164	0.85	1.18603	1.01631	0.85690	0.97952	1.01091
0.36	1.96837	0.50213	0.25510	0.44512	1.16766	0.86	1.17258	1.01692	0.86724	0.98238	1.00951
0.37	1.95251	0.52190	0.26730	0.46315	1.16365	0.87	1.15929	1.01724	0.87747	0.98499	1.00821
0.38	1.93648	0.54150	0.27963	0.48106	1.15962	0.88	1.14615	1.01730	0.88758	0.98736	1.00701
0.39	1.92030	0.56088	0.29208	0.49885	1.15558	0.89	1.13316	1.01709	0.89757	0.98951	1.00589
0.40	1.90397	0.58002	0.30464	0.51647	1.15152	0.90	1.12031	1.01663	0.90745	0.99143	1.00487
0.41	1.88752	0.59890	0.31729	0.53391	1.14745	0.91	1.10762	1.01593	0.91722	0.99315	1.00395
0.42	1.87095	0.61748	0.33004	0.55115	1.14337	0.92	1.09507	1.01499	0.92687	0.99465	1.00313
0.43	1.85429	0.63576	0.34286	0.56816	1.13930	0.93	1.08267	1.01382	0.93640	0.99596	1.00240
0.44	1.83753	0.65369	0.35575	0.58494	1.13523	0.94	1.07042	1.01244	0.94583	0.99707	1.00176
0.45	1.82070	0.67128	0.36869	0.60145	1.13117	0.95	1.05832	1.01084	0.95514	0.99799	1.00122
0.46	1.80381	0.68849	0.38169	0.61769	1.12712	0.96	1.04637	1.00905	0.96433	0.99873	1.00078
0.47	1.78687	0.70531	0.39472	0.63363	1.12309	0.97	1.03456	1.00706	0.97342	0.99929	1.00044
0.48	1.76988	0.72173	0.40778	0.64928	1.11908	0.98	1.02290	1.00488	0.98239	0.99969	1.00020
0.49	1.75288	0.73772	0.42087	0.66460	1.11510	0.99	1.01138	1.00253	0.99125	0.99992	1.00005
0.50	1.73585	0.75329	0.43396	0.67960	1.11114	1.00	1.00000	1.00000	1.00000	1.00000	1.00000

(Continued)

TABLE A.7.2 (*Continued*)

Rayleigh line flow table, k = 1.3

M	P/P*	T/T*	V/V*	T₀/T₀*	P₀/P₀*	M	P/P*	T/T*	V/V*	T₀/T₀*	P₀/P₀*
1.01	0.98877	0.99731	1.00864	0.99993	1.00005	1.51	0.58020	0.76756	1.32292	0.89572	1.13285
1.02	0.97768	0.99446	1.01717	0.99971	1.00020	1.52	0.57449	0.76253	1.32731	0.89287	1.13826
1.03	0.96672	0.99147	1.02560	0.99934	1.00044	1.53	0.56886	0.75752	1.33165	0.89001	1.14378
1.04	0.95591	0.98833	1.03391	0.99885	1.00079	1.54	0.56330	0.75253	1.33592	0.88716	1.14942
1.05	0.94524	0.98506	1.04212	0.99823	1.00123	1.55	0.55781	0.74755	1.34014	0.88430	1.15519
1.06	0.93470	0.98165	1.05023	0.99748	1.00178	1.56	0.55240	0.74259	1.34431	0.88145	1.16106
1.07	0.92430	0.97812	1.05823	0.99661	1.00242	1.57	0.54705	0.73765	1.34842	0.87860	1.16706
1.08	0.91403	0.97448	1.06613	0.99563	1.00316	1.58	0.54177	0.73274	1.35248	0.87576	1.17318
1.09	0.90390	0.97072	1.07392	0.99454	1.00401	1.59	0.53656	0.72784	1.35649	0.87292	1.17942
1.10	0.89390	0.96686	1.08162	0.99334	1.00495	1.60	0.53142	0.72297	1.36044	0.87008	1.18579
1.11	0.88403	0.96289	1.08921	0.99204	1.00599	1.61	0.52635	0.71812	1.36435	0.86725	1.19227
1.12	0.87429	0.95883	1.09670	0.99065	1.00713	1.62	0.52134	0.71330	1.36820	0.86443	1.19888
1.13	0.86467	0.95468	1.10410	0.98916	1.00838	1.63	0.51639	0.70849	1.37201	0.86161	1.20561
1.14	0.85518	0.95045	1.11140	0.98759	1.00972	1.64	0.51151	0.70372	1.37576	0.85880	1.21247
1.15	0.84582	0.94613	1.11860	0.98593	1.01117	1.65	0.50669	0.69896	1.37947	0.85600	1.21945
1.16	0.83658	0.94175	1.12571	0.98420	1.01271	1.66	0.50193	0.69424	1.38313	0.85321	1.22657
1.17	0.82747	0.93729	1.13272	0.98239	1.01436	1.67	0.49724	0.68954	1.38674	0.85043	1.23381
1.18	0.81847	0.93276	1.13964	0.98050	1.01611	1.68	0.49260	0.68486	1.39031	0.84766	1.24117
1.19	0.80959	0.92817	1.14647	0.97855	1.01796	1.69	0.48802	0.68022	1.39383	0.84490	1.24867
1.20	0.80084	0.92353	1.15320	0.97653	1.01992	1.70	0.48350	0.67560	1.39731	0.84215	1.25630
1.21	0.79219	0.91883	1.15985	0.97445	1.02197	1.71	0.47903	0.67100	1.40074	0.83940	1.26406
1.22	0.78367	0.91408	1.16641	0.97231	1.02413	1.72	0.47463	0.66644	1.40413	0.83668	1.27196
1.23	0.77525	0.90928	1.17288	0.97011	1.02639	1.73	0.47027	0.66190	1.40748	0.83396	1.27999
1.24	0.76695	0.90444	1.17927	0.96786	1.02876	1.74	0.46598	0.65739	1.41079	0.83125	1.28815
1.25	0.75876	0.89956	1.18557	0.96556	1.03122	1.75	0.46173	0.65291	1.41405	0.82856	1.29645
1.26	0.75068	0.89465	1.19178	0.96322	1.03379	1.76	0.45754	0.64846	1.41728	0.82588	1.30489
1.27	0.74271	0.88970	1.19792	0.96083	1.03647	1.77	0.45340	0.64404	1.42046	0.82321	1.31346
1.28	0.73484	0.88473	1.20397	0.95840	1.03925	1.78	0.44931	0.63964	1.42360	0.82056	1.32218
1.29	0.72708	0.87972	1.20994	0.95593	1.04213	1.79	0.44528	0.63528	1.42671	0.81792	1.33103
1.30	0.71942	0.87470	1.21583	0.95342	1.04512	1.80	0.44129	0.63095	1.42978	0.81529	1.34003
1.31	0.71187	0.86965	1.22164	0.95088	1.04822	1.81	0.43735	0.62664	1.43281	0.81268	1.34917
1.32	0.70442	0.86458	1.22737	0.94830	1.05141	1.82	0.43346	0.62236	1.43580	0.81008	1.35845
1.33	0.69706	0.85950	1.23303	0.94570	1.05472	1.83	0.42962	0.61812	1.43875	0.80749	1.36789
1.34	0.68980	0.85440	1.23861	0.94306	1.05813	1.84	0.42582	0.61390	1.44167	0.80492	1.37746
1.35	0.68264	0.84929	1.24412	0.94041	1.06165	1.85	0.42208	0.60971	1.44456	0.80237	1.38719
1.36	0.67558	0.84417	1.24955	0.93772	1.06528	1.86	0.41837	0.60556	1.44740	0.79983	1.39706
1.37	0.66861	0.83905	1.25492	0.93502	1.06901	1.87	0.41472	0.60143	1.45022	0.79730	1.40709
1.38	0.66173	0.83392	1.26021	0.93229	1.07285	1.88	0.41110	0.59733	1.45300	0.79479	1.41727
1.39	0.65495	0.82879	1.26542	0.92955	1.07680	1.89	0.40753	0.59326	1.45574	0.79230	1.42760
1.40	0.64825	0.82365	1.27057	0.92679	1.08086	1.90	0.40400	0.58922	1.45846	0.78982	1.43809
1.41	0.64165	0.81852	1.27566	0.92401	1.08503	1.91	0.40052	0.58522	1.46114	0.78735	1.44873
1.42	0.63513	0.81339	1.28067	0.92122	1.08930	1.92	0.39708	0.58124	1.46379	0.78490	1.45953
1.43	0.62870	0.80826	1.28562	0.91842	1.09369	1.93	0.39368	0.57729	1.46640	0.78247	1.47049
1.44	0.62235	0.80314	1.29050	0.91561	1.09819	1.94	0.39031	0.57337	1.46899	0.78005	1.48161
1.45	0.61609	0.79803	1.29532	0.91279	1.10280	1.95	0.38699	0.56948	1.47154	0.77765	1.49289
1.46	0.60990	0.79292	1.30007	0.90996	1.10753	1.96	0.38371	0.56562	1.47407	0.77526	1.50434
1.47	0.60381	0.78782	1.30476	0.90712	1.11236	1.97	0.38047	0.56179	1.47656	0.77289	1.51595
1.48	0.59779	0.78274	1.30939	0.90428	1.11731	1.98	0.37726	0.55798	1.47903	0.77053	1.52772
1.49	0.59185	0.77767	1.31396	0.90143	1.12238	1.99	0.37410	0.55421	1.48146	0.76819	1.53967
1.50	0.58599	0.77261	1.31847	0.89858	1.12755	2.00	0.37097	0.55047	1.48387	0.76587	1.55179

(Continued)

TABLE A.7.2 (*Continued*)

Rayleigh line flow table, k = 1.3

M	P/P*	T/T*	V/V*	T₀/T₀*	P₀/P₀*	M	P/P*	T/T*	V/V*	T₀/T₀*	P₀/P₀*
2.01	0.36787	0.54675	1.48625	0.76356	1.56407	**2.51**	0.25027	0.39460	1.57672	0.66740	2.43995
2.02	0.36482	0.54307	1.48860	0.76127	1.57654	**2.52**	0.24850	0.39215	1.57808	0.66583	2.46371
2.03	0.36180	0.53941	1.49093	0.75899	1.58917	**2.53**	0.24675	0.38972	1.57942	0.66427	2.48775
2.04	0.35881	0.53578	1.49322	0.75673	1.60198	**2.54**	0.24502	0.38731	1.58076	0.66272	2.51208
2.05	0.35586	0.53218	1.49549	0.75449	1.61497	**2.55**	0.24330	0.38492	1.58207	0.66119	2.53671
2.06	0.35294	0.52861	1.49774	0.75226	1.62814	**2.56**	0.24160	0.38255	1.58338	0.65967	2.56164
2.07	0.35006	0.52507	1.49996	0.75004	1.64150	**2.57**	0.23992	0.38020	1.58467	0.65816	2.58687
2.08	0.34721	0.52155	1.50215	0.74785	1.65503	**2.58**	0.23826	0.37787	1.58595	0.65666	2.61241
2.09	0.34439	0.51807	1.50432	0.74566	1.66875	**2.59**	0.23661	0.37556	1.58722	0.65517	2.63825
2.10	0.34160	0.51461	1.50646	0.74350	1.68266	**2.60**	0.23498	0.37326	1.58848	0.65370	2.66441
2.11	0.33885	0.51118	1.50858	0.74135	1.69676	**2.61**	0.23337	0.37099	1.58972	0.65223	2.69088
2.12	0.33612	0.50777	1.51067	0.73921	1.71105	**2.62**	0.23177	0.36873	1.59095	0.65078	2.71767
2.13	0.33343	0.50440	1.51275	0.73709	1.72553	**2.63**	0.23018	0.36649	1.59217	0.64934	2.74478
2.14	0.33077	0.50105	1.51479	0.73499	1.74020	**2.64**	0.22862	0.36427	1.59337	0.64791	2.77222
2.15	0.32814	0.49773	1.51682	0.73290	1.75508	**2.65**	0.22707	0.36207	1.59457	0.64649	2.79999
2.16	0.32554	0.49443	1.51882	0.73083	1.77015	**2.66**	0.22553	0.35989	1.59575	0.64508	2.82808
2.17	0.32296	0.49116	1.52080	0.72877	1.78542	**2.67**	0.22401	0.35772	1.59692	0.64369	2.85651
2.18	0.32042	0.48792	1.52276	0.72673	1.80089	**2.68**	0.22250	0.35557	1.59808	0.64230	2.88528
2.19	0.31790	0.48470	1.52469	0.72470	1.81657	**2.69**	0.22101	0.35344	1.59923	0.64093	2.91440
2.20	0.31541	0.48151	1.52660	0.72269	1.83246	**2.70**	0.21953	0.35133	1.60036	0.63956	2.94386
2.21	0.31295	0.47835	1.52850	0.72069	1.84855	**2.71**	0.21806	0.34923	1.60149	0.63821	2.97366
2.22	0.31052	0.47521	1.53037	0.71871	1.86486	**2.72**	0.21661	0.34715	1.60260	0.63687	3.00382
2.23	0.30811	0.47210	1.53222	0.71674	1.88138	**2.73**	0.21518	0.34508	1.60371	0.63554	3.03434
2.24	0.30573	0.46901	1.53405	0.71479	1.89811	**2.74**	0.21376	0.34304	1.60480	0.63421	3.06522
2.25	0.30338	0.46595	1.53586	0.71285	1.91506	**2.75**	0.21235	0.34101	1.60589	0.63290	3.09646
2.26	0.30105	0.46291	1.53765	0.71093	1.93223	**2.76**	0.21095	0.33899	1.60696	0.63160	3.12807
2.27	0.29875	0.45990	1.53942	0.70902	1.94963	**2.77**	0.20957	0.33700	1.60802	0.63031	3.16005
2.28	0.29647	0.45691	1.54118	0.70713	1.96724	**2.78**	0.20820	0.33501	1.60907	0.62903	3.19240
2.29	0.29422	0.45395	1.54291	0.70525	1.98509	**2.79**	0.20685	0.33305	1.61012	0.62776	3.22513
2.30	0.29199	0.45101	1.54462	0.70339	2.00316	**2.80**	0.20550	0.33110	1.61115	0.62649	3.25825
2.31	0.28978	0.44810	1.54632	0.70153	2.02146	**2.81**	0.20417	0.32916	1.61217	0.62524	3.29176
2.32	0.28760	0.44521	1.54800	0.69970	2.04000	**2.82**	0.20286	0.32724	1.61319	0.62400	3.32565
2.33	0.28545	0.44234	1.54966	0.69788	2.05877	**2.83**	0.20155	0.32534	1.61419	0.62277	3.35994
2.34	0.28331	0.43950	1.55130	0.69607	2.07779	**2.84**	0.20026	0.32345	1.61519	0.62154	3.39463
2.35	0.28120	0.43668	1.55292	0.69428	2.09704	**2.85**	0.19897	0.32158	1.61617	0.62033	3.42972
2.36	0.27911	0.43389	1.55453	0.69250	2.11654	**2.86**	0.19771	0.31972	1.61715	0.61913	3.46522
2.37	0.27704	0.43111	1.55612	0.69073	2.13628	**2.87**	0.19645	0.31787	1.61812	0.61793	3.50113
2.38	0.27500	0.42836	1.55769	0.68898	2.15627	**2.88**	0.19520	0.31605	1.61908	0.61674	3.53746
2.39	0.27297	0.42563	1.55925	0.68724	2.17651	**2.89**	0.19397	0.31423	1.62003	0.61557	3.57421
2.40	0.27097	0.42293	1.56079	0.68551	2.19700	**2.90**	0.19274	0.31243	1.62097	0.61440	3.61138
2.41	0.26899	0.42025	1.56232	0.68380	2.21775	**2.91**	0.19153	0.31064	1.62190	0.61324	3.64898
2.42	0.26703	0.41759	1.56382	0.68210	2.23876	**2.92**	0.19033	0.30887	1.62282	0.61209	3.68701
2.43	0.26509	0.41495	1.56532	0.68042	2.26004	**2.93**	0.18914	0.30711	1.62374	0.61095	3.72548
2.44	0.26317	0.41233	1.56679	0.67874	2.28157	**2.94**	0.18796	0.30537	1.62465	0.60982	3.76439
2.45	0.26127	0.40973	1.56826	0.67709	2.30337	**2.95**	0.18679	0.30364	1.62555	0.60869	3.80374
2.46	0.25939	0.40716	1.56970	0.67544	2.32545	**2.96**	0.18563	0.30192	1.62644	0.60758	3.84355
2.47	0.25753	0.40461	1.57113	0.67381	2.34779	**2.97**	0.18448	0.30022	1.62732	0.60647	3.88382
2.48	0.25568	0.40207	1.57255	0.67218	2.37041	**2.98**	0.18335	0.29852	1.62819	0.60537	3.92454
2.49	0.25386	0.39956	1.57395	0.67058	2.39331	**2.99**	0.18222	0.29685	1.62906	0.60428	3.96573
2.50	0.25205	0.39707	1.57534	0.66898	2.41649	**3.00**	0.18110	0.29518	1.62992	0.60320	4.00738

(*Continued*)

TABLE A.7.2 (*Continued*)

Rayleigh line flow table, k = 1.3

M	P/P*	T/T*	V/V*	T$_t$/T$_t$*	P$_t$/P$_t$*
3.10	0.17046	0.27923	1.63811	0.59282	4.45083
3.20	0.16070	0.26446	1.64561	0.58319	4.94672
3.30	0.15175	0.25076	1.65250	0.57424	5.50049
3.40	0.14350	0.23804	1.65885	0.56592	6.11812
3.50	0.13589	0.22622	1.66470	0.55818	6.80608
3.60	0.12887	0.21522	1.67010	0.55096	7.57143
3.70	0.12236	0.20497	1.67511	0.54423	8.42183
3.80	0.11633	0.19540	1.67975	0.53794	9.36556
3.90	0.11072	0.18646	1.68406	0.53206	10.4116
4.00	0.10550	0.17810	1.68807	0.52656	11.5697
4.10	0.10064	0.17027	1.69181	0.52139	12.8503
4.20	0.09611	0.16293	1.69530	0.51655	14.2647
4.30	0.09186	0.15604	1.69857	0.51201	15.8251
4.40	0.08789	0.14956	1.70162	0.50773	17.5446
4.50	0.08417	0.14347	1.70448	0.50370	19.4371
4.60	0.08068	0.13773	1.70717	0.49991	21.5178
4.70	0.07740	0.13232	1.70969	0.49633	23.8028
4.80	0.07431	0.12722	1.71207	0.49296	26.3094
4.90	0.07140	0.12240	1.71431	0.48976	29.0561
5.00	0.06866	0.11784	1.71642	0.48675	32.0625
5.10	0.06607	0.11353	1.71841	0.48389	35.3499
5.20	0.06362	0.10945	1.72029	0.48118	38.9406
5.30	0.06131	0.10557	1.72207	0.47861	42.8584
5.40	0.05911	0.10190	1.72376	0.47617	47.1290
5.50	0.05704	0.09841	1.72536	0.47386	51.7792
5.60	0.05507	0.09509	1.72687	0.47166	56.8379
5.70	0.05320	0.09194	1.72831	0.46956	62.3355
5.80	0.05142	0.08894	1.72968	0.46757	68.3042
5.90	0.04973	0.08608	1.73098	0.46567	74.7784
6.00	0.04812	0.08335	1.73222	0.46386	81.7942
6.20	0.04512	0.07827	1.73452	0.46048	97.6062
6.40	0.04240	0.07363	1.73662	0.45740	116.073
6.60	0.03991	0.06939	1.73853	0.45457	137.567
6.80	0.03764	0.06550	1.74028	0.45199	162.499
7.00	0.03555	0.06192	1.74189	0.44961	191.326
7.20	0.03363	0.05863	1.74336	0.44741	224.552
7.40	0.03186	0.05559	1.74472	0.44539	262.733
7.60	0.03023	0.05278	1.74598	0.44352	306.479
7.80	0.02872	0.05017	1.74714	0.44178	356.461
8.00	0.02732	0.04775	1.74822	0.44017	413.413
8.20	0.02601	0.04551	1.74922	0.43867	478.135
8.40	0.02480	0.04341	1.75015	0.43727	551.501
8.60	0.02368	0.04146	1.75102	0.43597	634.463
8.80	0.02262	0.03963	1.75183	0.43475	728.053
9.00	0.02164	0.03792	1.75259	0.43361	833.391
9.20	0.02071	0.03632	1.75330	0.43254	951.689
9.40	0.01985	0.03482	1.75396	0.43154	1084.26
9.60	0.01904	0.03340	1.75459	0.43060	1232.51
9.80	0.01828	0.03208	1.75517	0.42971	1397.98
10.0	0.01756	0.03083	1.75573	0.42888	1582.29

Index

Printed in the United States
by Baker & Taylor Publisher Services